Balancing Development and Sustainability in Tourism Destinations

Akhmad Saufi · Imanuella R. Andilolo
Norain Othman · Alan A. Lew
Editors

# Balancing Development and Sustainability in Tourism Destinations

## Proceedings of the Tourism Outlook Conference 2015

Springer

*Editors*
Akhmad Saufi
Economics and Business Faculty
Mataram University
Mataram
Indonesia

Imanuella R. Andilolo
Economics and Business Faculty
Mataram University
Mataram
Indonesia

Norain Othman
Universiti Teknologi MARA
Shah Alam
Malaysia

Alan A. Lew
Northern Arizona University
Flagstaff, AZ
USA

ISBN 978-981-10-1716-2 ISBN 978-981-10-1718-6 (eBook)
DOI 10.1007/978-981-10-1718-6

Library of Congress Control Number: 2016943455



Printed on acid-free paper

This Springer imprint is published by Springer Nature
The registered company is Springer Science+Business Media Singapore Pte Ltd.

# Preface

Sustainable development has been at the core of place and environmental management since it was introduced through the UN's World Commission on the Environment and Development (WCED) report, *Our Common Future*, in 1987. For many places, however, creating sustainable communities and environments seems to be continually beyond the reach of our achievements. Part of this challenge is because the concept of sustainable development includes both the goals of economic development and the often contradictory goals of environmental and cultural conservation. As such, sustainable development in general, and sustainable tourism in particular, is inherently contested and has outcomes that typically result in winners and losers.

The papers in this volume explored these issues from a global south perspective. They were selected from the highly successful Topical Tourism Outlook Conference, held on the island of Lombok in Indonesia, during July 29–31, 2015. The conference built on the first Sustainable Tourism Symposium organized by the University of Mataram (UNRAM) in July 2014 to initiate the expansion of their tourism studies program on the island of Lombok. It also incorporated the eighth conference in the Tourism Outlook Conference series, organized by the Universiti Teknologi Mara (UiTM), Malaysia, and the fourth conference in the Tropical Coastal and Island Tourism series, sponsored by Northern Arizona University and the International Geographical Union's Commission on Tourism, Leisure and Global Change.

The Lombok conference focused on the challenge of growth and conservation in communities seeking a better future that includes tourism. The island of Lombok, Indonesia, with its diversity of natural environments and social contexts, provided an ideal venue to address these issues.

The editors of this volume appreciate the efforts of the conference organizers and sponsors, which included:

- University of Mataram (UNRAM), Faculty of Economics, Indonesia
- Universiti Teknologi Mara, Shah Alam (UiTM), Faculty of Hotel and Tourism Management, Malaysia

- Northern Arizona University (NAU), Department of Geography, Planning and Recreation, USA
- James Cook University (JCU), School of Business, Australia and Singapore
- International Geographical Union (IGU), Commission on the Geography of Tourism, Leisure and Global Change
- *Tourism Geographies* journal, Routledge, UK
- Sabaragamuwa University, Department of Tourism Management, Sri Lanka

The selected chapters provide a platform for tourism scholars, hospitality and tourism industry practitioners, public and private land managers, community development workers, and others interested in tourism and development issues to share, exchange and debate ideas and knowledge related to our understanding and management of tourism and conservation. In addition, the papers have a special focus on tropical environments in South and Southeast Asia. The tropics comprise some of the richest and most complex ecosystems on the planet. The diversity and density of resources that exist in tropical environments means they are increasingly impacted by global economic and social forces, including economic development pressures that are often tourism and recreation related. In addition, they are being impacts rapid population growth that is driven by economic migrants, natural resource extraction, and the relentless onslaught of industrial fishing, global climate change, and rising sea levels. These pressures create enormous challenges for tropical communities due to the global scope of the problems, an inadequate understanding of the issues, and a paucity of effective management tools.

The authors and researchers who contributed to this book have a passion for understanding tourism in the context of these global issues. They, and the editors of this volume, appreciate the opportunity to contribute to this global discussion to help create a better future for our communities and for our world.

Flagstaff, AZ, USA Alan A. Lew, Ph.D., AICP

# Contents

# Editors and Contributors

## About the Editors

**Akhmad Saufi** is Senior Lecture in the Faculty of Economic and Busines, Mataram University Lombok Indonesia. He is a Doctor of Philosophy of International Tourism and Hospitality Management, and teaches Tourism Management. He is also Director of Tourism Study Centre of Economic and Business Faculty of Mataram University.

**Imanuella R. Andilolo** is Lecturer at the Management Department in Faculty of Economics and Business, University of Mataram, in Lombok, Indonesia. She obtained her master degree from University of Tilburg in the Netherlands. Her main research interest is tourism.

**Norain Othman** is Associate Professor in the Faculty of Hotel and Tourism Management, Universiti Teknologi MARA Puncak Alam, Selangor, Malaysia, and her areas of specialization are tourism development and tourism marketing and also event management. Her research focuses on sustainable community based on tourism and also Islamic Tourism.

**Dr. Alan A. Lew** is a Professor in the Department of Geography, Planning, and Recreation at Northern Arizona University where he teaches courses in geography, urban planning and tourism development. Dr. Lew's research interests focus on tourism in the Asia-Pacific region, particularly in Southeast Asia and East Asia, with topics covering community tourism planning, tourist movements, sustainable tourism, heritage tourism, tourism geography. He has been a Fulbright Scholar at the Universiti Teknologi MARA in Sabah, Malaysia, a Research Fellow at National Donghua University in Taiwan, a Concurrent Professor at Nanjing University, and an External Examiner at the Chinese University of Hong Kong. Professor Lew has authored over 80 articles and book chapters and has published several books, including *Understanding and Managing Tourism Impacts: An Integrated Approach* (2009), *Tourism Geography*, 3rd edition (2014), and *World Regional Geography*,

2nd edition (2015). He is the founding editor-in-chief of the journal, *Tourism Geographies* and is a Fellow of the International Academy for the Study of Tourism.

## Contributors

**Adam Abidin** is a working professional focusing on hospitality management. He has worked in the food and beverage industry for over a decade and holds a Master's degree in International Tourism and Hospitality Management from James Cook University. He currently oversees a restaurant and bar business as an entrepreneur.

**Fuadi Afif** is a Masters student in the Tourism Studies Program, UGM, Yogyakarta, Indonesia. His research interest is photography, animals and the environment.

**Johanna Adlin Ahmad** is Lecturer of Faculty of Hotel and Tourism Management in Universiti Teknologi MARA in Pulau Pinang, Malaysia. She has been serving the university for the past 8 years teaching Diploma in Tourism Management students. She has previously produced papers on community-based tourism, cultural tourism, and tourism interests in Sarawak of Borneo.

**Ismail Ahmad Ph.D.** is Professor in the Faculty of Business Management, Universiti Teknologi MARA, Malaysia. He obtained his Ph.D. in Economics from Universiti Kebangsaan Malaysia, UKM (2005), M.B.A. (Finance) from State University, Kentucky, USA (1987), and B.Sc. (Finance) Minor (Econs) from Indiana State University, Indiana, USA (1985). Currently, he is attached to Arshad Ayob Graduate Business School. He is also serving as a visiting professor at University of Sistan and Baluchistan, USB, Zahidan, Iran.

**Semra Günay Aktaş** is Professor in the Tourism Faculty, Department of Tour Guiding, Anadolu University. She received her Ph.D. from the Ankara University, Geography Department. Semra Günay Aktaşis is Dean of the Tourism Faculty, Anadolu University. Her research focuses on tourism geography.

**N. Aminudin** is Senior Lecturer and Head of the Department of Tourism Studies, Faculty of Hotel and Tourism Management, University of Technology MARA, Malaysia at Puncak Alam Campus where she teaches courses on tourism development and tourism economics. Her research focuses on sustainable community based tourism in Southeast Asia.

**Luh Micke Anggraini** is Lecturer of Hospitality and Tourism at Sekolah Tinggi Pariwisata Nusa Dua Bali Tourism Institute, a state-owned Tourism College coordinated by the Ministry of Tourism of the Republic of Indonesia. Her research focuses on destination planning and marketing, sustainable tourism, and tourism planning and policy in Bali and Indonesia. Her Ph.D. dissertation at the University of Western Sydney Australia explores place attachment, place identity, and tourism in Bali.

**Faiz I. Anuar** is a Senior Lecturer in the Department of Tourism Management, Universiti Teknologi MARA (UiTM), where he teaches courses on tourism innovation, technology, and event management. He received his Ph.D. from Texas A&M University, US. His research focuses on technology in tourism, augmented reality technology, and social media development in Southeast Asia.

**Khairil Wahidin Awang** is currently Associate Professor at the Department of Management and Marketing, Faculty of Economics and Management Universiti Putra Malaysia. With a Ph.D. from the University of Wales, Aberystwyth, he immersed himself researching in the arena of tourism geography. With an interest in sustainable tourism development, he seeks to explore the challenges posed by tourism-related SMEs along the lines considered conforming to the virtues of sustainability.

**Azila Azmi** is Lecturer of Tourism in the Faculty of Hotel and Tourism Management, Universiti Teknologi MARA Pulau Pinang. Her research interests are cross-border tourism, shopping tourism, and tourism supply chain management.

**Abu Bakar Webb** is a practicing lawyer with his own legal firm in Miri, Sarawak, Malaysia. He holds a Masters in Environmental Law. He is active in giving talks and discussions on environmental legal issues in several universities, departments and the social media.

**Ugra Bawono** is from Indonesia. He completed his undergraduate studies in Indonesia and pursued his studies Masters in Business Administration at James Cook University in Singapore.

**Hamid Edruce** are students in the Department of Tourism Management, Faculty of Hotel and Tourism Management, Universiti Teknologi MARA (UiTM) Kampus Puncak Alam, Selangor.

**Arni Abdul Gani** is Lecturer at the Department of Tourism Management Universiti Teknologi MARA. She received her Master of Tourism from James Cook University and currently a Ph.D. candidate at Universiti Putra Malaysia. Her research interests include tourism in protected areas, sustainable tourism practices and community participation in tourism.

**Myra P. Gunawan** is Researcher at the Tourism Planning and Development Center, Bandung Institute of Technology and leader of consultant team in the development of the Detailed Plan for Toraja. Myra P. Gunawan had also served as the Deputy Minister of Culture and Tourism, Republic of Indonesia.

**Nur Farihin Abd. Hadi Khan** obtained her Masters in Tourism Management from the Department of Hotel and Tourism Management at Universiti Teknologi MARA, Malaysia. She is currently Lecturer at the Department of Entrepreneurship and Business at Universiti Malaysia Kelantan (UMK). Her research interests are in the area of tourism management and hospitality management.

**Suhaida Halamy** is Lecturer in the Faculty of Information Management, UiTM Sarawak, Malaysia. She teaches courses on records management and information management. She has a great passion for the study of oral history and archives as a part of her teaching field.

**Ibrahim Abdul Hamid** currently holds the position as Chief Executive Officer and Dean at Kulliyyah of Hospitality and Halal Services, Insaniah University College. He has a proffessional qualification from Corporate Member of Hotel Catering Institutional Management Association (MHCIMA-UK).

**MaisarahAbd Hamid** is Lecturer of Faculty Hotel and Tourism Management in UniversitiTeknologi MARA Terengganu, Malaysia. She currently a Ph.D. candidate at Graduate School of Business, UniversitiSains Malaysia, Penang Malaysia. Her interest is on Tour operator involvement in sustainable tourism and tourism marketing.

**Mohd Hafiz Hanafiah** obtained his Master of Economics from Universiti Malaya in 2007. He is currently working as Senior Lecturer at the Department of Hotel and Tourism Management at Universiti Teknologi MARA, Malaysia. He is currently attending the Institute of Postgraduate Studies at Universiti Teknologi MARA Malaysia, undertaking his Ph.D. in Tourism Economics. He has won several awards in innovation competitions, presented numerous papers at academic conferences, and continues to be an active researcher in the field of tourism economics, marketing, and tourism planning.

**Norha Abu Hanifah** is Associate Professor in the Faculty of Law, Universiti Teknologi MARA, Shah Alam, Malaysia. She holds a Doctorate in Biodiversity Management. She teaches environmental law and is active in promoting environmental awareness.

**Mahesh Hapugoda MA** is attached to the Department of Languages, Sabaragamuwa University of Sri Lanka and currently serves there as Senior Lecturer in English. His research interests include Existential Literature, Postcolonial Theory, Frankfurt School Thoughts, Modern Literary Criticism and Zizek Studies and his interest in tourism literature derives from its early literary-sociological foundations initiated by Dean MacCannell, Daniel J. Boorstin, Jonathan Culler and others.

**Mohamad Abdullah Hemdi Ph.D.** is Associate Professor at the Faculty of Hotel and Tourism Management, Universiti Teknologi MARA, Malaysia. His research interest is in the area of human resources management, particularly in the aspects of hotel employees' attitudes and work-related outcomes. He has published his works in some international journals and conference proceedings. He is also the editorial board members for the Journal of Hospitality, Tourism, and Culinary Arts and Journal of Management and Science.

**Sanerya Hendrawan** is Lecturer in the Department of Business Administration, Parahyangan Catholic University, Bandung, West Java, Indonesia where he teaches courses on strategic and business development. His research focuses on the strategic development of business and entrepreneurship.

**Syamsul Herman Ph.D.** is Senior Lecturer in the Department of Recreation and Ecotourism, Faculty of Forestry, Universiti Putra Malaysia, 43400 UPM Serdang, Selangor, Malaysia where he teaches courses on wildlife recreation, introduction to outdoor recreation and tourism, principles of economics and recreation economic. His research focuses on responsible rural tourism network in Malaysia.

**Mohd Salmi Isa** is a Senior Lecturer at Graduate School of Business. She gained a PhD from the University of Hull in the United Kingdom. Having worked for past years on the service marketing, recently she has been researchon sustainable marketing surrounding Corporate Social Responsibility (CSR), neuromarketing, marketing innovation, green marketing and tourism marketing.

**Mohd Noor Ismawi Ismail** is Lecturer from the Department of Hotel and Tourism Management at Universiti Teknologi MARA, Malaysia. He obtained his Master of Science in International Hotel Management from the University of Surrey, United Kingdom in 2009. His research interests are in the area of hospitality management, boutique hotels, online bookings, customer satisfaction, and tourist's intention behavior.

**Orpha Jane** is Lecturer in the Department of Business Administration, Parahyangan Catholic University, Bandung, West Java, Indonesia where she teaches courses on strategic and business development. Her research focuses on the strategic development of business and entrepreneurship.

**Yan Jiang** is Associate Professor in the Department of Tourism, Zhejiang International Studies University in Hangzhou, Zhejiang Province, China, where she teaches courses on tourism psychology and tourism culture. Her research focuses on urban leisure and tourism management. Yan Jiang received her Ph. D. from the Graduate School of Chinese Academy of Social Sciences (CASS).

**Shazali Johari Ph.D.** is Senior Lecturer, associated with the Faculty of Forestry, Universiti Putra Malaysia, 43400 UPM Serdang, Selangor, Malaysia. His research focuses on community development and sustainable indigenous based tourism in Malaysia.

**Regina Jokom** is Lecturer in Hotel Management Program, Petra Christian University, Surabaya, Indonesia. She received her Master of Science (M.Sc.) in Innovative Hospitality Management from Ramon Llull University, Spain. She teaches on innovation and strategic management, and she focuses her research on customer behavior at hotel and restaurant, and innovative in hospitality industry.

**Nia Juliawati** is Lecturer in the Department of Business Administration, Parahyangan Catholic University, Bandung, West Java, Indonesia where she teaches courses on organizational behavior. Her research focuses on business organizational development and entrepreneurship.

**Yeliz Mert Kantar** is Associate Professor in the Faculty of Science. She received her Ph.D. from the Anadolu University, Department of Statistics.

**Devi Roza K. Kausar** is Assistant Professor at the Faculty of Tourism, Universitas Pancasila Jakarta. She was involved in the preparation of the Detailed Plan for National Strategic Tourism Area of Toraja as one of the tourism experts.

**Silverina Annabelle Kibat** is Lecturer in the Faculty of Hotel and Tourism Management, UiTM Sabah, Malaysia. She teaches courses on tourism management, specializing on tour planning, culture and tourism economics. Her previous researches were mostly on cultural heritage.

**Monika Kristanti** is Lecturer in Hotel Management Program, Petra Christian University, Surabaya, Indonesia. She got her Master of Art (M.A.) in International Hospitality Management from London Metropolitan University, London. She focuses her teaching on food and beverage service. She likes to do research on hospitality industry topics, especially related to food and beverage service and green hotel or restaurant.

**Puvaneswaran Kunasekaran Ph.D.** is a Postdoctoral Fellow in the Institute of Agricultural and Food Policy Studies, Universiti Putra Malaysia, 43400 UPM Serdang, Selangor, Malaysia where his research focuses on indigenous tourism and community development studies.

**Bet El Silisna Lagarense** is Senior Lecturer at Tourism Department, Manado State Polytechnic where she teaches courses on Hospitality and Tourism research, Tourism Information System, and the director of Manado Tourism Institute where she manages to develop courses on geography of tourism, tourism planning, tourism environment and marine tourism. Her research focuses on ecotourism, urban tourism and waterfront tourism. She received her Ph.D. from University of Waterloo, Canada and recently published in Asia Pacific Journal of Tourism Research published by Routledge, 2014.

**Resti Meilani** is Junior Lecturer in the Department of Forest Resources Conservation and Ecotourism, Faculty of Forestry, Bogor Agricultural University (IPB). She is assigned to the Outdoor Recreation and Ecotourism Division, and is a member of the teaching team for several undergraduate program courses, i.e., Outdoor Recreation and Ecotourism, Conservation Education, and Nature Interpretation. Her research focuses on the educational aspects of tourism development in Indonesia.

**Abdullah Mohamad** is currently Associate Professor at the Department of Recreation and Tourism, Faculty of Forestry Universiti Putra Malaysia. He obtained his Ph.D. from University of Aberdeen and has since pursue his career and research interests in the area of park and recreation management. He has conducted and managed various researches, concentrated in park management, the relationship between tourism and local community, including native people and tourism in protected areas.

**Endang Kustati Sri Harini Muntasib** is Professor in the Department of Forest Resources Conservation and Ecotourism, Faculty of Forestry, Bogor Agricultural University (IPB). She teaches various undergraduate and graduate program courses related to tourism, such as Outdoor Recreation and Ecotourism, Nature Interpretation, Conservation Education, and Ecotourism. She is the Head of the Outdoor Recreation and Ecotourism Division, and her research focuses on tourism development in Indonesia.

**Perunjodi Naidoo** is a Senior Lecturer at the School of Sustainable Development and Tourism, University of Technology, Mauritius and was previously Head of the Tourism, Leisure and Services Department. She was awarded a PhD in tourism in 2014. She is particularly intrested in tourism development in small island development states, the well-being of destination communities, tourism in emerging markets, and service management.

**Jordan Napitupulu** hails from Indonesia. He did his Bachelors in Business and Masters in Business Administration from James Cook University. He is currently based in Jakarta.

**Arissetyanto Nugroho** is the Rector of the Mercu Buana University, Jakarta, Indonesia where he teach management. His research focuses on hospitality and tourism marketing.

**Aknolt Kristian Pakpahan** is Lecturer in the Department of International Relations, Parahyangan Catholic University, Bandung, West Java, Indonesia, where he teaches courses on international politics and economics. His research focuses on the strategic development of community-based small business and entrepreneurship.

**Jenny Panchal** is Lecturer in tourism at James Cook University in Singapore. She worked in the resort industry before obtaining her Masters in Tourism Management at Victoria University of Wellington, New Zealand and Ph.D. in Tourism at James Cook University in Townsville, Australia. Her research interests lie in Asian spa and wellness tourism, tourist motivation and positive psychology.

**Gandhi Pawitan** is Lecturer in the Department of Business Administration, Parahyangan Catholic University, Bandung, West Java, Indonesia, where he teaches courses on Social and Business Statistics. His research focuses on sustainable development of community-based small business and entrepreneurship.

**Tri Kuntoro Priyambodo** is the Secretary of the Program of Postgraduate Tourism Studies, Universitas Gadjah Mada, Daerah Istimewa Yogyakarta, Indonesia. He is also Associate Professor of Computer Science and Electronics, Universitas Gadjah Mada, Daerah Istimewa Yogyakarta, Indonesia.

**Mohit Raghav** is an entrepreneur and CEO of World Lifestyle Pte. Ltd. He has been in the luxury travel trade for over 5 years after receiving his Bachelor's degree from James Cook University, Australia. His research interests are in lifestyle and travel experiences.

**Handayani Rahayuningsih** is a lecturer at the Tourism Diploma Program, Vocational School, Universitas Gadjah Mada, Daerah Istimewa Yogyakarta, Indonesia. She is interested in tourists behavior, photography, and food tourism researches.

**Sridar Ramachandran Ph.D.** is Associate Professor in the Department of Recreation and Ecotourism, Faculty of Forestry, Universiti Putra Malaysia, 43400 UPM Serdang, Selangor, Malaysia, where he teaches courses on parks, recreation, tourism development and tourism product and marketing. His research focuses on responsible rural tourism network in Malaysia.

**Rafidah Aida Ramli** is Senior Lecturer in the Faculty of Hotel and Tourism Management, Universiti Teknologi MARA Pulau Pinang, where she teaches courses in the foodservice department. Her research interest includes community nutrition, product development and food safety.

**Ikma Citra Ranteallo** is Lecturer in the Department of Sociology, Faculty of Social and Political Sciences, Universitas Udayana, in Denpasar, Indonesia. Her main research focuses on sociology of food, media, and new social movements.

**Iraj Ratnayake Ph.D.** is currently attached to the Department of Tourism Management, Sabaragamuwa University of Sri Lanka, as Senior Lecturer. His literal presentations mainly encompasses tourism planning and development. Besides research and writing, he also participates actively in training and consultation projects include tourism information systems, site planning, special interest tourism, visitor management and community tourism.

**Rusdan** is Senior Lecturer in the Faculty of Economic and Business, Mataram University Lombok Indonesia. He is a Master of Science in Marketing, and teaches Marketing Management. His research interest includes consumer behavior, tourism marketing and Islamic marketing.

**Sabrina Samsudin** is a master student from Faculty of Hotel and Tourism, Universiti Teknologi MARA, Puncak Alam, Selangor, Malaysia. Currently she is pursuing Master in Tourism Management majoring in marketing.

**Ahmad Shuib Ph.D.** is Professor in the Institute of Agricultural and Food Policy Studies, Universiti Putra Malaysia, 43400 UPM Serdang, Selangor, Malaysia, and his research focus includes responsible rural tourism network in Malaysia.

**Janfry Sihite** is the Head of Future Lab Indonesia where he is developing the capabilities of data analysis on social network application. He is currently a Ph.D. candidate at the University of Indonesia.

**Sulhaini** is Associate Professor in Faculty of Economic and Business, Mataram University Lombok Indonesia. She is a Doctor of Philosophy of Marketing, and teaches International Marketing. Her marketing research encompases relationship marketing, export behavior, brand origin knowledge tourism marketing, and Islamic marketing.

**Norzuwana Sumarjan Ph.D** is Senior Lecturer and the Head of Graduate Studies at the Department of Hotel and Tourism Management at Universiti Teknologi MARA, Malaysia. She graduated with her doctoral degree (Ph.D.) from Iowa State University, United States. Her research interests are in the area of hospitality quality management. She has published her work in several international journals and conference proceedings.

**Arzyana Sunkar** is Assistant Professor in the Department of Forest Resources Conservation and Ecotourism, Faculty of Forestry, Bogor Agricultural University (IPB). She teaches courses on Outdoor Recreation and Ecotourism, and Protected Area Management. Her research is on protected area management, conservation socio-psychology, community empowerment, and karst and cave management.

**K. Thirumaran** is Senior Lecturer specializing in tropical tourism issues. He worked in the tourism and hospitality industry before pursuing his doctorate at the National University of Singapore. His research interests include cultural tourism, policy and destination marketing. He is currently a Senior Lecturer at James Cook University, Singapore.

**Trianasari** is Lecturer at Diploma 3 Hotel Program at the Faculty of Economics and Business of Ganesha University of Education. Her research interests are tourism and hospitality service management, CSR, and business ethics. She teaches business ethics, management, and other hotel/hospitality related subjects.

**Agustinus Walansendow** is Senior Lecturer at Business Administration Department, Manado State Polytechnic, where he teaches business project, business statistics, Organisational Behaviour, Operational Management. He is a candidate of doctor of Economics at Sam Ratulangi University, Manado, Indonesia.

**H. Wee** currently served as Senior Lecturer at Faculty of Hotel and Tourism Management, University of Technology MARA, Malaysia at Puncak Alam Campus. Her major field of study is tourism management. She is currently at the final stage of completing her doctoral study.

**Yulia A. Widyaningsih** is Lecturer in the Faculty of Economics and Business UGM. Her research interest is in tourism marketing, cultural heritage tourism, and sustainability.

**Nuri Wulandari** is Lecturer and staff at the Department of Research, Indonesia Banking School at Jakarta, Indonesia where she teaches courses on Marketing. She received her Doctoral degree from University of Indonesia. Her research interests are consumer behavior and brand experience related to tourism and banking industry.

**Astri Yulia** is Lecturer in the Faculty of Business, Universiti Selangor. She holds a Ph.D. from Texas A&M University, US, and her research focuses on teaching, learning and culture, and research methodology.

**Deniz Yüncüis** is Assistant Professor in the Tourism Faculty, Department of Tourism Management Tourism, Anadolu University. Her research focuses on destination and event marketing and marketing research in tourism.

**Adi Yuniarta** is Lecturer at Diploma 3 Accounting Program at the Faculty of Economics and Business of Ganesha University of Education. He is an active researcher in the field of CSR, finance, and accounting. Besides research and teaching, he has involved in a great amount of social service activities.

**Kwartarini Wahyu Yuniarti** is Professor of Clinical Psychology at the Faculty of Psychology, Universitas Gadjah Mada, Daerah Istimewa Yogyakarta, Indonesia. Her researches focuses on health psychology, cross-cultural psychology, and indigeneous psychology.

**N.A.A. Zahid** is a Master of Science in Tourism Management graduate from University of Technology MARA, Malaysia. She is currently looking for a teaching opportunity in local universities. Her area of interests is in the field of event tourism.

# Chapter 1
# Introduction: Balancing Sustainability and Development in Tropical Tourist Destinations

**Akhmad Saufi**

**Abstract** The issues of conservation, restoration and sustainability in tropical environments are critical for the health of the world. As tourism industry continues to grow, it plays important role in the development of tropical countries, and benefits from the conservation and sustainability of the cultural and environmental resources. The issues of managing the balance between the development of tourism in tropical environments and the sustainability of traditional and sensitive places have inspired the conduct of an international tourism conference on 29–31 July 2015 in Lombok, Indonesia. A number of authors looked at various constituents of tourism sector and analyzed each of those from a sustainability standpoint. The tourism development and its sustainability were addressed in three main parts: (1) ecotourism, local culture, and marketing issues, (2) local community participation and tourist behavior issues; and, (3) sustainability and economic development in tourism.

**Keywords** Tourism development · Sustainability · Tropical countries · Conservation · Community participation

Tropical environments comprise some of the richest and most complex ecosystems on the planet. The diversity and density of resources that exist in tropical environments means they are increasingly impacted by global economic and social forces, including economic development pressures that are often tourism and recreation related. They are also being impacted by rapid population growth that is driven by economic migrants, major resource extraction in the form of oil and gas drilling and relentless industrial fishing, and global climate change and steadily rising sea levels. These pressures create enormous challenges for tropical communities due to the global scope of the problems, an inadequate understanding of the issues, and a paucity of effective management tools.

A. Saufi (✉)
University of Mataram, Mataram, Indonesia
e-mail: akh.saufi72@gmail.com



A. Saufi et al. (eds.), *Balancing Development and Sustainability in Tourism Destinations*, DOI 10.1007/978-981-10-1718-6_1

Given this context, the issues of conservation, restoration and sustainability in tropical environments are an enormously pressing concern for the health of the world. Tourism has an important role to play in addressing these issues because it is both an agent of development and a cause of change, while at the same time it benefits from the conservation and sustainability of the tropical world's cultural and environmental resources. How can we keep the balance between the development of tourism in tropical environments and the sustainability of those traditional and sensitive places?

This was the broad question that drove the undertaking of an international tourism conference that was held 29–31 July 2015 on the island of Lombok, Indonesia. This is also the question that the authors who have contributed to this book are trying to answer. The book takes tourism sustainability beyond the realms of external factors that matter to a single industry. The authors look at various constituents of the tourism sector and analyze each of those from a sustainability standpoint. The book includes articles and case studies that are global in nature, though with an emphasis on Asia, and that show how sustainable applications can be used and how concerns can be addressed.

The book consists of 36 chapters which are grouped into three parts. The first 12 chapters in part one discuss issues of ecotourism, local culture, and the marketing of natural and cultural products of tourism. There are linkages between ecotourism and biopiracy, and the need to protect indigenous and local communities from the practices of biopiracy (Chap. 2). Related to this, eco-friendly attitude are increasingly influencing tourist preferences to stay in green hotels during their holidays (Chap. 3). The conservation of culture through tourism in the context of Malaysia and several tourist destinations in Indonesia is discussed in Chaps. 4–7. Furthermore, various marketing issues related to *halal* tourism, tourist buying behaviour, and packaging are discussed in Chaps. 8–10. Movie touring and culinary tourism are issues discussed in Chaps. 11–13.

Part two includes 11 chapters that cover various issues on local community participation in tourism developmentand tourist behavior. Leisure activities and leisure skills of residents are discussed in the context of Hangzhou, China (Chap. 14 ).Demographic issues (Chaps. 15 and 16), the roles of community social capital in the development of ecotourism (Chap. 17), and the contribution of tourism to local wellbeing (Chaps. 18 and 20) provide insight into tourism and community development. Other chapters discuss the community participation in tourism planning (Chap. 21), dark tourism and its impacts on local economies and a community's memory (Chap. 22), cluster-based development models (Chap. 23), and Islamic tourism (Chap. 24).

Part three comprises 12 chapter that present a number of topics related to the sustainability and economic development in tourism. The chapters focus on issues related to economic impacts, such as tourism products and supply chains in the tourism industry (Chaps. 26 and 27), the competitiveness of tourism destinations (Chaps. 34 and 35), industry contributions to tourism sustainability (Chap. 28), the distribution of tourists in a destination (Chaps. 32 and 33), and marketing issues in tourism (Chaps. 25, 29, 30, 31 and 36).

Although much of evidence presented in this book is from tourism destinations in Asia, particularly China, Indonesia, Malaysia, Singapore, and Sri Lanka, many of lessons learned in these case studies apply to other tourism destinations in the world. The adoption of technology in tourism is overshadowing differences in management style and marketing strategies in developed and developing world tourism destinations. The ever increasing demand for travel across the globe is making the industry more sophisticated and is intensifying competition among the many places seeking tourist expenditures. Therefore, it is more important now than ever to understand the dynamic development of the tourism industry, which can only be accomplished through comprehensive studies by many different researchers in many different countries, as provided in this book.

# Part I
# Balancing Sustainability in Environmental and Cultural Tourism

**Norain Othman**

University of Technology MARA, Shah Alam, Malaysia

Concerns about the impact of economic activity on the environment have been an issue of importance since at least the dawn of the Industrial Revolution in Europe over 200 years ago. The wave of environmental concerns in the late twentieth century led to new environmental legislation, consumers demanding greener products, and competitors seizing competitive advantage by improving their environmental performance. Today, most companies are forced to face the challenge of going green. Environmental crises, such as the climate change, the discovery of the hole in the ozone layer, the rise of ocean temperatures causing massive coral reef destruction, and the sudden disruptions of melting glaciers, are some of the global concerns our contemporary world. Many of the answers to these problems have been grounded in the principles of sustainability, although that also worries many interested parties, including governments and big businesses.

The quest for sustainability has been difficult and achieving it may be too little and too late. Some suggest that we need to move beyond sustaining environments and supporting human development, and instead we should focus on restoring the environment for the future generations. How does one balance sustainability within an environment and culture in this context? Who is responsible for protecting or restoring the environment, heritage and culture within the local community at a tourist destination? All the stakeholders in tourism (including governments, tour and transport operators, community service providers, NGOs, tourists, local residents, and tourism associations) are responsible for achieving the goals of sustaining natural habitats, heritage sites, scenic landscapes and local cultures.

Balancing sustainability in tourism means to maintain a low impact on the natural environment and local culture, while helping to generate income and employments for the local community through the development and promotion of the tourism industry. A holistic and sustainability approach to this would involve strategies for a 'green economy' that are built on the three principles of low carbon, resource efficiency, and social inclusion. Sustainability indicators must involve notions of humanity, equity and social inclusion. Sustainable development is understood here to imply sustainable human development because development is ultimately about people and human societies are ultimately dependent upon their natural environment.

Hanifah and Webb (Chap. 2) discuss various legal issues and regulations about ecotourism and biopiracy in connection with the sustainability of tourism destination. International ecotourism guidelines started in the early 1990s with the establishment of The International Ecotourism Society (TIES), and were followed in 1992 by the United Nations Rio Earth Summit's environmental initiatives. Various guidelines and legal framework on ecotourism were introduced by The Convention on Biological Diversity (CBD) 1992, the Berlin Declaration on Sustainable Tourism 1997, and international bodies such as United Nations Environment Programme (UNEP) and the World Travel Tourism Council (WTTC). The chapter further examines the Malaysian Ecotourism Policy and the implementation of guidelines and strategies for sustainable tourism development in Malaysia.

Kristanti and Jokom (Chap. 3) cover the eco-friendly attitude and intention of guests to visit green hotels. The increasing consumer awareness of environmental problems has created a more positive attitude towards green activities and more involvement in daily eco-friendly behavior. Cultural heritage includes tangible elements, such as buildings, rural landscapes and villages, cities, art collections artifacts in museums, historic gardens, handicrafts and antiques. But it also includes intangible culture, such as music, dance, beliefs, ceremonies, rituals and folklore. Halamy and Kibat (Chap. 4) discuss on the importance of oral history in preserving intangible cultural heritage for future generation.

Kausar and Gunawan (Chap. 5) examine efforts to revitalize tourism planning efforts in the National Tourism Strategic Area of Toraja, which is a strategic tourism development and management area in Indonesia. The revitalization process is shown to include the diversification of tourist attractions beyond traditional elements and settlement, setting quality standards and improving information and communication technology (ICT) capacities among tourism service providers, and enhancing interpretations and the visitor management systems at Toraja's heritage sites. Anggraini (Chap. 6), Hillman et al. (Chap. 7), Anuar et al. (Chap. 9) and Johari et al. (Chap. 10) each discuss the restoration of place identity among residents and indigenous peoples, thus promote the culture and tradition of the local communities within tourism destinations. This often takes advantage of the interests of international tourists in purchasing souvenirs as gifts, evidence of having been there, and as memories of visiting a destination. Local community uses this opportunity to produces local handicraft and traditional crafts through cottage industries targeted to the international tourists.

As one of the more dynamic regions of global tourism growth, many new and emerging tourism research areas are developing in Asia. Lombok, Indonesia, for example, is known as the 'island of a thousand mosques' and Sulhaini et al. (Chap. 8) suggest that it can be marketed as a significant Islamic and *halal*-friendly tourism destination. Islamic tourism marketing is a new concept that needs further understanding as to how the process can develop under guidelines aligned with the Holy Quran and *Hadith*. In a different vein, Ismail et al. (Chap. 11) demonstrate how screen and film tourism plays a significant role in attracting and motivate tourist to visit movie locations that they are familiar with. Abidin et al. (Chap. 12) show how culinary tourism is being made increasingly popular through its promotion by celebrity chefs who market signature restaurants, cuisines, locations, menus and ambiance. Celebrity chefs enhance the tourist experience, with food as the focus of the attraction. On the other hand, Ranteallo and Andilolo (Chap. 13) show how every day tourists can also promote food tourism through social media through use of the hashtags:#foodgasm and #foodporn. By tracing the use of these terms, the impact of culinary tourism can be seen to increase food innovation and the culinary arts, along with cultural identities, which in the long run is another example of how tourism can contribute to balancing development and cultural sustainability in a destination.

# Chapter 2
# Ecotourism and Biopiracy: A Legal Perspective on the Sustainability of Tourism Destination in Malaysia

**Norha Abu Hanifah and Abu Bakar Webb**

**Abstract** Malaysia has the potential to become one of the top destinations for ecotourism, because it is one of the world's 12 mega diversity hotspots. Ecotourism is a growing niche market within the larger travel industry, with the potential of being an important sustainable development tools. However, without proper regulatory control, ecotourism may facilitate biopiracy. The issue of ecotourism and biopiracy has been the subject of major debates. Biopiracy, if not properly remedied, is detrimental to the interest of Malaysia over its genetic resources and traditional knowledge. The main purpose of this study is to examine the various legal issues pertaining to ecotourism and biopiracy in connection with sustainability of tourism destination in Malaysia, looking in particular at the roles played by the international and national laws.

**Keywords** Ecotourism · Biopiracy · Genetic resources · Traditional knowledge

## Introduction

The United Nation World Tourism Organisation (UNWTO) defines sustainable tourism as "Tourism which leads to management of all resources in such a way that economic, social and aesthetic needs can be fulfilled while maintaining cultural integrity, essential ecological processes, biological diversity and life support systems" (UNWTO 2002). Ecotourism is a subcategory of sustainable tourism. However, there is no single definition of ecotourism. According to The International Ecotourism Society (TIES), ecotourism covers all travels to natural areas that conserve the environment, contributing at the same time to the welfare of

N. Abu Hanifah (✉)
Faculty of Law, Universiti Teknologi MARA, Selangor, Malaysia
e-mail: norha99@yahoo.com

A.B. Webb
Messrs Webb & Co Lutong, Miri Sarawak, Malaysia
e-mail: bhipni@yahoo.com

 
A. Saufi et al. (eds.), *Balancing Development and Sustainability in Tourism Destinations*, DOI 10.1007/978-981-10-1718-6_2

the local people. Ceballos-Lascurain (1996) describes ecotourism as "traveling to relatively undisturbed or uncontaminated natural areas with the specific objectives of studying, admiring and enjoying the scenery and its wild plants and animals, as well as any existing cultural manifestations (both past and present) found in these areas".

Third World Network (TWN) points out that in any event, governments and other concerned parties should be alerted and seriously ponder the question whether it is wise to indiscriminately promote tourism forms that facilitate the stealing and smuggling of local biological resources and traditional knowledge (TK), before necessary legal frameworks and administrative mechanisms are in place to effectively combat abuses and exploitation (TWN 2002). Since Malaysia is positioning itself as a globally competitive ecotourism destination, this inevitably calls for an urgent review of its policies and environmental legislation pertaining to sustainable ecotourism and biopiracy.

## Ecotourism in Malaysia

### *An Overview*

Malaysia is identified as one of the world's 12 mega biological diversity areas. Covering almost 60 % of land mass, Malaysia's tropical rainforests are millions of years old and they are home to an incredibly diverse array of flora and fauna. This tropical forest is a wealth of genetic resources, with an estimated of 15,000 species of flowering plants, 286 species of mammals, 150,000 species of invertebrates and 4000 species of fishes (Tourism Malaysia (TDC) 2008).

Malaysia offers tourists a range of activities in which they may engage to experience and fully appreciate the ecotourism experience such as caving, hiking, jungle trekking, white water rafting, rock climbing, bird watching, diving and river cruising. The Malaysian National Tourism Policy (NTP) was formulated in 1992 by the then Ministry of Culture, Arts and Tourism (MOCAT) to develop the tourism industry. The NTP incorporated necessary guidelines for tourism destination development and ecotourism, has been identified as one of the sustainable tourism form in this plan.

Since ecotourism supports local economies, the National Ecotourism Plan was prepared in 1996 by the Worldwide Fund for Nature Malaysia (WWF Malaysia) for MOCAT. It contains strategies and guidelines for the development and management of ecotourism in Malaysia. In 2004, the MOCAT was split to facilitate the establishment of a separate ministry responsible solely for matters related to tourism, i.e. the Ministry of Tourism (Hamzah 2004).

Malaysia has a significant stake for harnessing the potential of biotechnology and bioprospecting for achieving sustainable economic development. However, countries with valuable assets of genetic resources and TK are at risk because there are concerns that the substances from the genetic resources and TK are being used and patented by third parties, with few or none of the benefits being shared with the original TK-holders and without their prior informed consent (Twarog and Kapoor 2004).

## Regulating Ecotourism

### *The International Regime*

The ecotourism international regime gained momentum in the early 1990s due to the establishment of TIES in 1990. This was followed by the 1992 United Nations Rio Earth Summit (Rio Declaration on Environment and Development) where 178 countries signed a number of environmental initiatives including Agenda 21 which is a comprehensive plan of action to be taken globally, nationally and locally. The Convention on Biological Diversity (CBD) 1992 offers decision-makers guidance on ecotourism where its three main objectives are the conservation of biological diversity, the sustainable use of its components of biological diversity and the fair and equitable sharing of the benefits arising out of utilisation of genetic resources.

In 1997 the Berlin Declaration on Sustainable Tourism gave an important baseline for ecotourism and point No. 16 of this declaration stated that:

> Tourism should be restricted, and where necessary prevented, in ecologically and culturally sensitive areas.

In response to concerns that many species were becoming endangered because of international trade, hence, the Convention on International Trade in Endangered Species of Wild Fauna and Flora (CITES) was entered into force in 1975. This convention is particularly important to ecotourism because it has made the industry much more sustainable.

Principle 4 of the UN Rio Declaration on Environment and Development mentioned that "in order to achieve sustainable development, environmental protection shall constitute an integral part of the development process and cannot be considered in isolation from it". The Rio Declaration indicates that the sustainable development of the environment shall be integrated with other development processes. Most importantly, based on the concept of sustainable development, ecotourism should not only protect genetic resources but it should also accommodate and meets the human rights aspect to promote and to protect all resources of human being such as economic, social and cultural integrity.

However, according to Simon (1999), the current legal framework is a patchwork of agreements and treaties that concern trade more than tourism and are often in conflicts. She argues that although international bodies such as United Nations Environment Programme (UNEP) and the World Travel and Tourism Council (WTTC) are moving towards a unified set of guidelines, their implementation will remain problematic due to a lack of systematic measurement and enforcement.

The UN had declared 2002 as the International Year of Ecotourism (IYE). However, critics argue that the UN proclamation of 2002 as the IYE has created a major debate because as already noted, the growing awareness that the ecotourism industry is not as benign as initially believed (TWN 2002). It has been argued that in many studies conducted around the world, ecotourism falls short of the ideals inherent in the principles it promotes conservation of nature and cultures, benefits to local people and local participation.

## *The Malaysian Ecotourism Policy and Legal Regime*

The tourism planning organisation in Malaysia is complex and influenced by the three-tier form of government, i.e. Federal government, State government and Local Authorities (Hamzah 2004). Malaysia is one of the implementer countries of Agenda 21. Chapter 28 of the Agenda 21 clearly binds the local authority to take lead in the implementation of the guidelines and strategies for stakeholders' participation in sustainable tourism development.

Since tourism is a Federal affair hence the overall policy planning is carried out by the Ministry of Tourism. It is important to note that the Secretary General of the Ministry of Tourism is given the responsibility for licensing and enforcement of all matters relating to tourism under the Tourism Industry Act 1992. Section 34 of the Act allows the Minister to make such regulations as he/she may consider expedient for the purpose of the Act. Under this section the following regulations have been made:

- Tourism Industry (Compounding of Offences) Regulations 1992;
- Tourism Industry (Tour Operating Business and Travel Agency Business) Regulations 1991;
- Tourism Industry (Licensing and Control of Tourist Guides) Regulations 1991; and
- Tourism Industry (Licensing of Tourism Training Institution) Regulations 1994.

In East Malaysia, the Sabah Tourism Promotion Corporation is established under the Sabah Tourism Promotion Corporation Enactment 1981. Unfortunately, this legislation has no specific reference to the role of ecotourism or nature-based tourism. In Sarawak, the Sarawak Tourism Board Ordinance was enacted in 1994. In 1995 the Sarawak Tourism Board was established whose functions and responsibilities are similar to those of the Malaysia Tourism Promotion Board and Sabah Tourism Promotion Corporation.

Malaysian government has taken various actions to protect the natural environment and its indigenous people. The Aboriginal Peoples Act 1954 is one of the earliest legislation that has a definite impact on ecotourism. It basically takes the international concept of sustainable development with human rights and nationalises it. Under the National Land Code 1965 it is possible for State governments to reserve land for many different purposes, and this power has been exercised for matters relevant to ecotourism.

A prominent law passed in 1974 to prevent and control pollution and land degradation is the Environmental Quality Act. This law indirectly supports the government's policy on sustainable tourism and legality of ecotourism by way of setting down policies through which the environment can be developed and protected. In Peninsular Malaysia, the National Parks Act 1980 strengthens sustainable ecotourism by providing for the establishment and control of National Parks to preserve and protect wild life, plant life and objects of geological, archaeological, historical and ethnological and other scientific and scenic interest.

The Fisheries Act 1985 provides for the conservation, management and development of maritime and estuarine fishing and fisheries, marine turtles and riverine fishing. This Act is particularly important to ecotourism as activities such as scuba diving, underwater photography and snorkelling are often carried out in marine parks. In addition, the Fisheries (Prohibited Areas) Regulations 1994 declares five areas to be fisheries prohibited areas where no collection of shells, mollusks and corals is allowed, and where a license must be obtained in order to enter the area to kill or capture any fish.

Under its obligation as a signatory to the CBD, Malaysia launched its National Policy on Biological Diversity on 16 April 1998 with the aim to conserve Malaysia's biological diversity and to ensure that its components are utilised in a sustainable manner for the continued progress and socio-economic development of the nation. In 2010, the Wildlife Conservation Act was enacted to replace the 38-year-old Protection of Wild Life Act 1972. This latest Wildlife Act provides significantly higher penalties and mandatory jail terms for wildlife crime. The Act widens the list of agencies empowered to enforce wildlife laws by including Police and Customs officers and it protects more species of wildlife. Unfortunately, this Act is only applicable for the protection and conservation of wildlife in Peninsular Malaysia and the Federal Territory of Labuan. Also in 2010, the International Trade in Endangered Species Act 2008 came into force which has a positive impact on sustainable ecotourism in Malaysia.

## *Ecotourism or Biopiracy?*

There is evidence that a growing number of the Northern-based pharmaceutical corporations, biotechnology companies and their intermediaries are stalking the forests, fields and waters of the developing world often posing as tourists for the purpose of developing patented rare genetic traits of biological riches and indigenous knowledge (ETC Group 1995).

However, proponents for biotechnology argue that the contention that biopiracy is a problem that is rather unsound. Previous research by the Australian APEC Study Centre revealed no instances of the forcible, illegal removal of genetic resources in any jurisdiction. In addition, it did not reveal substantial cases of biopiracy nor "any instance of highly profitable returns from a product developed via the acquisition of genetic resources from developing countries" (Asia-Pacific Economic Cooperation (APEC), 2009). It has been argued that most allegations of biopiracy are so thoroughly riddled with inconsistencies and outright lies that the entire genre, pending further clarification, must be consigned to the realm of "rural" legend (Chen 2006).

Nevertheless, at the 1998 World Travel Mart in London, the World Customs Organization (WCO) warned that this unprecedented illegal global trade in flora and fauna has resulted in vast damages and economic losses (TWN 1999). The WCO in cooperation with the CITES recognised the linkages between tourism and the escalating theft and smuggling in flora and fauna and produced a brochure to raise awareness on this matter (TWN 2002). WCO officials pointed out that customs authorities as well as the travel and tourism industry need to be fully educated on how much damage the illicit biotrade can do to societies, cultures and the environment. According to TWN, despite all warnings, it has been argued that tourism policy-makers, the industry and large conservation organisations tend to ignore this burning issue, probably out of fear that the exposure of these illegal activities in connection with (eco) tourism can create image problems and hurt the funding of projects.

Some famous examples of biopiracy cases which operate through application of patents to genetic resources and TK are as follows:

1. Rosy Periwinkle, a plant native to Madagascar but widely introduced to other tropical countries has long been cultivated for herbal and traditional medicine to treat diabetes, malaria and Hodgkin's disease. Western pharmaceutical companies have begun extracting substances from the plant to treat leukaemia. Accusations of biopiracy have arisen because patents have been placed on these substances, with no compensation paid to indigenous communities.
2. The Neem tree is a major component of traditional Hindu medicine (Ayurveda), and is often prescribed for skin-related conditions. In 1995, US patent office granted patent on an antifungal agent from the seeds of the Neem Tree. The patent was eventually overturned in 2005 after the Indian government responded to a widespread outcry and initiated legal action.

It has been argued that between 25 % and 50 % of current prescription pharmaceuticals come from plants, either directly or through modifications by biochemical methods, and the value of drugs to the U.S. pharmaceutical industry coming from plant species is estimated at over 30 billion USD per year (Zakrzewski 2002).

## Regulating Biopiracy

### *The International Regime*

The three conventions that are most relevant when speaking of biopiracy are the CBD, Nagoya Protocol and Trade-Related Aspects of Intellectual Property Rights

(TRIPS). The preamble of the CBD states that traditional knowledge, innovations and practices are of importance to the conservation of biological diversity and that indigenous and local communities have a close and traditional dependence on biological resources. The two most important articles in the CBD that are relevant in controlling biopiracy are Article 3 and Article 8(j). Article 3 recognises the sovereign rights states have in accordance with the Charter of the United Nations and the principle of international law, "the sovereign right to exploit their own resources pursuant to their own environmental policies, and the responsibility to ensure that activities within their jurisdiction or control do not cause damage to the environment of other States or of area beyond the limits of national jurisdiction". Article 8(j) of the CBD states that signatory states should respect, preserve and maintain knowledge, innovations and practices of indigenous and local communities.

The adoption of Nagoya Protocol on Access and Benefit Sharing (ABS) in October 2010 reaffirms that genetic resources are subject to national sovereignty and offers the opportunity to recognise the rights of indigenous and local communities over their genetic resources and associated TK.

As regards the TRIPS Agreement 1994, its main aim is to give intellectual property rights (IPRs) to inventors through patents. Article 27.1 provides that "patents shall be available for any inventions, whether products or processes, in all fields of technology, provided that they are new, involve an inventive step and are capable of industrial application". Article 27.2 allows Members to exclude from patentability inventions in order to "protect human, animal or plant life or health or to avoid serious prejudices to the environment". However, developing countries are apprehensive that the TRIPS is merely an exploitative mechanism employed to patent indigenous biological material (Abott 1996–1997).

## *Towards Biopiracy Law in Malaysia*

Since it has been argued that there are linkages between ecotourism and biopiracy, the big question is whether the Malaysian environmental laws are adequate in protecting biological resources and TK from biopiracy? Indeed, Malaysia is actively involved in negotiations related to the development of the international regime on ABS. Malaysia aims to have a national law on ABS in place to implement the provisions of Article 15 of the CBD (which deals with access to genetic resources) and the related provisions in the Nagoya Protocol. This will further strengthen Malaysia's ability to deal with emerging issues in biodiversity, alongside the Biosafety Act which was passed in 2007. However, the Biosafety Act which seeks to protect biological diversity from the potential risks posed by living modified organisms is silent on the subject of ABS and TK.

Although no national ABS law currently exists, however, under Section 4 of the Forestry Act 1984, the State Forestry Director is empowered by the State Authority to control the removal of plants or resources from the forest. In addition, under

Section 34 of the Act, researchers have to apply for a license in order to carry out research activities in a permanent forest reserve. The Sarawak state government is the first country in the Federation of Malaysia to adopt access legislation of plant genetic resources in the form of the Sarawak Biodiversity (Access, Collection and Research) Regulations 1998, which is in line with the CBD. Interest and value in having ABS in Sabah are somewhat evident in the provisions of the Sabah Biodiversity Enactment 2000 (SBE 2000). Available resources to formulate and implement its ABS system lie in the SBE 2000, its institutional elements (Sabah Biodiversity Council/Sabah Biodiversity Centre) and Biodiversity Centre Fund. Their proactive positions determined by the fact that some of Malaysia's richest biodiversity are found in these two states.

## *Shortcomings*

The CBD is the key mechanism for promoting international cooperation on protecting nature. It is the first convention to establish the sovereign right of a state over its natural resources and access to those resources (Bautista 2007). Yet the US with its dynamic leadership in conserving global biodiversity has yet to ratify the treaty. Snape (2010) argues that now more than ever, the engagement and leadership of the US is necessary to protect biological diversity and the natural services enjoyed by Americans and others throughout the world. Indeed, the CBD's effectiveness and the urgent cause of stemming the ongoing high rate of global biodiversity loss both suffer from the lack of official involvement and support from the US (Snape 2010).

Some critics argue that there is a potential conflict between the CBD and the TRIPS Agreement (McManis 1998; Tejera 1999). Indeed the TRIPS Agreement is not only in conflict with the CBD but also with the Nagoya Protocol. The CBD and Nagoya Protocol are concerning environment and peoples' right and the TRIPS Agreement are concerning commercial rights. According to the CBD and Nagoya Protocol, countries have the right to regulate access to biological resources and TK, and to determine benefit sharing arrangements. The TRIPS Agreement has no provision for patent holder to share benefits with communities in countries of origin. TRIPS enables persons or institutions to patent a country's biological resources outside the country of origin and arguably this facilitates the conditions for misappropriation of ownership or rights over living organism. In TRIPS there is no provision to obtain prior informed consent unlike the CBD and the Nagoya Protocol to check against misappropriation or biopiracy. According to Shiva, the TRIPS Agreement is a globalisation of U.S. style patent laws which encourage the patenting of centuries old indigenous knowledge as "novel invention" (Raj 1999).

In Malaysia, the preventive environmental measures in the forms of laws and regulations have occasionally been said to be sufficient in regulating sustainable ecotourism. However, sustainability criteria in ecotourism activities, which are instrumental for containing its potential adverse impacts, are not clearly defined at the national levels. According to some critics, because such criteria are not legally established, it may be more difficulty to (i) guarantee the optimal conservation of the natural environment, which is a precondition for the development of ecotourism and to (ii) assess the liability of ecotourism operators in cases of environmental and social damages (Garcia et al. 2004).

Malaysia is not a party to the Nagoya Protocol. Its ABS Bill is still in a draft form and it has not yet become law. The above-mentioned Sabah Biodiversity Enactment 2000 and the Sarawak Biodiversity (Access, Collection and Research) Regulations 1998 standing alone may not be sufficient to effectively regulate biological diversity prospecting and exploitation of genetic resources and TK. This is regrettable as there are significant benefits to the harmonisation of access legislation among the other states in the federation. Without a proper national ABS legislation that clearly spells out the right to decide who gains access to the country's genetic resources and TK, the monetary and non-monetary terms of benefit sharing, stake in ownership over final product, inclusion of Certificate of Origin by User and certificate of having obtained PIC from local authorities, illicit bioprospecting or biopiracy would not be easily countered (Sham 2010).

However, the greatest challenge in implementing environmental regulations as regards to ecotourism and biopiracy in Malaysia is stemming from the complexities of the Federal Constitution. Regulatory control on ecotourism and biopiracy may not be effective with autonomous states like Sabah and Sarawak in East Malaysia where, the relationship between the powers and responsibilities of the Federal and State government in environmental matters is already a very loose and complex one. Indeed, the *Bakun Dam* case places significant restriction on the federal government to regulate environmental matters in the various Malaysian states in future. In the same context, another challenging factor is to create legally protected areas and ensure proper supervision of the existing ones to guarantee that the limits and specific uses of such areas are respected.

Untrained, inefficient and insufficient personnel are obstacles for the adequate supervision and management of natural areas where ecotourism can ideally operate. Since ecotourism in Malaysia operates in remote and rural places, the areas not properly supervised or under any form of legal protection, are likely to be negatively impacted by activity such as biopiracy.

## Conclusion

As already noted, according to some critics, there are serious grounds for concern over the accelerating process of ecotourism and its implication on the environment especially in relation to biopiracy. Indeed, there are some valid reasons why the Malaysian Government should be interested in these recent developments and trends and to come out with some strategic planning policy to face these challenges.

Currently Malaysia does not have national law on ABS. Without a comprehensive ABS regulation, biopiracy can be threatening to the conservation of genetic resources and TK. In answering the above question of whether Malaysia has adequate law to protect its biological resources and TK from biopiracy, the answer is a NO. Hence, specific regulation dealing with the diverse economic aspects of biopiracy must be clearly established, at regional and national levels before indiscriminately promoting ecotourism.

## References

Abott FM (1996–1997) The WTO TRIPS agreement and global economic development. CHL *Kent* Law Rev 72

Australian Asia Pacific Economic Cooperation (APEC) Study Centre (2009) Genetic resources and growth—TRIPs and CBD—how to improve access and benefit sharing without stifling agricultural development and innovation. Melbourne

Bautista LB (2007) Bioprospecting or biopiracy: does the trips agreement undermines the interests of developing countries. Philippine Law J 82(1)

Ceballos-Lascurain H (1996) Tourism, ecotourism and protected areas, International union for conservation of nature (IUCN)

Chen J (2006) There's no such thing as biopiracy…And it's a good thing too. McGeorge Law Rev 37:23

ETC Group (1995) Bio-prospectors hall of shame… or guess who's coming to pirate your plants?! Pros and cons of bilateral bioprospecting agreements, December, USA

Garcia B, Gasques JG, Bastos ET (2004) Ecotourism in the Amazon. IEMSS 2004 Conference. Available at http://www.iemss.org/iemss2004/pdf/ecotourism/garcecol.pdf

Hamzah A (2004) Policy and planning of the tourism industry in Malaysia. In: Proceedings of the 6th ADRF general meeting, Bangkok, Thailand

McManis CR (1998) The interface between international intellectual property and environmental protection: biodiversity and biotechnology. *WASH. U.L.Q* 76:255

Ministry of Science, Technology and the Environment (1997) Assessment of biological diversity in Malaysia, Kuala Lumpur

Shiva V (1999) Promoting biopiracy, blocking TRIPs reform: seattle and the politics of the W.T.O/TRIPs review

Robinson DF (2011) Confronting biopiracy, challenges, cases and international debates. Rev European Commun Int Environ Law 20(2)

Sham MA (2010) ABS and Traditional Knowledge in Sabah. Biocultural diversity and conservation learning platform briefing notes, vol 2

Simon F (1999) Regulating Ecotourism: legal frameworks and market guidelines. Bulletin, Yale F&ES, p 197

Snape WJ III (2010) Joining the convention on biological diversity: a legal and scientific overview of why the united states must wake up. Sustain Develop Law Policy, Spring, 6

Tejera VM (1999) Tripping our property rights: is it possible to reconcile the convention on biological diversity with Article 27 of the TRIPS agreement. New England L. Rev 33:967

Third World Network (1999) Briefing on tourism, development and environment issues in the Mekong sub region, vol. 5, no. 2 March–April

Third World Network (2002) International year of reviewing ecotourism. Penang, Malaysia

Tourism Malaysia (2008) Media info: ecotourism in malaysia, communications and publicity division, tourism Malaysia, Putrajaya. Available at http://corporatetourism.gov.my

Twarog S, Kapoor P (eds) (2004) Protecting and promoting traditional knowledge: systems, national experiences and international dimensions. UN conference on trade and development, United Nations, New York and Geneva, iii

Zakrzewski PA (2002) Bioprospecting or biopiracy? The pharmaceutical industry's use of indigenous medicinal plants as a source of potential drug candidates. University of Toronto Medical Journal 79(3):252–254

# Chapter 3
# The Influence of Eco-friendly Attitudes on Tourists' Intention Toward Green Hotels

**Monika Kristanti and Regina Jokom**

**Abstract** The increasing awareness in environmental issue drives companies to implement green practice in the business. Therefore, some hotels name themselves as green hotels in order to attract potential customers. This study examines tourists' eco-friendly attitudes and the effect on their intention toward green hotels. The eco-friendly attitudes were measured by tourists' perception based on the attributes, such as, severity of environmental problems, inconvenience of being environmental friendly, importance of being environmental friendly, and level of responsibility of business corporations. Therefore, tourists were asked about their intentions to visit, to engage in word-of-mouth behaviors, and to pay more for a green hotel. Questionnaires were distributed to the domestic and international tourists in Juanda International Airport Surabaya using convenience sampling technique. The study shows that tourists have positive responses in environmental issues; however, they tend not to separate piles of garbage for recycling reason. The result reveals that tourists' eco-friendly attitudes significantly influence their intention to visit, to engage in word-of-mouth behaviors, and to pay more for green hotels. Partially, attitudes in being environmental friendly make the largest unique contribution in intention to visit while perception of environmental problem contributes the most in word-of-mouth intention. In addition, intention to pay more is highly affected by tourists' inconvenience of being environmental friendly.

**Keywords** Eco-friendly attitude · Intention to visit · Intention to pay more · Intention to spread word-of-mouth · Green hotels

M. Kristanti (✉) · R. Jokom
Hotel Management Program, Petra Christian University, Surabaya, Indonesia
e-mail: mkrist@petra.ac.id

R. Jokom
e-mail: regina@petra.ac.id



A. Saufi et al. (eds.), *Balancing Development and Sustainability in Tourism Destinations*, DOI 10.1007/978-981-10-1718-6_3

## Introduction

The increasing awareness in environmental issue drives companies to implement green practice in the business. Most of the companies make this environmental issue as a part of their social responsibility (Lee 2009).

Lodging industry is one of the main industries that creates environmental problems. The industries consume lots of water, energy, and unrecyclable products that in the end endanger the environment. Therefore, some hotels start to implement innovative methods and name themselves as green hotels in order to attract potential customer, especially customers who have eco-friendly intention toward hotels that operate in a green way (Han et al, 2011). The green behavior of customers could be affected by knowledge, awareness, and attitudes to the environment (Aman et al. 2012).

In Indonesia, there are some hotels that have green hotel labels, such as Borobudur Hotel Jakarta, St. Regis Resort Bali, Grand Melia Hotel Jakarta, Nusa Dua Beach Hotel Bali, and others (Indonesia Official Tourism Website, 2013). However, there are not too many researches in Indonesia that have been conducted on the study of the correlations between hotel guests' eco-friendly attitudes in everyday lives and their eco-friendly intentions to stay, to talk about, and to pay extra for a green hotel.

This study helps hotel management to grasp current and potential customers' green purchasing behavior. This study tries to answer some research questions:

1. Does eco-friendly attitude influence tourists' eco-friendly intention toward green hotel?
2. Which eco-friendly attitude has the most influence to eco-friendly intention of tourists?

## Literature Review

### *Green Hotel*

The hotel industry has the purpose of providing comfortable service or supplies, such as hot and cold water, food and drinks, towels, lighting, air-conditioning, etc. In providing these services, hotels obviously use lots of water, energy, unrecyclable products, and natural resources that can bring negative impact for the environment. As environmental awareness of individual increases, people are increasingly trying to find eco-friendly hotels. Consequently, many hotels are starting to apply some innovative ways to escalate green operations (Han et al 2011).

Wolfe and Shanklin (2001) define green as the action to preserve the environment. Green hotel is an eco-friendly hotel operation that performs or follows various environmentally friendly practices or programs such as saving water and energy, using eco-friendly purchasing policies, and reducing waste disposals to protect the natural environment and reduce operational costs (Green Hotel Association 2008). Green hotel follows environmentally friendly guidance and implements environmental management actively, and green hotel is committed to protect the environment by having eco-labels. Implementing green practice in hotel not only does the responsibility to the environment, but also enables to bring a huge impact to the operational cost (Han et al 2011).

Companies who sell green products can attract more customers. 67 % of American people are willing to pay around 5–10 % higher for green products (Han et al 2011). Mensah (2004) states 90 % of hotel customers choose to stay the night at green hotel. It is supported by Bulter (2008) that 16 % customers of Kimpton Hotel and Restaurant like to stay there because of the implementation of green practice; for example, the use of nontoxic cleaning agents and the availability of dust bin in the room.

## *Eco-friendly Attitudes and Intention*

The increasing awareness in environmental problems creates positive attitude toward green activities and more involvement in daily eco-friendly behavior. This attitude is called eco-friendly attitude that consists of four (4) dimensions: (1) perceived severity of environmental problems, (2) inconvenience of being environmental friendly, (3) importance of being environmental friendly, and (4) level of responsibility of business corporations. In other words, individuals who have eco-friendly attitudes generally notice the seriousness of ecological problems, argue the statement that being environmentally friendly is inconvenient, realize the importance of being ecologically friendly, and strongly believe that hospitality operations should be responsible for the environment (Han et al. 2011).

Eco-friendly intention is the intention of customers to visit green hotel, to take part in positive word-of-mouth and to pay extra for green hotel (Han et al. 2009). It is supported by Manaktola and Jauhari (2007) who state that the customers having awareness to environment choose to visit green hotel; 22 % customers always search information about green hotel before they stay in a hotel, but positive attitude to green practices does not direct the customers to pay more for green hotel. There are only 15 % customers approving to pay more, and other customers state that hotel should be responsible for the cost. Furthermore, 40 % customers are willing to pay 4–6 % higher to stay at green hotel.

## *Research Methods*

The type of the research is causal quantitative aiming to identify the influence of eco-friendly attitude to eco-friendly intention of tourists toward green hotel.

The population of the research is domestic and international tourists at the age of 17–60 years and they have the experience of staying in a hotel. The sampling method is convenience sampling technique, and 240 questionnaires were distributed at Juanda International Airport Surabaya. Hair et al (1998) mention that the number of sample used in multiple regression analysis varies from 20 to 1000 respondents depending on the number of independent variables. The questionnaire used was 7-point Likert-type scale ranging from strongly disagree to strongly agree. The methods used for the data analysis were descriptive statistics (mean), multiple linear regression, and *t*-test.

The questionnaire consists of four parts. The first part is definition of green hotel to help participants know what green hotel is. The second part is about demographic information of participants, such as age, gender, knowledge about green hotel, and previous experience with a green hotel. The third part is about eco-friendly attitudes in their everyday life, and the fourth section consists of questions about eco-friendly intentions to visit, to spread word-of-mouth, and to pay more for a green hotel.

## Findings

240 questionnaires were distributed, and 207 questionnaires could be used for further analysis. Thus, the response rate was 86.25 %. The majority of respondents in this research were men (55.1 %), at the age of 17–35 years (58.9 %), coming from Indonesia (50.7 %) and Asia, such as China, Hong Kong, Thailand, India, etc (41.1 %), having undergraduate study (42.7 %), and working as students, entrepreneurs and private sector employees (73.5 %). Most of the tourists doubted whether they have knowledge about green hotel (37.7 %) and they did not know whether they have ever stayed at green hotel (44.9 %).

Multiple linear regressions were employed to examine the influence of attitude components on intentions to stay at a green hotel, to say positive word-of-mouth and to pay extra. Standardized coefficients and *t*-values were applied to determine which component of attitudes has dominant impact to intention toward green hotel.

## The Mean of Attitudes and Intentions

Based on Table 1, it can be seen that tourists have good environmentally friendly attitudes. The tourists strongly disagree (mean = 2.12) that people do not have to worry about conservation of electricity, water, and trees; and disagree (mean = 2.24) that people do not have to worry about the earth's present condition. The tourists realize that severity of environmental problems is a serious matter. Moreover, the tourists disagree (mean = 3.00) doing recycle and separating waste bring too much

**Table 1** Mean of eco-friendly attitude

| Variable | Measurement items | Mean |
|---|---|---|
| Eco-friendly attitudes | Severity of environments problems (SEP) | |
| | 1. The earth has enough electricity, water, and trees that we do not have to worry about conservation | 2.12 |
| | 2. The earth is a closed system where everything eventually returns to normal, so I see no need to worry about its present state | 2.24 |
| | Mean of SEP | 2.18 |
| | Inconvenience of being environmentally friendly (INEF) | |
| | 1. Recycling is too much trouble | 2.83 |
| | 2. Keeping separate piles of garbage for recycling is too much trouble | 3.17 |
| | Mean of INEF | 3.00 |
| | Importance of being environmentally friendly (IMEF) | |
| | 1. Recycling will reduce pollution | 4.78 |
| | 2. Recycling is important to save natural resources | 5.48 |
| | Mean of IMEF | 5.13 |
| | Level of responsibility of business corporations (LRBC) | |
| | 1. Hospitality operations (e.g., hotels and restaurants) are concerned about the environment | 5.48 |
| | 2. Packaged food or paper companies are concerned about the environment | 5.55 |
| | Mean of LRBC | 5.52 |

difficulty. They agree (mean = 5.13) that recycle is really important to save natural resources and reduce pollution. The tourists aware that environmental problems are not only individual's responsibility, but also the responsibility of hospitality operations (hotel and restaurant), and industries that support hospitality operations such as packaged food and paper companies. These industries should concern about the environment (mean = 5.52).

The intentions of tourists toward green hotel can be seen in Table 2. It shows that tourists have the intention to visit green hotel (mean = 5.03). Tourists will stay, always try to stay and have the willingness to stay the night at a green hotel while traveling. Furthermore, tourists have the intention to talk about green hotel (mean = 5.08). Tourists will recommend and suggest friends and families to stay at green hotel while traveling, and also tourists will say positive things about environmentally friendly hotel. In addition, tourists have intention to pay additional charge for a green hotel (mean = 4.86). Tourists will spend more money to stay the night at a green hotel, and it is acceptable if they have to pay more and would like to have additional charge for a green hotel. The highest intention of tourists toward green hotel is intention to say positive things about, and the lowest is intention to pay more for a green hotel.

**Table 2** Mean of Eco-Friendly Intention

| Variable | Measurement items | Mean |
|---|---|---|
| Intention to visit (IV) | 1. I will stay at a green hotel when traveling | 4.90 |
| | 2. I will make an effort to stay at a green hotel when traveling | 4.94 |
| | 3. I am willing to stay at a green hotel when traveling | 5.24 |
| | Total mean | 5.03 |
| Intention to spread word-of-mouth (ISWOM) | 1. I will encourage my friends and relatives to stay at a green hotel when traveling | 5.01 |
| | 2. If someone is looking for a hotel, I will suggest to him/ her to stay at a green hotel | 4.98 |
| | 3. I will say positive things about an environmentally friendly hotel | 5.26 |
| | Total mean | 5.08 |
| Intention to pay more (IPM) | 1. I will spend extra in order to stay at an environmentally friendly hotel | 4.87 |
| | 2. It is acceptable to pay more for a hotel that engages in green practices | 4.87 |
| | 3. I am willing to pay more for a green hotel | 4.85 |
| | Total mean | 4.86 |

**Table 3** Determination of eco-friendly intentions

| Eco-friendly attitudes | Intention to visit | | Intention to spread WOM | | Intention to pay more | |
|---|---|---|---|---|---|---|
| | Beta | *t*-values | Beta | *t*-values | Beta | *t*-values |
| Severity of environmental problems | −0.107 | −1.362 | −0.234** | −3.046 | −0.151 | −1.947 |
| Inconvenience of being environmentally friendly | 0.084 | 1.058 | 0.196* | 2.546 | 0.313** | 4.025 |
| Importance of being environmentally friendly | 0.209** | 2.945 | 0.112 | 1.620 | −0.061 | −0.875 |
| Level of responsibility of business corporations | 0.055 | 0.745 | 0.152* | 2.135 | 0.115 | 1.596 |
| **Adjusted R2** | | | | | | |
| Intention to visit | 0.052 | | | | | |
| Intention to spread WOM | 0.102 | | | | | |
| Intention to pay more | 0.083 | | | | | |

* significant at 0.05
** significant at 0.01

## *The Impact of Eco-friendly Attitudes*

The findings of multiple regression analysis are presented in Table 3. It is shown that components of eco-friendly attitudes in most cases are related with eco-friendly intentions toward green hotel. Severity of environmental problems (SEP) was significantly correlates only with intention to spread word-of-mouth (ßSEP → IV = −0.107, $t$ = −1.362; ßSEP → ISWOM = −0.234, $t$ = −3.046; ßSEP → IPM = −0.151, $t$ = −1.947). Inconvenience of being environmentally friendly (INEF) significantly strengthened tourists' intentions to have positive word-of-mouth and to pay extra charge for a green hotel (ßINEF → IV = 0.084, $t$ = 1.058; ßINEF → ISWOM = 0.196, $t$ = 2.546; ßINEF → IPM = 0.313, $t$ = 4.025). Importance of being environmentally friendly (IMEF) was significantly correlates with intention to visit a green hotel (ßIMEF → IV = 0.209, $t$ = 2.945; ßIMEF → ISWOM = 0.112, $t$ = 1.620; ßIMEF → IPM = −0.061, $t$ = −0.875), and level of responsibility of business corporations (LRBC) was significantly related to intention to talk about green hotel (ßLRBC → IV = 0.055, $t$ = 0.745; ßLRBC → ISWOM = 0.152, $t$ = 2.135; ßLRBC → IPM = 0.115, $t$ = 1.596).

Individuals see that severity of environmental problems is not likely to induce tourists' willingness to visit a green hotel or to pay more for a green hotel. Inconvenience of being ecologically friendly is not likely to create tourists' willingness to visit a green hotel. The perceived importance of being environmentally friendly does not cause the willingness to talk about green hotel or to pay extra charge for a green hotel. Moreover, individual's perceived level of responsibility of business corporations is not likely to induce the willingness to visit a green hotel or to pay more for a green hotel.

Furthermore, the results of the comparisons among the beta coefficients along with $t$-values reveal that severity of environmental problem has a greater impact on intentions to spread word-of-mouth, inconvenience of being environmentally friendly has a greater impact on intention to pay more, and importance of being environmentally friendly has a greater impact on interest to visit green hotel. The four dimensions of eco-friendly attitudes clarifies approximately 5.2 % of the total variance in the intention to visit, 10.2 % in the intention to spread word-of-mouth, and 8.3 % in the intention to pay more.

## Discussion

The findings, in general, show that hotel customers' eco-friendly attitudes positively affect tourists' intention to visit green hotel, to spread word-of-mouth, and to pay more for a green hotel. However, partially, it is only some dimensions that significantly affect the intention. The multiple regression analysis states that the importance of being environmentally friendly best clarifies tourists' interest to visit green hotel. These results are in line with the previous findings by Han et al (2011). The findings suggest that among the four (4) dimensions of attitudes, the perceived importance of being environmentally friendly is the best way to encourage intentions to visit a green hotel. Hence, green hotel management should educate guests through environmental campaigns or informal education, such as brochures, seminars or exhibitions. If the individuals have knowledge for environmentally friendly, individuals can perform eco-friendly practices or being environmentally friendly in everyday life.

In addition, inconvenience of being environmentally friendly has the greatest impact on the intention to pay more for a green hotel. This result is supported by the mean of inconvenience of environmentally friendly that tourists support the recycling and separate piles of garbage. Even though tourists tend to be unwilling to separate piles of garbage, they still realize that it is important. Moreover, severity of environmental problems has the largest contribution on the intention to spread word-of-mouth. Based on these two dimensions, it shows that tourists need action that can be recognized to inspire them to have the intention toward green hotel, to pay more and to spread favorable word-of-mouth. Tourists will appreciate if hotels have done the recycling and separating garbage program, and reduced the use of electricity, water and energy. It generates green hotels to apply their environmental commitment in their facilities and makes the guests know the hotels' concerns for being environmentally friendly.

In addition, the four dimensions of eco-friendly attitudes could define approximately 5.2 % of the total variance in intention to stay the night, 10.2 % in intention to have positive word-of-mouth, and 8.3 % in intention to pay additional charge. These dimensions merely have small contribution to the intention. This result is supported by Han et al (2011) who mention that tourists' behavior or intentions are strongly correlated with tourist's previous experience or knowledge. In this study, 35.7 % of the participants did not have knowledge about green hotel, and 37.7 % of the participants were not sure if they have knowledge of green hotel. In addition, 41.5 % of the respondents never stayed at a green hotel, and 44.9 % said that they were not sure whether they had ever stayed at green hotel. Given this fact, it can be considered that many tourists do not have knowledge about green hotels and do not even know the presence of green hotels that in the end affect their intention toward green hotels. Obviously, green hotel management has to be more active in notifying current and potential customers of hotel's eco-friendly facilities, programs and practices by using eco-labels or green hotel labels. Another reason for small contribution to the intention itself is the lack of environmental awareness. It is related to the fact that majority of respondents for this study were Indonesian (50.7 %) and Asian (41.1 %). They do understand the eco-friendly attitudes but they do not implement it properly because of the personal inconvenience and additional cost, for example, they still use unrecyclable products and have difficulties in separating piles of garbage. Environmental awareness instills a positive attitude toward eco-friendly activities, encourages people to engage more in eco-friendly behavior in everyday lives, and in the end, encourages eco-friendly intention (Han et al. 2011). It is also explained by Manaktola and Jauhari (2007) that guests who are conscious of hotel's green programs show their preference to visit a green hotel.

## Conclusion

Based on the findings, tourists' eco-friendly attitudes influence their intention to visit, to engage in word-of-mouth behaviors, and to pay more for green hotels. Partially, attitudes in being environmental friendly make the largest unique contribution in intention to visit while perception of environmental problem contributes the most in word-of-mouth intention. In addition, intention to pay more is highly affected by tourist's inconvenience of being environmental friendly. There were some limitations which need to be addressed for the future research. The first limitation is related to the sample composition, in which the majority respondents were Indonesian and Asian. Hence, further research on western travelers should be conducted to give broader view. Furthermore, this study only used quantitative approach which does not reveal the reason behind the answer. Therefore, it is suggested that future studies should focus on investigating the reasons for the eco-friendly attitude and intention toward green hotels using qualitative approach for better result and understanding.

## References

Aman AHL, Harun A, Husein Z (2012) The influence of environmental knowledge and concern on green purchase intention the role of attitude as a mediating variable. Br J Arts Social Sci 7(II)

Bulter J (2008) The compelling "hard case" for "green hotel" development. Cornell Hotel Restaurant Adm Q 49:234–244

Green Hotel Association (2008) What are green hotels? Retrieved from http://www.greenhotels.com/whatare.htm

Hair JF, Anderson RE, Tatham RL, Black WC (1998) Multivariate data analysis. Prentice Hall, Upper Saddle River, New York

Han H, Hsu L, Lee J (2009) Empirical investigation of the roles of attitudes toward green behaviors, overall image, gender, and age in hotel customer's eco-friendly decision making process. Int J Hospitality Manage 28(4):519–528

Han H, Hsu L, Lee J, Sheu C (2011) Are lodging customers ready to go green? An examination of attitudes, demographics, and eco-friendly intentions. Int J Hospitality Manage 30:345–355

Indonesia Official Tourism Website (2013) Indonesia awards top 10 green hotels 2013. At http://indonesia.travel/en/news/detail/1069/indonesia-awards-top-10-green-hotels-2013

Lee KH (2009) Why and how to adopt green management into business organizations? The case study of Korean SMEs in manufacturing industry. Manag Decis 47(7):1101–1121

Manaktola K, Jauhari V (2007) Exploring consumer attitude and behavior towards green practices in the lodging industry in India. Int J Contem Hospitality Manage 19(5):364–377

Mensah I (2004) Environmental management practices in US hotels. Retrieved from http://www.hotel-online.com/news/PR2004.2nd/May04EnvironmentalPractices.html on 18 April 2008

Wolfe KL, Shanklin CW (2001) Environmental practices and management concerns of conference center administrations. J Hospitality Tourism Res 25(2):209–216

# Chapter 4
# Reconnecting the Past Through Oral History: An Effort in Preserving Malaysians' Intangible Cultural Heritage

**Suhaida Halamy and Silverina Anabelle Kibat**

**Abstract** Malaysian are among the societies in the world who started late in documenting their history. Most stories and cultural heritage were commonly shared orally from one generation to another. Hence, the lack of original historical resources has been a major problem in reconstructing Malaysian history. Without oral history being recorded and written, and with the loss of the elders who were willing to preserve and pass along this history, cultural memories began to wither, along with many of the colorful cultures and heritage. Malaysia, known as a melting pot of race and culture, has a society who had long discovered their culture and heritage through oral history as a part of their rich historical heritage. In view of that, there is a need of encouragement and effort on historical research and oral history writing for the continued preservation and appreciation of oral history among its people. This paper provides an overview on the importance of oral history in preserving the valuable oral treasures among the society, and to further enrich the local culture and heritage for the future generations. The findings of this paper suggest that there is a need for requirement of improved level of awareness on the value of oral history, and as well as a greater exposure through education for the younger generations and tourists, in order to ensure that oral history is appropriately understood and accepted at all levels.

**Keywords** Oral history · Cultural and heritage · Preserve · Tourist awareness

S. Halamy (✉)
Universiti Teknologi MARA Sarawak, Sarawak, Malaysia
e-mail: dida@sarawak.uitm.edu.my; dasecret_82@yahoo.com

S.A. Kibat
Universiti Teknologi MARA Sabah, Sabah, Malaysia
e-mail: silve598@sabah.uitm.edu.my; silverinakibat@gmail.com



A. Saufi et al. (eds.), *Balancing Development and Sustainability in Tourism Destinations*, DOI 10.1007/978-981-10-1718-6_4

## Introduction

The National Committee on Oral History National Archive of Malaysia (1991) defines oral history as "the technique of eliciting the reminiscences of selected individuals through recorded interview sessions." The committee further elaborates that the individuals selected are those considered the best able to provide such information derived from their personal involvement and experience of historical events, from their special relationship with a particular personality, or the bearing that a specific period in history may have had on their lives. The recorded interview, when transcribed, allows for the convenient use of oral history as both research as well as teaching material. In this way, oral history can be made to contribute significantly to the existing resources on local history.

In an effort to promote the writing of the history of Malaysia, it is not necessarily devoted to writing about well-known figures and major events that happened in the past. Surely these are important, but we should not forget that history includes the story and events of ordinary people. Hence, patterned topics of local history and the history of the country should be acknowledged appropriately and the opportunity to express this heritage can be done through oral history. Oral history is considered as living memories which have great importance in the history of mankind. It has also become a key resource for reconstructing, reconnecting, recreating, and recording the past and present knowledge of a society.

In short, oral history can be defined as an activity, a detached academic process of inquiry into the memories of people who have experienced the recent past directly. This inquiry and the responses it generates are recorded to supplement written records that have been found wanting in some areas for historical or aesthetic analysis. It is a studied abstract and analytic practice of historians and other social scientist, and it relies heavily either on manual, mechanical, or electronic recording devices (Rais 2005).

## Definition of Cultural Heritage

In Malaysia, National Heritage Act 2005 (under Sect. 67) [2] has categorized heritage into two; cultural heritage and nature heritage. Cultural heritage is further subdivided into two—"unfixed heritage" and "fixed heritage." Under "fixed heritage," there are tangible and intangible heritage.

According to the Act, cultural heritage is divided into two aspects: tangible and intangible heritage which differ in nature, although they belong to the same category. "Tangible Heritage" refers to something that is permanently being seen and it is touchable. It can be categorized as "Static" and "Movable." The term "Static" refers to anything that is permanently there such as historical sites of Lembah Bujang, Langkawi, Melaka; monument or building, for example Tomb, Palace,

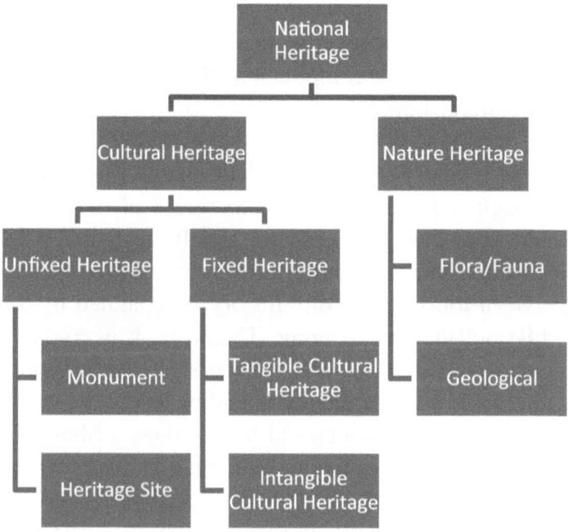


**Fig. 4.1** Categories of Malaysia National Heritage, National Heritage Act 2005 (2008)

Fortress or Minarets; and forest, cave, mountain, river, flora, and fauna are considered as nature. Meanwhile, "Movable" refers to cultural artifacts that can be moved such as manuscript and textile.

The National Heritage Act 2005 (2008) defines intangible cultural heritage as "...a phrase, language, pronunciation of the tongue, musical notes, audible lyrics, songs, folk songs, oral traditions, music, dance as generated by the performing arts, theatre, musical compositions and sound, martial arts that has existed or exists in relation to the heritage of Malaysia or any part of Malaysia or the Malaysian community in relation to heritage." Intangible cultural heritage is important not only for its cultural manifestation itself, but rather the wealth of knowledge and skills that are transmitted through it from one generation to the next. The social and economic value of this transmission of knowledge is relevant for the minorities and the mainstream social groups within a state, and it is as important for the developing states as it is for the developed ones (UNESCO) (Fig. 4.1).

## Intangible Cultural Heritage as Tourism Product

Developing historical legacy does have its own advantages. This is because, in comparison to the authenticity of culture, historical heritage is easier to preserve or rebuild because they consist of buildings or monuments formed in the past. Regardless if the buildings or monuments were built thousands or hundreds of years ago, they have the potential to be redeveloped. Buildings can be restored and if they are damaged, they can be reconstructed based on past records and photos. This is because the development of historical events has often been recorded either in the form of historical records, or oral history. Therefore, in developing historical heritage, it requires a combination of historians, experts in architecture, archeologists, and cultural figures.

Tourism activities in Malaysia have started since the eighteenth century with the arrival of more foreign travelers, mainly the Europeans and Arabians, to Malaya. Through their journey and experience of the past, they came up with oral documentation that has become widely recognized as an indispensable means of recording and preserving information. In the Malaysian's context, the correlation between tourism and oral history is separated by a huge gap. Tourism is thriving, while oral history is dying. Therefore, it is crucial to record reminiscences, experiences, testimonies, and reflections from individuals when they are still alive. Otherwise, the information would be lost to the future generation. As an old African proverb says, "when an old person dies, a library burns to the ground."

From the perspective of tourism, oral history of historical events and phenomena is important to be strengthened. Such information can be raised as a powerful element that could attract the attention of tourists. Additionally, the element of oral history has its own motives, as well as a unique attraction and potential to become the object of attention of tourists. This is because the motive of tourists can be stimulated either by real or fictional events that are to be explored, appreciated or be associated with a phenomenon, either through reading or storytelling. Despite limited available research, the needs for tourism associated with well-known people, places, and events can be identified based on the roles and effort taken by the government, tourism industry, researchers, activists, local community, and students (Richard and Wilson 2007). These needs appear to be based on these three reasons

(a) *See and be where they have been.* This is about a desire to have a physical and concrete experience which links the tourists to past events and people. For instance, a strong factor is the desire to tread in the footsteps of famous people, or to stand where battles have taken place.
(b) *Connecting with what they have read.* Following from the above, tourists are motivated by a desire to better understand and picture the past created for them by famous writers, either real or imagined. For example, a landscape inspired by a scene from a novel or a piece of music, which are portrayed as real events. Thus, for some people, it is important to gain insights into what motivates creative geniuses.
(c) *Connecting with other like-minded people.* Connecting with other people with similar interests is a key reason why some will embark on a literary tour or attend a commemorative event overseas. It provides an opportunity to discuss and develop an often long held interest in a group environment.

Oral history in Sarawak is handed down from one generation to another. Most of oral history in Sarawak has been documented and published such as the Sarawak Museum Journal and Sarawak Gazette. These continued efforts are carried out by

*Dewan Bahasa dan Pustaka* and *Pustaka* Sarawak Branch through the collection, translation, and publication in English and Malay Language. These activities are noted as increasingly important in the future as they have now become uncommon, and those who have valuable experiences have gone without their stories being documented.

## Challenges

Malaysia, like any other countries, may experience the loss of its stories and experiences of the past very quickly. This is expedited by the advancement of technology whereby people rely mostly on telephone, fax, emails, and face-to-face interactions to communicate. The traditions such as writing letters, diaries, biographies, and memoirs have become less important and uncommon. As a result, one of the major weaknesses identified in the effort to rewrite the history of Malaysia is the lack of historical sources, for example the history of the states in Malaysia. Only few states possess the natural resources associated with the historical development of their states, but even then, it is only up to the nineteenth century. For instance, the Sarawak Museum was the earliest institution in the country to carry out oral history and started recording materials in 1957.

In other situations, the lack of awareness among many Malaysians has contributed to the loss of cultural heritage. For instance, Sabah was once an equal if not superior counterpart to Malaya as a former colony of the British. The obliteration of Sabah townships has been recorded as the most devastating in the history of the British Empire during WWII. Sadly, the documentation on Sabah's architectural legacy is greatly lacking. Neither the archives nor libraries in both Sabah and peninsular Malaysia had kept much information on Sabah's traditional houses or pre-war buildings (Richard 2014). Today, Sabah does not have many architectural remnants left from this period of history that accurately portray its past glory as a former colony of the British.

Therefore, in circumstances where documenting stories and past events is no longer a common practice, it is the oral history that may help determine that not all development is lost or forgotten as historical evidence. Oral history can play a big role in maintaining and increasing resources available with regard to Malaysian history in this century. Additionally, oral history can bring the history in and out of the community and help the less privileged toward self-dignity and self-confidence. It makes contact and creates understanding between social classes and different generations. With mutual understanding between both individuals and historians, it can give a sense of belonging to a place or time. Oral history offers challenges to the acknowledged myths of history and as a means for an essential transformation of the social meaning of history.

## Reevaluating the Value of Oral History

Some scholars in Malaysia have expressed concerns over the development of historical stories in an increasingly less interested society due to the lack of efforts to collect, document, and highlight past events or stories. There are several factors contributing to this phenomenon. The first factor is the lack of exposure among the public, which includes the authorities, practitioners, or educators at all levels. Practitioners who are too old can no longer afford to continue the effort and struggle. Meanwhile, most of the younger generations do not have much interest in acquiring the skill in oral history from the elders. Thus, the skill on oral history continues to disappear, and is not shared and passed on to the younger generations. Resulting from this, much of the personal history of the late twentieth and twenty-first centuries would be lost to future historians.

Another factor is the lack of energy and expertise of young people, mainly academicians who are interested in finding, collecting, and documenting historical development. Those who are interested in the field of oral history are not many. This may be due to the negative perception that oral history does not have much commercial value. In fact, oral history allows one to create valuable sources or materials by talking to people in the community where questions can be answered about the past, which may help in the process of documentation. Besides, hopes, feelings, aspirations, disappointments, family histories, and personal experiences can be learnt through oral history.

Apart from that, the lack of research done by scholars or students in higher institutions on the use of oral history is another factor contributing to the concerns over the development of historical stories in Malaysia. For students, available documented materials or resources are commonly used in the process of learning as opposed to using oral history as the primary source of knowledge. Therefore, oral history is commonly treated as reading materials instead of the source of knowledge to be researched further. If this continues to occur, the community or the public may experience a great loss of their cultural heritage because oral history can contribute greatly to the development of the local community, and tourists alike. History after all is all about human experience. Through oral history, researchers and interviewees can come together in conversation about a commonly shared interest as with all human interactions and this has the potential to be tremendously rewarding for both parties.

## Recommendation

In general, the community displays the lack of awareness on the purpose of oral history, except for those who have the interest to do study this field. The position of oral history is much marginalized in the society. In educational settings, not many students recognize and appreciate oral history. This could be due to the perception that oral history has no commercial value compared to other subjects, and that oral history promises limited employment opportunities for the students. Such a response should be observed and reviewed in order for oral history to be duly acknowledged and developed among students and the society at large. Hence, oral history should be disclosed as a subject in schools either in the secondary or primary school to raise awareness on the importance of oral history among these young generations.

In the context of tourism, respective ministries such as the Ministry of Tourism, and the Ministry of Information, Communication and Culture may collaborate with other agencies in displaying oral history collection at places of tourist attraction such as the Cultural Village, museum, archives, and some other places. This would encourage tourists to visit for the purpose of enriching their knowledge about the country and its history.

## Conclusion

Information collected and gathered through oral history should be treated as guidance, not as the ultimate source because memories often fade and facts get confused with other facts. Information obtained through oral history, however, exists nowhere else and must be taken at face value. Actual historical stories never fade as long as people continue to tell the stories. Hence, we need to mobilize efforts to document the stories more meaningfully through oral history. This is because oral history collection gives us an opportunity to review the sequence of events that happened in the past for us to understand the legacy of the past in order to move into the future. Besides, exposure to oral history is important because we can evaluate the history for a better life at present and in the future. For example, oral history collection on the reforms brought about by the influence of the West colonial power, particularly Britain and the Netherlands, began to change the mentality, philosophy, and the social community in Malaysia at the threshold of the eighteenth century. In short, history is taken as a point of fact that cannot be denied; forming and shaping movements of the human mind. Oral history has been an important method to assist researchers in discovering and using the materials in order to sustain the intangible cultural heritage in Malaysia. Oral history is a valuable research tool with historical value and nature.

The discussion above is one small aspect in the field of tourism and oral history. With a strong growth in tourism industry, oral tradition as a form of culture should be in line with the development. Nevertheless, cultural factors remain important because humans do naturally have an instinctive curiosity to witness the splendor of the cultural heritage of a race for knowledge expansion. Since Malaysia does not have Angkor Wat, Taj Mahal, and the Great Wall of China, proper attention should be given to stories through oral history such as ethnic diversity and cultural heritage, food heritage, performing arts heritage, and others in Malaysia that could become tourist attractions too.

Oral history is a priceless cultural heritage. Although it is only a piece of heritage, its role and its contribution toward constructing and shaping thought and taste is immense. Its role is not only as a source of information but also as a vehicle for education, sciences, culture, and history. Hence, it needs to be maintained, improved, excavated, searched, and documented in variety forms of media. Oral history can serve as a tourism product and can contribute to a country's economy. It should be seen as more than a mere subject that can provide knowledge to the society, readers, tourists, and researchers.

## References

Azhar R (2008, June 26). Mengenali Sastera Rakyat. *Utusan Sarawak*. Retrieved from http://cetasasbuya.blogspot.com/

George WE (2010) Intangible cultural heritage, ownership, copyrights and tourism. Int J Culture, Tourism Hospitality Res 40(4):376–388

Jusoh J, Masron T, Hamid NFA, Shahrin N (2013) Tourist expectation and satisfaction towards physical infrastructure and heritage elements in Melaka UNESCO World Heritage Site. Acad J Interdisc Stud MCSER Publ 2(8):734–739. Retrieved from doi:10.5901/ajis.2013.v2n8p733

Malaysia Government (2008) National heritage act 2005 (Act 645)

National Committee on Oral History (1991) Handbook on oral history. National Archives of Malaysia, Kuala Lumpur

Norgrove K, Mirchandani S (2007) The role of archives within tourism: a review of current archives tourism activity and opportunities for tourism development involving archives in North East England. Retrieved from http://webarchive.nationalarchives.gov.uk/

Rais Y (2005). Keynote address in working papers of the Seminar on preservation and dissemination of oral history as national heritage, Kuala Lumpur, pp 7–9 April

Richard NS (2014) Fighting to preserve our heritage. Retrieved from www.thestar.com.my/opinion/columnist/ok-bah/

Richard G, Wilson J (2007) Tourism, creativity and development. Retrieved from http://liepajaskultura.lv/uploads/files/4169.pdf

Samsudin M, Mohamad S (2013) Pengaruh Warisan Sejarah dalam Industri Pelancongan Langkawi. Int J Malay World Civil (Iman) 1(1):99–109. Retrieved from http://journalarticle.ukm.my

Singer GG (2011) The importance of the intangible cultural heritage. Retrieved from http://www.academia.edu/

Sokial RN (2014, January 22) Fighting to preserve our heritage. The Star Online. Retrieved from http://www.thestar.com.my

Swain ED (2003) Oral history in the archives: its documentary role in the twenty-first century. Am Archivsit 66(1):139–158. Retrieved from http://www.columbia.edu/

Trapp-Fallon JM (2003) Searching for rich narratives of tourism and leisure experience: how oral history could provide an answer. JSTOR: Tour Hospitality Res 4(4), 297–305. Retrieved from http://www.jstor.org/stable/23745914

UNESCO. Intangible Cultural Heritage.Retrieved from www.unesco.org/culture/ich/

# Chapter 5
# Revitalizing Tourism in the Heritage Land of Toraja: Planning as a Process

**Devi Roza K. Kausar and Myra P. Gunawan**

**Abstract** Toraja in South Sulawesi, Indonesia, is a tourist destination known for its ethnicity as the main attraction. It is located in the northern part of South Sulawesi Province, or about eight hours drive from the province capital, Makassar. The traditional houses of Toraja with their distinctive shape, the ancestral house known as Tongkonan, cliff-hanging, and stone-carved burial sites, and traditional ceremonies such as the death ritual, RambuSolok, were considered interesting for the taste of ethnic tourists. International tourist arrival to Toraja reached its peak in the 1997 but plunged to a low number from 1998 to 2003 when a series of unfortunate situations ranging from economic crises, the fall of New Order regime, to safety issues hit Indonesian tourism sector. International tourist arrival to Toraja continued to fall from 2004 to 2009. In 2013, the number of tourist arrival finally picked up and the revitalization of tourism in Toraja is underway. Through the Indonesian Government Regulation Number 50, 2011 on the National Master Plan for Tourism Development, the government has again designated Toraja as one of national strategic tourism areas to be prioritized in its tourism development plan. This paper will describe the tourism revitalization planning process that was undertaken from July 2014 to December 2014 involving teams of academics from various background, consultants, and local stakeholders, using data collection technics such as observations and site visits, market survey, and focus group discussions (FGD). Some of the objectives identified from this process include diversification of visitor attractions beyond traditional settlements—as expected by the emerging Asian

D.R.K. Kausar (✉)
Faculty of Tourism, Universitas Pancasila, Jakarta, Indonesia
e-mail: devikausar@univpancasila.ac.id

M.P. Gunawan
Tourism Planning and Development Center, Institute of Technology Bandung, West Java, Indonesia
e-mail: myrapg@yahoo.com



A. Saufi et al. (eds.), *Balancing Development and Sustainability in Tourism Destinations*, DOI 10.1007/978-981-10-1718-6_5

tourists to Toraja; setting quality standards and improving ICT capacity among tourism service providers; and enhancing interpretations and visitor management systems within heritage sites.

**Keywords** Toraja · Heritage · Tourism planning · Revitalization

## Introduction

Toraja in South Sulawesi, Indonesia, is a tourist destination known for its ethnicity as the main attraction. It is located in the northern part of South Sulawesi Province, or about eight hours drive from the province capital, Makassar. Administratively, destination Toraja encompasses two regencies, namely, Tana Toraja and North Toraja. The two regencies are led by two Regents.

Earlier tourism related researches on Toraja examined Toraja from an anthropological perspectives (Adams 1984, 2003; McGregor 2000). Toraja is also renown as a producer of specialty arabica coffee (Neilson and Shonk 2014). Tourism in Toraja began in 1972, when it was included in the master plan for tourism development in Indonesia (Adams 1984). Toraja, along with Nias and West Sumatra, were three areas beside Bali and several districts in Java that were designated as tourism development area. The traditional houses of Toraja with their distinctive shape, the ancestral house known as Tongkonan, cliff-hanging, and stone-carved burial sites, and traditional ceremonies such as the death ritual, Rambu Solok, were considered interesting for the taste of ethnic tourists (Adams 1984). International tourists arrivals to Toraja reached its peak in the 1997 with 41,856 visits by foreign tourists, dominated by Europeans.

From 1998 to 2003, a series of unfortunate situations hit Indonesian tourism sector. From monetary crisis in 1998 that led to the fall of Indonesia's second President, Soeharto, a fragile political situation in the beginning of reformation era in 2000, to terrorist attack on the tourist island of Bali in 2002 and the bombing of Marriott Hotel in Jakarta in 2003. In 2003, the number of foreign tourists visiting Toraja reached only 15,385. As many international visitors to Toraja used Bali as their point of entry to Indonesia, the "Bali Bombing" had impacted severely on the number of incoming tourists to Bali and also to Toraja as a consequence. International tourist arrival to Toraja continued to fall from 2004 to 2009, with an average number of 5500 foreign tourists, but picked up to 43,149 foreign tourists and 56,099 domestic tourists in 2013.

European countries such as France, Italy, Spain, Germany, and the Netherlands have long been the main tourist generating countries for Toraja (PT LAPI ITB 2014). However, these last few years have seen an increasing number of tourists coming from East Asian countries such as Japan, China, and South Korea to Toraja (PT LAPI ITB, 2014). International tourists are either coming through Makassar (an international hub for the eastern part of Indonesia) directly from overseas or entering through Bali and continuing to Makassar, before proceeding with ground transportation to Toraja for about 8 h.

As for domestic tourists, the number has been increasing with the rising trend to explore the vast country further beyond Bali among younger and highly educated Indonesians, and the growing fashion for backpacking.

The increasing number of tourist arrival is in line with the government intention toward the revitalization of tourism in Toraja. Through the Indonesian Government Regulation Number 50 year 2011 on the National Master Plan for Tourism Development, the government has again designated Toraja as one of 88 National Strategic Tourism Areas (KSPN) to be prioritized in its tourism development plan. The Indonesian Government through Government Regulation Number 50 year 2011, define KSPN as the area that has the primary function of tourism or have the potential to support national tourism development.

In 2012, the Indonesian government facilitates the formation of destination management organization (DMO) in Toraja with local working group, consisting of industry stakeholders and community representatives, becoming the partner of Government Office of Tourism and Culture in managing destination Toraja. The establishment of DMO Toraja was also aimed at facilitating coordination between stakeholders. In 2014, the planning process is underway for the revitalization of Toraja as one of the national strategic tourism areas.

This paper will discuss the tourism revitalization planning process that was undertaken from July 2014 to December 2014 involving teams of academics from various background, consultants, and local stakeholders, using data collection technics such as observations and site visits, market survey, and focus group discussions (FGD). Some of the objectives identified from this process include diversification of visitor attractions beyond traditional settlements—as expected by the emerging Asian tourists to Toraja; setting quality standards for tourism service providers such as hotels, car rentals, and restaurants; improving information and communication technology capacity among tourism service providers; and enhancing interpretations and visitor management systems within traditional settlements that were nominated as World Heritage Site. Besides formulating plans for the realization of these objectives, the detailed tourism planning document also include designs for new facilities to improve tourism services and boost Toraja's overall ambience as a destination.

## Tourism Planning Process Within the National Strategic Tourism Area of Toraja

The aim of modern planning, according to Williams cited in Mason (2003) is to seek optimal solutions to problems, increase and maximize development benefit. In the field of tourism, Gunn (1994) stated that tourism planning should aim at least on the following goals, i.e., enhance visitor satisfactions, increase rewards to ownership and development, resource protection, and integration of tourism into the social and economic life of communities and areas. In a nutshell, tourism planning can help to shape and control physical patterns of development, conserve scarce resources, provide framework for active promotion and marketing of destinations and serve as a mechanism to integrate tourism with other sectors (Williams cited in Mason 2003).

Gunn (1994), however, reminds that the challenge of tourism planning is the fact that tourism is a fragmented industry, in which coordination among government agencies, between public and private sectors, and among private enterprises is often a challenging task (Jamal and Getz 1995). Thus, within this planning process, there are four major areas in which the revitalization of tourism in Toraja is set to progress, i.e., destination (and attractions), tourism industry, market, and institutional development. This is in line with the main legal framework for tourism sector in the national level, Law No. 10 of 2009 on Tourism, which includes aspects of tourism destinations, tourism industry, tourism marketing and tourism institutions. The tourism revitalization plan which is officially titled "Detailed Plan for National Tourism Strategic Area of Toraja" is to provide direction for the development and management of strategic tourism area to support the realization of tourism development that is sustainable and contributes to improved quality of life.

## Study Methods

Overall, the study that was conducted in relation to preparing the detailed plan for tourism revitalization in Toraja involves four major phase: preparation phase, data collection and field survey phase, analysis and synthesis, and finally, planning and design. In the preparatory stage, secondary data as well as related literatures were collected; several experts were dispatched to Toraja for preliminary studies, and local working group (Toraja DMO) was consulted.

Data collection and field survey phase involved field observation, interviews, and meetings with experts for consultation and confirmation; interviews with related agencies, community leaders, tourism entrepreneurs, and stakeholders including tourists, tour operators and guides; and focus groups discussions (FGD) with actors of tourism industry (hotel owners, tour operators, and tourism industry association), FGD with youth groups and FGD with local authorities such as Regents, head of regency tourism office, and head of other related offices, e.g., public works and regional planning office. FGDs were conducted to ensure greater community involvement as suggested by Jamal and Getz (1995) and to achieve better coordination with other sectors (Gunn 1994).

The third phase is analysis and synthesis of all data and information collected. The last phase is planning and design, which include preparation of vision and mission of Toraja KSPN, defining plans for destination, tourism industry, market, and institutional development, and preparing detailed plans and design for selected areas.

## The Heritage Land of Toraja

Toraja's traditional settlements that are centered on the "tongkonan" are the very attraction of tourism in Toraja. They are also the cultural heritage of the Torajans that is nominated as World Heritage Site (WHS) under the name of Toraja Traditional Settlements. The first nomination dossier was actually submitted in 2005 by the Department of Culture and Tourism of the Republic of Indonesia to UNESCO, consisting of 10 properties scattered throughout the region. The nomination was a series nomination in which 10 different settlements were nominated under the same title. It is worth noting that at first the nomination intended by the Indonesian Government only included the settlement of Ke'te Ke'su (also very popular among tourists). However, a UNESCO meeting for nomination of South-East Asian WHS that was conducted in Toraja in 2001 recommended that the nomination should cover more settlements (Adams 2003).

The fact that these settlements are private properties presents a specific challenge on how to manage these properties once they become World Heritage Site. On the ground, these settlements are private properties owned by families, while the institution that is responsible for coordinating area management and site preservation is The Office of Archaeological Heritage Conservation in Makassar (the province capital) and The Regency Office of Tourism and Culture (Department of Culture and Tourism 2005). While waiting for UNESCO's decision, a lot of changes happened in these settlements, many of them are not really in line with the spirit of preservation that the WHS is all about. New houses and buildings with modern design emerged in the settlements, sometimes too close to the tongkonan. The bamboo roofs were replaced by zinc roofs and in settlements that were frequently visited by tourists, souvenir shops and small eateries emerged almost sporadically.

From the 2005 nomination dossier, UNESCO saw a potential problem in the coordination and management of the sites. Coordination mechanism between institutions and between these institutions with the local foundations representing owners of the settlement was not very clear. The Government of Indonesia then was requested to submit another dossier, which was submitted by the Ministry of Education and Culture in 2013 that consisted of 13 sites which were scattered within the North Toraja and Tana Toraja Regency. The North Toraja Regency is a new regency that was formed in 2008 when the previous Tana Toraja Regency was divided into two regencies.

## Setting Priorities

As mentioned before in the earlier part of this paper, there are four major areas in which the revitalization of tourism in Toraja is set to progress, i.e., destination (and attractions), tourism industry, market, and institutional development. This paper will limit itself to describe some of the objectives that will be prioritized in the tourism planning process.

## *Diversification of Tourists' Attractions Beyond Traditional Settlements*

From FGD that was conducted with tourism industry, it was known that beside European tourists that were considered as Toraja's long time market, tourists from East Asian countries, especially China had started to come to Toraja. In contrast to European tourists, however, Asian tourists expect some diversification beyond traditional settlements. Unlike their Western counterpart, Asian tourists are not really keen on spending their holiday visiting burial site after burial site, no matter how unique they are. According to several tour guides that attended the FGD, this is because East Asian tourists have different perception about going to graves or burial sites, in which they usually do it at a specific time of the year. While most European tourists are interested in visiting as many settlements (most include cliff-hanging graves, stone-carved burial sites, or burial caves), Asian tourists like to spend more time for leisure sight-seeing and shopping.

In addition, these last few years have seen the emergence of tourists (mostly Westerners) who come to Toraja for trekking. Toraja's karsts and hills were explored by Lonely Planet team a few years ago and Toraja geo-trekking route was published in the Lonely Planet. From then on, quite a number of tourists come to Toraja for trekking. In order to assist the local government in developing Toraja's geo tourism, the Detailed Plan for National Strategic Tourism Area of Toraja recommends several trekking routes that can be developed.

Another potential product that still needs a lot of improvement is agrotourism, especially one that is based on coffee. Pango-pango coffee plantation in Tana Toraja Regency, for instance, need better visitor facilities. Moreover, Toraja coffee is popular internationally among coffee lovers, but on the contrary, many tourists that are interviewed in this study said that coffee houses were still very rare in Toraja. Cooperation between coffee growers, coffee production companies, and hospitality entrepreneur is needed to link coffee as an agricultural commodity with tourism.

## *Setting Quality Standards and Improving ICT Capacity Among Tourism Services Providers*

Observations, survey, and interviews to tourism service providers that were carried out as part of this study concluded that local ownership is still strong in Toraja. However, many cases showed that the quality of facilities and services needs to be improved. In many hotels, maintaining the quality of their facilities is a problem due to low occupancy rate. As for restaurants, most international tourists that were surveyed for this study stated that for an international destination, Toraja is still lacking in choices of restaurants as well as food variety. On the other hand, domestic tourists feel the need for more Halal food in Toraja. Unlike most areas in Indonesia, Toraja is a region where Islam is not the religion of majority, therefore affecting the availability of Halal food outlets. Although in reality Halal foods and restaurants are not that difficult to find, many domestic tourists still perceive Toraja as a destination lacking in conformity to Halal standard. The Government thus should try to increase the number of Halal food outlets, provide affordable Halal certification scheme, and increase the feasibility of Halal outlets in its tourism promotion.

Another issue within the tourism industry that needs to be dealt with immediately is improving ICT capacity among tourism services providers. While most tourism businesses in Toraja are small to medium enterprises, basic facilities and services that relies on ICT, such as internet booking or even credit card acceptance are still low. This tourism planning document recommends steps that can be taken in improving ICT capacity among stakeholders.

## *Enhancing Interpretations and Visitor Management Systems Within Toraja's Heritage Sites*

The objective of visitor management is to improve the quality of experience and minimize the negative impact of the visiting tourists (Mason and Kuo 2006). Visitor management is especially needed in cultural heritage sites, where sustainability and integrity of the site and the area became a priority. Through surveys to heritage sites in Toraja which comprise of traditional settlements and all of their elements (granaries, field, and burial sites), the team identified the following issues: the absence of visitor centers and adequate interpretation facilities in almost every heritage site, either in the form of trained and qualified local guides as well as in the form of interpretation boards, visitors booklet, etc.; the ambience of "living heritage" is lacking in several sites as occupants of tongkonan move to newer and more modern houses; the increase in the number of souvenir shops; and the lack of good and clean restrooms.

The detailed plan for tourism revitalization in Toraja thus suggested following steps: Building a Visitor Information Center that provide information about the heritage site as Toraja culture in general; utilize the Information Center as an area where new visitors wait before they can enter the main area; establish rules for visitors; monitor and regulate the number of visitors in sensitive and fragile area such as burial site.

## Conclusion

The preparation of Detailed Plan for National Strategic Tourism Area of Toraja as discussed and described in this paper is one of the steps taken in a bid to revitalize tourism in Toraja. There are four major areas in which the revitalization of tourism in Toraja is set to progress, namely destination (and attractions), tourism industry, market, and institutional development. The Plan among others recommends that the revitalization process should prioritize on the following matters, i.e., diversification of tourist attractions beyond traditional settlements, setting quality standards and improving ICT capacity among tourism service providers, and enhancing interpretations and visitor management systems within Toraja's heritage sites. To achieve these priorities, cooperation between various stakeholders is essential.

## References

Adams K (1984) Come to TanaToraja, "Land of the Heavenly Kings" travel agents as brokers in ethnicity. Ann Tourism Res 11:469–485

Adams K (2003) The Politics of Heritage in TanaToraja, Indonesia: interplaying the local and the global. Indonesia Malay World 31(89):91–107. doi:10.1080/13639810304444

Department of Culture and Tourism, The Republic of Indonesia (2005) Nomination of the Tana Toraja traditional settlement for the inscription in the world heritage list

Gunn CA (1994) Tourism planning: basics, concepts and cases, 3rd edn. Taylor & Francis, Philadelphia

Jamal T, Getz D (1995) Collaboration theory and community tourism planning. Ann Tourism Res 22(1):186–204

Mason P (2003) Tourism impacts, planning and management. Butterworth-Heinemann, Oxford

Mason P, Kuo IL (2006) Visitor management at Stonehenge, UK. In: Leask A, Fyall A (eds) Managing world heritage sites. Butterworth-Heinemann, Oxford

McGregor A (2000) Dynamic texts and tourist gaze: death, bones and buffalo. Ann Tourism Res 27(1):27–50

Neilson J, Shonk F (2014) Chained to development? livelihoods and global value chain in the coffee-producing Toraja Region. Aust Geogr 45(3):269–288. doi:10.1080/00049182.2014.929998

PT LAPI ITB (2014) Inception report: detailed plan for national strategic tourism area of Toraja

# Chapter 6
# A Local Sense of Place for Cultural Sustainability: Reconstruction of Place Identity in Kuta, Bali

**Luh Micke Anggraini**

**Abstract** Local sense of place refers to the emotional and reciprocal relationship between the residents of a community and the environment they inhabit. It is fundamental to cultural sustainability, especially after a place is exposed to tourism. This paper derives from a digital ethnography that examines the reconstruction of place identity by local community in *desa adat* Kuta, a Balinese customary village, which is also a popular beach destination in Bali, Indonesia. The local residents' contemporary expressions with regard to sense of place were analyzed from the local community's social media (Facebook group). This study adopts discourse analysis to identify the recovery of the indigenous characteristics of Kuta and Balinese values expressed by the inhabitants, despite the challenge of living in a tourist destination. The findings suggest that the indigenous community has a significant role in maintaining the traditions, identity, values, and esteem of their village, according to *desa adat* Kuta. Concomitantly, the identities of the people of Kuta are being (re)-constructed from their local emotional attachments with their village. Further, the residents also leverage tourism development to enhance their livelihood and to contribute to their identity formation. The study's conclusion has an important implication, particularly for advancing existing cultural tourism policy, by suggesting the incorporation of local sense of place in tourism planning and development in Bali.

**Keywords** Sense of place · Cultural sustainability · Place identity · Local community · Kuta · Bali · Indonesia

L.M. Anggraini (✉)
Sekolah Tinggi Pariwisata Nusa Dua Bali(Bali Tourism Institute),
Nusa Dua, Bali, Indonesia
e-mail: mickebali@yahoo.com



A. Saufi et al. (eds.), *Balancing Development and Sustainability in Tourism Destinations*, DOI 10.1007/978-981-10-1718-6_6

## Introduction

Along with many other factors, tourism has had a considerable influence, with sociocultural significance, on the lives of people who live in traditional places, in the sense that it affects their quality of life and puts pressure on local values. Studies have suggested that issues relating to place identity are overlooked by tourism industry planners and policy makers (e.g., Carter et al. 2007; Kerstetter and Bricker 2009). A sense of place, manifested as place identity, should therefore be revealed to strengthen the local character of an area while also enhancing the tourist experience.

This paper is drawn from ethnography about local sense of place in Bali after the local people had experienced considerable transformation due to tourism activities. The Balinese customary village of Kuta was observed to illustrate the way local identity there had been refabricated after the place had been extensively industrialized. From this perspective, this study aims to suggest an inclusive approach in illuminating place identity established through the emotional relationship between people and place, as performed by the indigenous residents living in a geographically bounded community, especially after the meanings of place have been altered after being exposed to tourism. This process is proposed in navigating the future development of tourism at the community level and to support cultural sustainability in affected areas.

## Place Identity and Cultural Sustainability in a Tourism Destination

Several definitions of 'sense of place' have denoted this construct as a reciprocal and emotional relationship between humans and their environments (Hubbard et al. 2004; Schofield and Szymanski 2011; Vanclay 2008). Sense of place has also been used to describe the distinctiveness or unique character of particular localities and regions (Foote and Azaryahu 2009). It can be inferred that sense of place is what people feel matters most regarding the place to which they belong. Sampson and Goodrich (2009) and Altman and Low (1992) argued that place attachment, as the core idea of sense of place, can foster and sustain group, community and cultural identity, with a unique cultural identity and a sense of community distinctiveness. Derived from the notion of sense of place, 'place identity' embodies the notion of a symbiotic relationship between person and place. Significant to both individuals and communities, place identity can be described as

> … a sub-structure of the self-identity of the person consisting of broadly conceived cognitions about the physical world in which the individual lives. These cognitions represent memories, ideas, feelings, attitudes values, preferences, meanings, and conceptions of behavior and experience, which relate to the variety and complexity of physical settings that define the day-to-day existence of every human being (Proshansky et al. 1983, p. 59).

This relationship involves perceptions of the qualities that imbue places with a distinctive character, as well as an individual's tie or association with place (Peterson 1988). At the center of such physical environment-related cognitions is the 'environmental past' of the person; a past consisting of places, spaces, and their properties, which have served instrumentally in the satisfaction of the person's biological, psychological, social, and cultural needs.

Thus, people understand places differently and distinguish between them by drawing on key somatic features, cultural elements, historical memories, experiential connections, and so on. Relph (1976) indicated that place identity is a product of an awareness of a place's 'deep symbolic significance' and an appreciation for its individual character. This identity also develops from recognition of a place's attributes drawn from its natural setting, cultural diversity, topographic conditions, and general quality of life. Any place might have special meaning to the local community, valued by the people as characterizing a local area and having a distinctive quality that sets it apart from other places (Schofield and Szymanski 2011). This study therefore adopts the above definitions by suggesting that place identity is a complex cognitive structure that is characterized by a host of attitudes, values, thoughts, beliefs, meanings, and behavior tendencies that go well beyond simple emotional attachments and a sense of belonging to a particular place.

Thorne (2012) argued that tourism experiences are located in a particular place. His proposition was based on Richard Florida's opinion that 'place is becoming the central organizing unit of our economy and society,' which implies that any place-based activities, including tourism, have 'place' as their product. As found in many studies, tourism brings modernity and multicultural society, and involves cultural commodification in destination communities through a complex series of production activities (e.g., Cohen 1988; Mowforth and Munt 2009; Urry and Larsen 2011). Lew (2014), for instance, acknowledged that tourism also has a role in producing local and tourist culture. However, the uncontrolled development of tourism has caused significant damage, as suggested by Salazar and Bushell (2013); tourism can disadvantage the local people when its implementation destroys the local culture and exploits the natural resources. In this sense, the development of tourism is associated with the production of 'constructed' space for representation (Frisvoll 2012) and is deemed a vehicle enabling modernity to disrupt a small-scale, static, and well-ordered locality (OECD 1994) leading to negative impacts on its traditional culture. Further, gentrification has also become an issue of practises in commodified places (Kennedy and Leonard 2001; Smith 2007).

According to Relph (1976, p. 20), the home is considered the place of greatest personal significance in someone's life, 'the central reference point of human existence.' After the modification of the physical environment of neighborhoods and the changes to place characteristics that tourism brings, a local community experiences a challenge in sustaining their own culture. Bali, as a tourism destination, anticipated the impact of tourism on local culture by promoting 'cultural tourism' and issuing a 'cultural tourism policy' to safeguard tourism development on the island. However, understanding the way this strategy can be implemented at the community level is more important than having a policy, particularly understanding the way cultural identity can be maintained in an environment in which the local residents live with tourism or share their place with tourists.

Cultural sustainability is arguably the most neglected dimension of sustainable development (Myllyviita et al. 2014). In UNESCO's (2001) declaration on the topic, culture is regarded as 'distinctive spiritual, material, intellectual and emotional features of society or a social group … that … encompasses … art and literature, lifetyles, ways of living together, value systems, tradition and beliefs.' According to the Sustainable Development Research Institute (1998), cultural sustainability is 'the ability to retain and to allow change to be guided in ways that are consistent with the cultural values of a people.'

Places or spaces are utilized and designed by people and thus provide the foundation for creativity and new cultural development. Even in small places, such as a town or village, significant cultural productions can be generated in small clusters of multi-ethnic urban communities that become known widely (Anheier and Raj Isar 2011). Therefore, this study considers that places should be perceived as the cultural production of local identities. In this case, places in Bali, as everywhere, have become centers of cultural activities to determine what is vital to sustainability. To understand the relationship between tourism, place identity, and sustainability, this study investigates (1) the impact of tourism on place identity; (2) the recovery of place identity after extensive tourism development; and (3) the implications for cultural sustainability.

## Study Area: Kuta, Bali, Indonesia

Kuta is a Balinese customary village (*desa adat*),[1] which is also known as a popular beach destination in the South of Bali Island in Indonesia. No longer a traditional fishing village, the current landscape of Kuta abounds with symbols of place identity related to Bali and tourism as the place has reached the 'stagnation' stage of area life cycle (Butler 1980). The distinctive attributes of Kuta reflect key physiographic, socioeconomic, historic, and cultural features such as surf beaches, beach culture, shops, restaurants and cafes, nightlife, traffic, and typical tourism activities in the tropical environment.

The relationship between the *Kutanese* (Balinese residents who are members of *desa adat* Kuta) and Kuta village is the focus of this case study. As observed in Kuta, the sense of place and place identity are very much associated with the history, culture, and tourism, and have become parts of the life course brought by the place to individuals and the memories people hold about the place (Altman and Low 1992; Proshansky et al. 1983; Rubinstein and Parmelee 1992). There have

[1]The village of *desa adat* Kuta covers 723 ha located on a 7 km beachside facing the Indian Ocean to the west. The area is relatively flat, less than 100 m above sea level. In 2010, around 2917 families were registered as residents, out of a population of 12,337 (Badan Pusat Statistik Kabupaten Badung 2011). The village consists of 13 traditional neighborhoods (*banjar*).

been enduring issues concerning place identity associated with historical background. The low status of Kuta's initial settlement was partly based on its past as a slave port as well as old Balinese attitudes that classified the beach environment in a lower social status. Kuta was associated with poor, low-class farmers and fishermen and was seen as a place of exile for those who had committed serious crimes or were social rejects (Picard 1996).

Over the last century, Kuta has been transformed for different purposes and has been growing in an uncontrolled way. Kuta has become an example of a market-driven, unplanned tourist area attracting temporary workers, and the villagers have experienced enduring social and environmental issues (Vickers 1989, 2012). It is important to note that as Kuta village is a *desa adat* (as are many other villages in Bali), the local residents are responsible for maintaining their sociocultural obligations in a Balinese village system, such as conducting temple festivals in their home village and participating in communal activities. Local community activities thus occur simultaneously with daily tourism activities in their village. These 'ways of operating' establish numerous practises through which the residents reappropriate the space through sociocultural productions (de Carteau 1984) and the socio-spatial interactions of everyday life, which help create a sense of place for themselves and others (Perkins and Thorns 2012). After long-term tourism development in this village, the maintenance of the social relationship of the local community in Kuta has been a challenge, especially since the village has been transformed into a tourist town.

## Methodology

For this research, the local people's contemporary expressions regarding sense of place were analyzed from online reference to the local community's social media (Facebook group).[2] This study adopted discourse analysis to identify the recovery of the indigenous characteristics of Kuta and Balinese values by the inhabitants who are experiencing the challenge of living in a tourist destination. Digital ethnography (Murthy 2008) was applied in this research to collect the data from contemporary conversations on the social network sites of Facebook groups (Boyd and Ellison 2007), and from relevant websites. This approach aimed to understand the way Balinese residents in Kuta signify their place after tourism in everyday life, revealing the place identity—the main questions of this research.

Social media was used because of its viability in obtaining the current data of the studied area. The Kuta Facebook Community was developed in early 2012, initiated by the local community of Kuta to foster community cohesion through online communication among the local residents of Kuta village. As with many other

[2]https://www.facebook.com/groups/kutacom/.

Facebook groups, it offers discussion boards and opportunities for members to find someone who can provide them with information.

Despite its exclusiveness (prioritizing local residents), membership of this group is voluntary, participatory, and noncommercial, and it is also open to those who feel obliged to contribute to community development in Kuta. The group started its online activities in March 2012 with 728 members. By July 2015, the Kuta Facebook Community had 2136 members, with more than 1000 uploaded photographs and enduring online commentaries and conversations. Most of the contents are about the sociocultural and tourism activities in Kuta, as reported by local residents.

The use of social network sites is not only a feasible method for documentary but also a tool for sharing stories, and creating and communicating histories with the wider community. In this regard, the Kuta Facebook Community may help to mediate the aspirations of locals for the future development of tourism in Kuta. The discourse analysis of identities (Silverman 2006) that was used in this research aims to reveal the sociopsychological characteristics of a person or community.

## Findings and Discussion: Kuta Identities After Tourism

The way the Balinese community has preserved their sense of place in Kuta, such as through place attachment and place identity, as well as the physical environment as resources of tourism development, could be observed through the Kuta Facebook Community. The discussion topics included tourism development issues, such as the results of place transformation (both desirable and undesirable), and the economic, sociocultural, and environmental impacts of tourism in Kuta. The conversations also revealed a lack of public consultation in tourism planning and development in this area, through their expressions of ignorance of the tangible and intangible local values and identities in Kuta. In relation to place identity issues, the results suggested that the local community had their own interpretations about Kuta after tourism, and they shared their thoughts on the impacts of tourism on the local identity.

A number of aspects of present-day Kuta are associated with tourism activities in this area, such as shopping districts with modern architecture, surfing, and traffic congestion. Compared to other places, beach surfing here is considered an acceptable activity by the locals and adopted as a desirable trait. In this case, the residents have embraced a nonindigenous cultural product that is practiced widely among the community members and they support fostering surfing as a local identity. As revealed in this study, once a tourism activity is believed to create a positive image for a place, there is enthusiasm in the local community to promote this interest and become attached to this identity. Surfing also maintains the people's connection with the beach environment and relieves tensions with the local values that are generated by tourism. It is argued that a sense of place can support a community's or culture's response to crisis, or when experiencing urban neighborhood change, racial and ethnic conflict, or national tensions (Sampson and Goodrich 2009). Thus, the positive emotional bonding between people and place that is found in surfing may help the people when they experience the undesirable impacts of tourism in their environment.

This study also found that the Balinese in Kuta adhere to the local principles of *Desa Kala Patra*,[3] applying this value both locally and contextually. It can be inferred that the local values of respecting the place–time context remain strong, as the people still refer to Kuta as a *desa adat*. However, the study also revealed that in Kuta, the Balinese identity of this village is not always deemed appropriate from the local perspective. In one case, locals criticized the Balinese statues of a Balinese dancer and a traditional musician that were placed at the entrance to their village, despite the representation of Balinese identities. They said that according to the principle of *Desa Kala Patra*, such a practice was an example of things being 'out of context,' which most Balinese would wish to avoid. In this case, the residents preferred that the statues would represent Kuta, rather than Bali, and argued that Kuta held a number of figures that would be better symbols of their village. They suggested alternative ideas for statues that would better represent Kuta, such as of the Hindu God Baruna,[4] because Kuta is a coastal village. They even endorsed a statue of a surfer, which is unrelated to Balinese culture. This recommendation is understandable because since the onset of tourism in Kuta in the mid-twentieth century, surfing has become a local subculture (Koke 1987; Mabbett 1987), and has recently been adopted as the identity of the village, with villagers admitting that 'Kuta and surfing are now inseparable.'

Other similar instances of the recovery of 'Kuta identity' were found in this study. Instead of promoting their Balinese identity, Kuta residents have attempted to promote local significance. Examples were public facilities, such as a passage-way and community art hall that were named after a Kuta-born Balinese musician and Legong dance creator, the late Mr. Lotring. Further, Lotring's creation, Legong dance, was used in the annual tourism festival 'Kuta Karnival' in 2011 and 2012, in addition to the use of surfing to symbolize Kuta.

This study also revealed that the local and national government programs applied in Kuta had neglected the crucial intermediary step of 'consulting with the community.' This case showed that the identity of Kuta had been reconstructed through social processes undertaken within the bounds of a certain landscape with particular aspects, which should be taken into consideration in the government's and industry's agendas in Kuta. Identifying Kuta simply as 'Bali' is no longer relevant, as the identity of Kuta is not derived solely from the Balinese culture.

The online community activities that were described by the Kuta Facebook Community showed that local people should be involved in the interpretation of local places and culture. With this approach, local people can take an active part in

---

[3]*Desa Kala Patra* is the place–time–occasion concept in Balinese Hinduism, which determines the character of a place.

[4]God of the sea in Balinese Hindu mythology.

all of the process of this interpretation, including the research and the presentation and celebration of people and place (Bramwell and Lane 1993), which is aligned with the principle of sustainable tourism of encouraging community participation. Although tourism is believed to have a considerable impact on local culture and to have a disruptive effect on place attachment, within the arrangement of a *desa adat* community, the local residents of Kuta have survived and are striving to reestablish a network of community members.

## Conclusions

This study has shown that the indigenous community in Bali who live within the *desa adat* customary village of Kuta has attempted to foster local culture, responding to the effects of tourism on place identity. The Kuta Facebook Community, as a social network site, can be considered a workable media for maintaining community cohesion in highly transformed places, to sustain the social and cultural life. This undertaking contributes notably to the persistence of Kuta as a representation of a customary Balinese village, not only fostering a sense of place but also helping to empower individuals and the wider community, and assisting in shaping personal and community identities. Further, it has encouraged the residents to understand, value, and maintain their environment, cultural resources, and traditions.

Alongside this process, Kuta identities are (re)-produced from local emotional attachments with their village after being exposed to tourism. As such, the local community of Kuta perceives tourism as part of their identity, as it has been practiced for a long time. Another important finding of this study has been that the local community of Kuta has attempted to (re)-construct the identity of their village in a way that is intrinsically 'Kuta' through the reinterpretation of people, place, and activities in their localities. The reinvention of 'Kutaness' has been a local initiative to distinguish themselves from other Balinese on the island. It is also obvious that the local community aims to (re)-shape the positive image of Kuta as a tropical beach destination according to cultural tourism principles.

As with all research of an exploratory nature, this study has some limitations. It lacks comprehensiveness due to the limited time available and types of social media observed. Moreover, it observes place identity-related discourse only in online communities, rather than in the actual setting. In future analyses of place identity in Bali, community perceptions expressed both online and offline should be integrated, to examine cultural sustainability of a place in a more far-reaching manner. Upcoming research should focus on a different geographic area of Bali (considering that the mountainous regions have been regarded as superior in Balinese culture), and investigate tourism customers' perceptions of place identity in other tourist areas of Bali.

Acknowledging its significant influence on people, place, and culture, local sense of place has been proved to be indispensible for the cultural sustainability of a tourism destination. As shown in this study, sense of place also has the capacity to foster community and cultural identity while developing tourism in Bali. Tourism development should seek to understand further the link between place, community, and identity. It is recommended that government at all level, planners, and developers should include the community's sense of place in tourism policy and practises, to safeguard local culture.

**Acknowledgments** This paper is derived from my Ph.D. thesis.

## References

Altman I, Low SM (1992) Place attachment, vol 12. Plenum Press, New York, London

Anheier H, Raj Isar Y (eds) (2011) Heritage, memory and identity. Sage, London

Badan Pusat Statistik Kabupaten Badung. (2011). Badung dalam angka (Badung in figures). Retrieved from http://www.badungkab.go.id/bda_2011/Tabel%203.1.2.pdf

Boyd DM, Ellison NB (2007) Social network sites: definition, history, and scholarship. J Computer-mediated Commun 13(1):210–230

Bramwell B, Lane B (1993) Interpretation and sustainable tourism: the potential and the pitfalls. J Sustain Tour 1(2):71–80

Butler R (1980) The concept of a tourist area cycle of evolution: Implications for management of resources. Can Geogr/Le Géographe Canadien 24(11):5–12

Carter J, Dyer P, Sharma B (2007) Dis-placed voices: sense of place and place identity on the Sunshine Coast. Soc Cult Geogr 8(5):755–773

Cohen E (1988) Authenticity and commodification in tourism. Ann Tourism Res 15:371–386

de Carteau M (1984) The practice of everyday life. Los Angeles, University of California Press, Berkeley

Foote KE, Azaryahu M (2009) Sense of place. In: Rob K, Nigel T (eds) International encyclopedia of human geography. Elsevier, Oxford, pp 96–100

Frisvoll S (2012) Power in the production of spaces transformed by rural tourism. J Rural Stud 28 (4):447–457

Hubbard P, Kitchin R, Valentine G (eds) (2004) Key thinkers on space and place. Sage, London

Kennedy M, Leonard P (2001) Dealing with neighborhood change: a primer on gentrification and policy choices. A discussion paper. The Brookings Institute Centre on Urban and Metropolitan Policy, Washington, DC

Kerstetter D, Bricker K (2009) Exploring Fijian's sense of place after exposure to tourism development. J Sustain Tour 17(6):691–708

Koke L (1987) Our hotel in Bali. January Books, Wellington

Lew AA (2014) Introduction to special issue—cultural geographies of tourism: image, identity and place. Tourism Geogr: Int J Tour Space, Place Environ, 16(2):171–173

Mabbett H (1987) In praise of Kuta: from slave port to fishing village to the most popular resort in Bali. January Books, Wellington

Mowforth M, Munt I (2009) Tourism and sustainability: development globalisation and new tourism in the third world, 3rd edn. Routledge, Oxon and New York

Murthy D (2008) Digital ethnography: an examination of the use of new technologies for social research. Sociology 42:837–855

Myllyviita T, La'htinen K, Hujala T, Leskinen LA, Sikanen L, Leskinen P (2014) Identifying and rating cultural sustainability indicators: a case study of wood-based bioenergy systems in Eastern Finland. Environ Develop Sustain, 16:287–304

OECD (1994) Tourism strategies and rural development. Organization for Economic Co-operation and Development, Paris. Retrieved from http://www.oecd.org/cfe/tourism/2755218.pdf
Perkins HC, Thorns DC (2012) Place, identity and everyday life in a globalizing world. Palgrave Macmillan, New York
Peterson G (1988) Local symbols and place identity: Tucson and Albuquerque. Social Sci J 25 (4):451–461
Picard M (1996) Bali: cultural tourism and touristic culture (D. Darling, Trans.). Archipelago Press, Singapore
Proshansky HM, Fabian AK, Kaminoff R (1983) Place–identity: physical world socialisation of the self. J Environ Psychol 3:57–83
Relph E (1976) Place and placelessness. Pion Limited, London
Rubinstein RL, Parmelee PA (1992) Attachment to place and the representation of the life course by the elderly. In: Altman I, Low SM (eds) Place attachment. Plenum Press, New York, NY, pp 139–163
Salazar JF, Bushell R (2013) Heritage for sale: Indigenous tourism and misinterpretations of voice in northern Chile. In: Staiff R, Bushell R, Watson S (eds) Heritage and tourism: Place, encounter, engagement. Routledge, London and New York, pp 187–212
Sampson KA, Goodrich CG (2009) Making place: identity construction and community formation through 'sense of place' in Westland, New Zealand. Soc Nat Res 22:901–915
Schofield J, Szymanski R (eds) (2011) Local heritage, global context: cultural perspectives on sense of place. Ashgate, Surrey
Silverman D (2006) Interpreting qualitative data. Sage, London
Smith M (2007) Space, place, placelessness in the culturally regenerated city. In: Richards G (ed) Cultural tourism: global and local perspectives. Haworth Press, Binghamton, NY, pp 91–111
Sustainable Development Research Institute (1998) Social capital formation and institutions for sustainability. Workshop proceedings prepared by Asoka Mendis. Sustainable Development Research Institute, Vancouver. Retrieved from www.williambowles.info/mimo/refs/soc_cap.html
Thorne S (2012) Place-based cultural tourism: a new planning paradigm. Retrieved from http://economicdevelopment.org/2012/10/place-based-cultural-tourism-a-new-planning-paradigm/
UNESCO (2001) UNESCO universal declaration of cultural diversity (UNCED). Retrieved from http://portal.unesco.org/en/ev.php-URL_ID=13179&URL_DO=DO_TOPIC&URL_SECTION=201.html
Urry J, Larsen J (2011). The tourist gaze 3.0. Sage, London
Vanclay F (2008) Place matters. In: Vanclay F, Higgins M, Blackshaw A (eds) Making sense of place: exploring concepts and expressions of place through different sense and lenses. National Museum of Australia, Canberra, pp 3–11
Vickers A (1989) Bali: a paradise created. Penguin Books Australia, Ringwood
Vickers A (2012) Bali: a paradise created, 2nd edn. Tuttle Publishing, Singapore

# Chapter 7
# Perceptions of Impacts and Development in a Cultural Tourism Hub: Ubud, Bali

**Peita Hillman, Brent D. Moyle and Betty Weiler**

**Abstract** Tourism provides a raft of economic benefits to destination communities, including employment, income, foreign investment, improved infrastructure and facilities. However, core benefits received are often juxtaposed with local concerns regarding economic, environmental and social impacts of tourism, such as seasonality, congestion, acculturation and foreign ownership. Much literature supports the notion that residents support tourism, as long as benefits derived outweigh any costs incurred. However, such studies have not specifically considered the perspectives of locals who work in the tourism industry. Consequently, the aim of this research is to explore the perceptions of locals employed in the tourism industry of Ubud, Bali, towards further development. To achieve this objective a series of 21 in-depth, semi-structured interviews with Balinese locals employed in the tourism industry were undertaken. The interviews were conducted with the help of a bilingual university-educated local "gatekeeper", and were audio recorded, transcribed and then analysed using open, axial and selective coding to identify common themes. Key findings revealed that respondents value their community and surroundings, due primarily to the village feel, scenery and climate. Respondents showed concern for the potential of tourism development changes to Ubud such as modern, western style accommodation, loss of traditional farming land and increased traffic. However, many were equally appreciative of the benefits received from tourism such as job opportunities, economic development, opportunities for cultural creativity and improvements to the local infrastructure and facilities. Respondents expressed a desire for more tourists to travel to Ubud whilst keeping

P. Hillman (✉)
School of International Tourism, International College of Management, Sydney, Australia
e-mail: phillman@icms.edu.au

B.D. Moyle
Griffith Business School, Griffith Institute for Tourism, Griffith University, Gold Coast Campus, Southport, QLD, Australia
e-mail: b.moyle@griffith.edu.au

B. Weiler
School of Business and Tourism, Southern Cross University, Coolangatta, QLD, Australia
e-mail: betty.weiler@scu.edu.au



A. Saufi et al. (eds.), *Balancing Development and Sustainability in Tourism Destinations*, DOI 10.1007/978-981-10-1718-6_7

development levels sustainable for future generations. While an increase in tourists and job opportunities in the future was welcomed by most respondents, some also stipulated that authorities must maintain control of tourism development, limiting the number and scale of new hotels and restaurants whilst preserving the physical environment and culture. Long-term tourism industry employees tended to express more caution about further tourism development in Ubud as opposed to recently employed locals. Distance was also a key factor, with locals with a workplace situated further away from the centre of Ubud more likely to support further development than those who worked in the centre of town. These findings suggest that locals employed in tourism in Ubud would likely support further development if located away from the centre of town and built according to Balinese tradition and local developmental regulations. Younger respondents also were more supportive of future development than older respondents citing job creation and career opportunities as incentives. Future research should consider perceptions of residents in destination communities not directly employed in the tourism industry to compare and contrast their views on the impacts of tourism and future development with those who are tourism industry employees. In addition, future research should enhance the conceptual clarity surrounding issues of how tourism and future development impacts destination communities of unique culture and traditions.

**Keywords** Bali · Tourism impacts · Employees' perceptions · Communities · Tourism development

## Introduction

The island of Bali is located between Lombok and Java in the Republic of Indonesia (Hitchcock and Putra 2007). Bali is considered a tourism paradise that boasts a range of accommodation choices from affordable resorts to luxury retreats, beautiful landscapes such as lush rice terraces and unique cultural offerings including dance, music, art and cuisine (Byczek 2011). In 2014, it was estimated that there were over 10 million visitors to Bali, of which approximately 3.7 million were international visitors (Bali Government Tourist Office 2015a, b, c).

Centrally located, 35 km from the capital of Denpasar, the regional centre of Ubud has been traditionally considered as a place to escape from the mass tourism and high level of development on the south side of the island (Harmini 2009). Visitors to Ubud can experience the famous Monkey Forest, Ubud Market, museums and walking tours of the rice paddies (Bali Tourism Board, n.d.)

However, Ubud has its share of tourism development issues, such as agricultural land being sold to developers to construct more hotels, villas and restaurants for visitors (Sutawa 2012). Ubud has also been reported to be displaying initial signs of acculturation, where one dominant culture, usually Western, infiltrates another culture (Hall 2007). One illustration of this is the strong presence of imported brands in the main tourist area including its own Starbucks coffee outlet and Pandora jewellery store.

Previous studies tend to focus on identifying variables associated with tourism impacts such as employment opportunities, standard of living, pollution and measure residents' attitudes around these (Kim et al. 2012; Nunkoo and Ramkissoon 2010). However, there is limited research exploring how locals who are actually employed in the tourism industry perceive the impacts of tourism on their surrounding environment and how this influences their support for future development.

Respondents in the present study were asked for their opinion on how tourism has affected Ubud's economic, social and cultural environments. Rather than suggesting what the impacts could be such as inflation, dependence on tourism and job creation, residents responded with their own suggestions so knowledge could be built entirely from their perspective.

## Literature Review

Tourism has been recognised by previous studies to generate economic benefits for the Balinese community such as employment, income, foreign investment, improved infrastructure and facilities (Sutawa 2012; Hampton 2005). However, concerns about rapid, unregulated development and the environmental and social impacts of tourism on the Balinese community have been raised by tourism scholars (Warren 2012; Cole 2008).

Cole (2012) identified that a key reason these concerns have been unearthed is because tourism in Bali lacks the capital or entrepreneurship for appropriate infrastructure and sustainable development. For instance, MacRae (2002) argues that the scale of development which includes unlicensed and illegal support facilities had left irreversible scars on the landscape as traditional rice paddies and farm land is sold to developers, 85 % of whom are non-Balinese. More recently, it has been reported that large numbers of tourists visiting the island are creating impacts for the environment including litter, waste and pollution (MacRae 2011; Gurtner 2007). Physical impacts from demands on the environment are increasingly attributed to the level of tourism development, in particular, the strains on the water supply and associated issues with water quality and saltwater intrusion (Cole 2012). In addition to pressure on scarce resources such as water, it has been identified that Bali has experienced several of the negative social consequences of mass tourism including traffic congestion, overcrowding at key tourist sites, commodification of culture and undesirable tourist behaviour (Deery et al. 2012). The positive and negative impacts that tourism development poses for a community have been demonstrated to influence the level of support that will be present (Andereck and Nyaupane 2011).

Involvement and participation in tourism from the host community in the region is recognised as an important requirement for sustainable tourism and associated development, without which support for the industry is unlikely (Lee 2013; Deery et al. 2005). Some argue that tourism should also reflect a single identity of the region, which does not conflict with local cultural and environmental values (Angelkova et al. 2012). However, Byczek (2011) identified that portraying a single identity poses a significant challenge for island planners, and that the traditional way of life is at odds with mass tourism and a westernised urban landscape.

Previous literature has identified a lack of meaningful involvement in the tourism industry by locals can lead to unfavourable attitudes towards development (Lee 2013), which is an obstacle in the pursuit of sustainable tourism (Waligo et al. 2012). As the density of tourism development increases so does the likeliness of negative attitudes towards it from the local community (Hunt and Stronza 2013). The most common form of involvement in the tourism industry is through direct employment of locals, in tourism-related businesses such as hotels, transport and tour operators (Collinson 2012). Previous studies have identified that if locals are not directly involved in the tourism industry, especially on islands, the propensity for negative social impacts of tourism increases (Moyle et al. 2010).

In addition, previous studies have identified a number of other factors which may influence residents' perceptions and attitudes towards tourism (Andereck et al. 2005; Nunkoo and Ramkissoon 2011). These include though are not limited to the type and length of tourist–host interaction, the importance of tourism to the community group, level of economic dependence on the tourism industry, the extent of tourism development and density, length of time living in the community and the location of the community in relation to the tourism zone (Deery et al. 2012; Teye et al. 2002). Other authors confirm that there are several community constructs to consider including housing, neighbourhood, family, satisfaction with surrounding community, social life and employment (Sirgy and Cornwell 2001; Grzeskowiak et al. 2003).

## Method

Twenty-one semi-structured interviews were conducted with local tourism industry employees in Ubud, Bali. Previous studies have recognised that local industry employees are well positioned to comment on tourism impacts and development, as they have a clear economic connection to the industry (Bohdanowicz and Zientara 2009). Building on an approach from Simmons and Fairweather (2005), three stages were used to recruit and select participants for the research. First, an online search was conducted to identify key tourism businesses in the region. Second, restaurant staff, hotel workers and taxi drivers were approached and invited to participate in the research at their place of employment. Third, purposive sampling was used via a Bahasa Indonesian speaking gatekeeper who approached potential respondents and set up interviews from referrals from initial participants and contacts in the local tourism industry (Altinay and Paraskevas 2008). All participants were required to be 18 years of age or older, be of Balinese descent, currently working in the tourism industry, have lived in the region for at least two years and have a competent level of English language skills.

Interviews explored a range of different issues relating to tourism including their perceptions of economic, social and environmental effects, level of involvement in the tourism industry, and support for current and future tourism development. Also discussed was the frequency of contact with tourists, development restrictions and what level of growth in tourism they would be willing to accept for Ubud. Interviews continued until the point of literal replication was reached, at which point the interviews were coded for emergent themes using open, axial and selective coding (Neuman 2011). Pseudonyms are used to protect the anonymity of respondents.

## Results

Preliminary findings indicated positive views towards tourism's contribution to the community of Ubud and specifically their way of life, yet concerns were also raised by several respondents about the region succumbing to mass tourism, as seen in the southern resort areas of Kuta and Nusa Dua. Ba'Mua observed; "*People come here because it's paradise, and it's cheap, tourism has brought bad changes to Bali, in Kuta the culture has moved already, it's like Australia now, but in Ubud we still keep the traditional culture*".

Respondents were asked for their opinion on how tourism has affected Ubud's economy. A wide variety of economic effects were raised and many were balanced in their views with both positive and negative comments. The job opportunities available through tourism were frequently mentioned, as Ba'Mua confirmed; "*When tourism or many tourists are coming then it is easy to get a job in tourism*" and Catur agreed; "*More hotels that means job vacancies, so all Ubud people and also from other parts of Bali and Indonesia, they can work in Ubud*". Tourism was viewed as an effective job creator for many local residents as Dede observed; "*If I talk about the economy, I think it's a good increase, we have a good economy here because there are a lot of work spaces (new jobs)*".

Tourism was viewed favourably for the economic development of the community, as respondents such as Jangha enthused; "*Tourism puts Bali into a more advanced state, it gives income to the local residents in Ubud*" and Uri, who has worked in tourism longer than any other respondents; "*Before tourism life was very poor, and with tourism, life was better because there is money to send children to school, not just the people from the hotel, the side effects like the farmers, transportation, you know the side effect, it's quite big the impact of tourism in Bali*".

Respondents identified seasonality of tourism was an issue for all of those working in the industry in Ubud and trying to provide for their families, Anbeh explained; "*We have a low season and high season, low season is January to April, May starts high season and then June, July, August, we have also the same for Christmas and New Year, we have high season so the economy will drop in low season, my salary is also down*". Uri, who owns a hotel in Ubud, was concerned about falling occupancy rates even though tourist numbers to Bali are increasing; "*This year occupancy is dropping, I don't know why it really puzzles me. I wonder if it was because of the election but it dropped. Some of the other hotels are empty and they're suffering too*". Overall, Bali's occupancy rates are indeed falling, in 2014 rates were 57 %, down from 60 % in 2013 and 63 % in 2012 (Bali Government Tourism Office, 2015).

Foreign ownership of land and development was an issue of concern for locals such as Ba'Mua who would like to see Ubud locals retain their ownership of tourism businesses; "*When the owner of the business is local people, like Balinese locals, they can keep it protected. But I worry when the owner is from foreigner, I worry they don't use like traditional style here in Ubud. They use…everything is modern, like the style of hotel is modern*". The gatekeeper observed from another resident Nyoman; *"he is actually worried about people in Ubud selling their land to foreign, outside investors. Then we have foreigners become our boss. He wants the local people to manage their own business to become the boss*". Issuance of building permits in Ubud and greater Bali was discussed by several respondents, suggesting perhaps the application process needs to be reviewed. Catur stated; "*Too easy, too easy to get a permit*" and Dimia agreed; "*They give easily to make a building, so I'm a little bit disappointed with the government*". Another resident, Uri suggested zoning rules can be disregarded by those in power; "*Everybody wants development, 'cause they say it will add to their cash… they don't do a real good blueprint of development. Even if they do, they just ignore it, if an area is restricted not for building and somebody speaks to the authority and then there's some bribery, they move the boundary*!"

In terms of sociocultural effects from tourism, respondents felt strongly that their Balinese Hindu culture has been preserved from any negative influences as Dimia explained; "*For the culture—no Ubud keep the culture very good, they still do what they have to do*". and Dewi supported this view; "*I think so far it is good, no nothing negative*". Locals also acknowledge that culture is an important tourism drawcard for Ubud with many visitors interested in viewing dance performances as Epana confirmed; "*Yes the culture is one of the things that attract tourists to come to Ubud"*. Keeping the culture traditional and avoiding cultural change was also of importance to respondents and many felt residents were in a strong position to continue their cultural practices intact, as Tanu stated; "*I want to keep the tradition, Bali's still like natural, still ceremony, if the King is strong, I think will not change, the King in Ubud is very strong*" and Catur observed; "*Some Balinese dance, they perform for tourism, so Balinese have not lost their culture*".

Respondents appeared to be pleased that Ubud has tight restrictions on bar and nightclub opening hours, as Anbeh affirmed; "*In Ubud, no way—no way pub, no way discotheque. Ubud is just a traditional village with culture*". and Tanu noted of Kuta; "*Like many bars, dance for music, 24 h a day. In Ubud, only until 12 o'clock, it's really quiet yeah*". Ultimately, many respondents acknowledge that if culture is negatively affected it will also impact the appeal of the region to potential tourists, as Anbeh declared; "*I think we are boring that we like traditional but others like modern, but it's important to protect because why many tourists come to Ubud, we must protect the question, the answer for the question*".

The effects of tourism on the physical environment were also deemed to be both positive and negative by the respondents. Tourism was identified to promote neighbourhood pride in keeping their area looking aesthetically appealing, as Catur proudly mentioned; "*In my village Penestanan, we started the project, Penestanan Clean and Green every two weeks, on Sunday, we do that, so not only local people join that but also the tourists, the guests who stay in our village, they join with us to clean the village*".

The respondents also expressed concern over crowding, as Ba'Mua lamented; "*During 10 years since I have been living back in Ubud—not too much change, only with traffic*" and Nina observed; "*As I see, it's still good I think, but I don't know whether (it will stay that way). I don't like the traffic jams, so many buildings*". Ubud is renowned for its scenic rice paddies throughout the region, the respondents frequently mentioned that the number of rice paddies appears to be declining, Anbeh, who has worked in a restaurant in the main street of Ubud for seventeen years; "*Before around Monkey Forest there is rice fields, but now there is many restaurants, there is businesses also, we people, we are here*".

A section of the interviews focused explicitly on uncovering respondents perceptions of proposed future development in the Ubud region. Whilst the respondents largely agreed that they would like more tourists to visit Ubud, some expressed they are worried about the density of tourism development in the centre of Ubud. Krishna observed; "*It seems like everything is getting smaller and smaller because of the buildings, if Ubud gets too busy it will become like Kuta*". Ba'Mua pointed out that if tourism development is excessive it may mean losing their visitor market; "*tourism is controlled here, we must keep it traditional, it's an asset for Ubud, when we cannot keep it traditional I worry that tourists will not come to Ubud again*".

Nonetheless, several residents like Epana did however express overt support for further development in Ubud; "*There could be more, further, yes.... as long as they build hotels based on the rules, it's ok*". Locals did mention that support would waiver if appropriate regulations were not put in place for future development in Ubud. Uri predicted; "*Yeah, I think if they don't control it, don't put restrictions, it will develop, so far, it's still controlled, but not really tight, not tight*". Respondents were probed on their ideal view of the development of Ubud in 10 years' time. Despite several noting that the only constant in life is change, the overwhelming view was strong, their views seemed to confirm their desire for Ubud to retain its unique cultural and historic integrity and identity as Dimia stated; "*I hope the rice paddies are still in front of our resort not buildings*". The vision of respondents like Catur was to mitigate the existence of key impacts reported to affect quality of life; "*I hope it's better than now, not more traffic*". Unequivocally the most common response when probed into the vision for tourism development over a 10 year period was that Ubud did not become like its southern tourism counterpart, as Nina declared; "*I think this is enough for Ubud, because otherwise it will become like Kuta*".

Perceptions of impacts such as job creation, career opportunities and support for future development were far more positive for younger respondents who had been working in tourism for less than one year. It was also apparent that the further away from the centre of Ubud the respondent's workplace was located, the more likely support was expressed for further development of hotels and attractions, along with more positive perceptions of impacts from tourism in the region.

## Conclusions

Tourism in Ubud was viewed favourably in terms of job creation and economic development. However concerns about the physical environment such as increased congestion and loss of traditional farming land, were also raised. Respondents largely felt that tourism has not affected Ubud's Balinese culture allowing it to be preserved for future generations.

At the same time respondents seemed supportive of more visitors coming to Ubud, most likely as it will benefit their careers and allow them to support their families financially. There was also a genuine sense that they like seeing tourists and interacting with them on a day-to-day basis in their jobs and if tourists enjoy Ubud, they are intrinsically more satisfied. Residents were divided over an increase in further tourism development, some were supportive as they felt in order to attract more visitors, more development must occur. The younger respondents were also more supportive of more tourism development, as were respondents whose workplace was located further away from the centre of Ubud.

Tourism and development is seen to be tightly controlled and monitored with strong policies to promote local ownership and employment. However, locals were concerned with congestion, development density and the rapid pace of tourism development in the area, most respondents felt there was enough hotels and restaurants already, however they would welcome an increase in visitors. When visioning how they would like Ubud to be in the future, all would like Ubud to remain exactly as it is, not wishing to become another Kuta, the mass tourism hub of Bali's south.

## References

Altinay L, Paraskevas A (2008) Planning research in hospitality and tourism. Routledge, Abingdon

Andereck KL, Nyaupane GP (2011) Exploring the nature of tourism and quality of life perceptions among residents. J Travel Res 50(3):248–260

Andereck KL, Valentine KM, Knopf RC, Vogt CA (2005) Residents' perceptions of community tourism impacts. Ann Tourism Res 32(4):1056–1076

Angelkova T, Koteski C, Jakovlev Z, Mitrevska E (2012) Sustainability and competitiveness of tourism. Procedia-Social Behav Sci 44:221–227

Bali Government Tourist Office (2015a) The number of domestic tourist arrivals to Bali by month 2004–2014 Retrieved 11 June 2015 from: http://www.disparda.baliprov.go.id/en/Statistics2
Bali Government Tourist Office (2015b) The number of foreign tourist arrivals to bali by month 2008–2015 Retrieved 11 June 2015 from: http://www.disparda.baliprov.go.id/en/Statistics2
Bali Government Tourist Office (2015c) Room occupancy rate of classified hotel in Bali 2011–2015 Retrieved 11 June 2015 from: http://www.disparda.baliprov.go.id/en/Statistics2
Bali Tourism Board (n.d.) Ecotourism (nature tourism) Ubud Retrieved 15 June 2015 from http://www.balitourismboard.org/ubud.html
Bohdanowicz P, Zientara P (2009) Hotel companies' contribution to improving the quality of life of local communities and the well-being of their employees. Tour Hospitality Res 9(2):147–158
Byczek C (2011) Blessings for all? community-based ecotourism in bali between global, national, and local interests–a case study. ASEAS—Austrian J South-East Asian Stud 4(1):81–106
Cole S (2012) A political ecology of water equity and tourism: a case study from Bali. Ann Tourism Res 39(2):1221–1241
Cole S (2008). Tourism, culture and development: hopes, dreams and realities in East Indonesia. Channel View Publications
Collinson L (2012) Is Bali doing as well as it should be? Retrieved January 6, 2014 from http://www.abc.net.au/foreign/content/2012/Bali%20Article.pdf
Deery M, Jago L, Fredline L (2005) A framework for the development of social and socioeconomic indicators for sustainable tourism in communities. Tourism Rev Int 9(1):69–77
Deery M, Jago L, Fredline L (2012) Rethinking social impacts of tourism research: a new research agenda. Tour Manag 33(1):64–73
Grzeskowiak S, Sirgy JM, Widgery R (2003) Residents' satisfaction with community services: predictors and outcomes. J Reg Anal Policy 33:1–36
Gurtner Y (2007) Tourism crisis: management and recovery in tourist-reliant destinations. Communities living with hazards. James Cook University, Australia, Centre for Disaster Studies, pp 82–101
Hall CM (2007) Introduction to tourism in Australia, development issues and change, 5th edn. Pearson Education, Australia
Hampton MP (2005) Heritage, local communities and economic development. Ann Tourism Res 32(3):735–759
Harmini AA (2009) Traditional knowledge based ecotourism in Bali. Retrieved January 20, 2014 from http://repository.rmutp.ac.th/handle/123456789/310
Hitchcock MP, Putra ND (2007) Tourism, development & terrorism in Bali. Ashgate Publishing Group, Great Britain
Hunt C, Stronza A (2013) Stage-based tourism models and resident attitudes towards tourism in an emerging destination in the developing world. J Sustain Tourism 22(2):279–298
Kim K, Uysal M, Sirgy MJ (2012) How does tourism in a community impact the quality of life of community residents? Tour Manag 36:527–540
Lee TH (2013) Influence analysis of community resident support for sustainable tourism development. Tour Manag 34:37–46
MacRae G (2011) Solid waste management in tropical Asia: what can we learn from Bali? Waste Manage Res 30(1):72–79
MacRae G (2002) Land tenure, land reform and commodification. In: Reuter T (ed) Inequality, Crisis and Social Change in Indonesia: The Muted Worlds of Bali (p 145). Routledge Curzon, London
Moyle B, Croy G, Weiler B (2010) Community perceptions of tourism: Bruny and Magnetic Islands, Australia. Asia Pacific J Tour Res 15(3):353–366
Neuman WL (2011) Social research methods: qualitative and quantitative approaches, 7th edn. Pearson, Allyn and Bacon, Boston
Nunkoo R, Ramkissoon H (2011) Residents' satisfaction with community attributes and support for tourism. J Hospitality Tour Res 35(2):171–190
Nunkoo R, Ramkissoon H (2010) Small island urban tourism: a residents' perspective. Curr Issues Tour 13(1):37–60

Simmons DG, Fairweather JR (2005) Understanding the tourism host-guest encounter in New Zealand: foundations for adaptive planning and management. EOS Ecology
Sirgy MJ, Cornwell T (2001) Further validation of the Sirgy et al.'s measure of community quality of life. Social Indicators Res, 56(2):125–143
Sutawa GK (2012) Issues on Bali tourism development and community empowerment to support sustainable tourism development. Procedia Econ Fin 4:413–422
Teye V, Sönmez S, Sirakaya E (2002) Resident attitudes toward tourism development. Ann Tour Res 29(3):668–688
Waligo VM, Clarke J, Hawkins R (2012) Implementing sustainable tourism: s multi-stakeholder involvement management framework. Tour Manag 36:342–353
Warren C (2012) Risk and the sacred: environment, media and public opinion in Bali. Oceania 82(3):294–307

# Chapter 8
# Developing *Halal* Tourist Destination: Investigating Lombok's Potentials from Destination Marketing Perspective

**Sulhaini, Akhmad Saufi and Rusdan**

**Abstract** Lombok is one of the most Muslim-populated Islands of Indonesia and is entitled 'island of a thousand mosques'. Such status may be one of the reasons behind the plan of Indonesian Government to develop Lombok as an Islamic *(halal-friendly)* tourist destination in Indonesia. This study is aimed at investigating the potentials of Lombok and its surroundings (a number of small islands or gilis around Lombok) to adopt the Islamic principles (followed to develop *halal-friendly* tourist destination) from the destination marketing perspectives. Data was collected from electronic and non-electronic promotional materials used by tourism agencies and private tourism enterprises (travel agents, hotels, and restaurants) in promoting Lombok and its surroundings. The data was then analyzed using content analysis approach. The study finds two main categories: (1) the *potential halal-friendly* tourist destinations, including all tourist destinations from within Lombok Island, and (2) the *moderate halal-friendly* tourist destinations, including the gilis. These findings suggest that the Islamic tourism principles can be adopted by all tourism enterprises in Lombok, and used to market Lombok as a *halal-friendly* tourist destination. However, such principles cannot be fully adopted by all tourism enterprises on the gilis. Instead, the *halal-friendly* tourism can be adopted by individual tourism enterprises and used as an individual market positioning on the gilis. The study brings implications to both tourism studies and management.

**Keywords** Islamic tourism · Halal-friendly tourism · Positioning · Promotion · Lombok · Three gilis

Sulhaini (✉) · A. Saufi · Rusdan
Department of Management, Faculty of Economics and Business, Mataram University, Mataram, Lombok, Indonesia
e-mail: niniys@yahoo.co.uk

A. Saufi
e-mail: Akh.saufi72@gmail.com

Rusdan
e-mail: erdanmuslim58@gmail.com

 
A. Saufi et al. (eds.), *Balancing Development and Sustainability in Tourism Destinations*, DOI 10.1007/978-981-10-1718-6_8

## Introduction

In Muslim world, tourism is about God's command to mankind, and is an important topic to study (Jafari and Scott 2014). This includes the marketing management of Islamic tourism (Haq and Wong 2010). Islamic tourism can be defined as tourism by Muslims and non-Muslims in the Muslim world. It facilitates non-Muslims to understand Muslims and creates greater market opportunities of Islamic destinations (Zamani-Farahani and Henderson 2010). Since Islamic principles govern much of business practices including tourism as acceptable or halal, it is also called 'halal tourism' (Battour et al. 2014). According to the concept, private tourism enterprises and tourism agencies' commitment in Islamic tourism may promote non-Muslims' understanding towards Muslim worlds while adopting Islamic principles. Then, their roles are not only involving economic but also, and perhaps more importantly, religious considerations.

Since Islamic tourism has religious motives, it can be seen as an exploration on all creations that have been created by God in order to obtain knowledge and increase faith (Jafari and Scott 2014) and thus it can also be called as 'purposeful tourism' (Elsamadicy and Negm 2013) or 'spiritual tourism' and people who carry out this have special interest to achieve specific objectives with specific needs (Haq and Wong 2010). Meanwhile, Islamic travelling that is compatible to Islamic principles is regarded as 'purposeful movement' (Zamani-Farahani and Henderson 2010). The core purpose is to conduct a spiritual journey to serve God and to get to know more of His creations including other people from other countries. This entails social and religious benefits of Islamic tourism. Muslims are then encouraged to take spiritual journeys. The holy book of Islam (Qur'an) endorses travelling in at least fourteen chapters/surah (Zamani-Farahani and Henderson 2010). Islam encourages people to travel by mean observing and learning God's creation to gain more understandings and thus travelling can also be seen as spiritual journeys since it can strengthen travellers' faith. This involves the principle of Tawhid, which is the oneness of God. By travelling, one Muslim is experiencing, seeing and learning the beauty of God's creations and knowing the greatness of God. Therefore, travelling must convey Islamic principles and adopting the principles can be viewed as worship of God or *ibadah* requiring the service providers to be good hosts. Put simply, Islamic tourism provides economic, social and religious benefits for the hosts/service providers, their activities are guided by the Islamic principles.

## Islamic Tourism Marketing

Islamic marketing is a very new concept and its emergence reflects trends in the Muslim world and beyond (Wilson 2012). For Muslims, Islam is not just a religion nor a part of culture but it is a way of life providing guidelines on how to conduct business, yet research on Islamic marketing is still very limited (Sandikci 2011; Hussnain 2011). Then, this study contributes to knowledge by studying Islamic tourism on Lombok and its surroundings through integration of Islamic marketing in the analysis.

Islamic marketing has been defined as the process of identification and implementation of value maximation strategies for the welfare of the stakeholders in particular and the society in general governed by the guidelines given in the Qur'an and Hadith (Hussnain 2011). This means Islamic marketing advances value creation in business for a wider society. This idea is transferable and relevance to a wider non-Muslim customers/community generating sustained advantages (Wilson 2012). Then, it is reasonable to assume that the adoption of Islamic principles will bring benefits to Muslims and non-Muslim consumers/visitors.

Islamic marketing implies that when targeting Muslims consumers, the practice of marketing needs to take on Islamic principles and thus it is different from marketing in general. All marketing activities must convey the Qur'an and Hadith eschewing selfish pursuit of profit maximation to suit individual marketer desires and ambitions. The guidance in the Qur'an and Hadith comes into individual conscience and influences their behaviour from within (Saeed et al. 2001). Wilson (2012) therefore suggests that Islamic marketing can be seen as an acknowledgement of a God-conscious approach to marketing, which tends towards the Islamic norms and values. Muslim consumers may have distinctive characters and behaviour. In targeting this segment, marketers offer a wide variety of products, which are positioned as Islamic or Halal including tourism products/destinations (Sandikci 2011). Islamic teachings influence Muslim tourists' behaviour. The consumers may evaluate a product or a destination's attributes such as: the inclusion of prayer/worship facilities, halal food, alcoholic and gambling free zones, Islamic entertainments, Islamic dress codes, general Islamic morality and the Islamic call to prayer (Battour et al. 2014). Muslim consumers/visitors choose the best destination, which is permissible according to Islamic law. Therefore, in targeting this segment, marketing strategies of an Islamic destination must be aligned with Islamic law.

Islamic marketing literature seems to emphasize segmentation where religion is the main factor explaining Muslim consumers' behaviour and ensuring products' halalness is the only objective. Nonetheless, it should refocus on the daily practices for which the product may be relevant and provide solutions that will assist Muslims be better Muslims and live proper Islamic lives. Also, one has to understand that Islamic marketing is unique since all commercial endeavours taken by a person are part of worship (ibadah) of God (Saeed et al. 2001). Service providers in the tourism industry can thus obtain religious benefits. However, this entails multiple and dynamic interpretations among Muslims. Understanding the variations will be fruitful as the basis of market segmentation (Sandikci 2011; Saeed and Azmi 2014).

Abdullah et al. (2013) identified factors (five Cs) affecting Islamic marketing mix. First, *commitment* stresses instilling assurance in business dealings via trustworthy marketing activities by implementing Islamic principles. Second, *character* describes the distinctive character of Islamic marketing. This can be seen in product design, specification and promotion, which are in accordance with Islamic principles. Third, *conformity* relies on compliance in terms of process involved in creating and delivering Islamic products/services. Fourth, *conscience* relates to ethical dimensions of Islamic marketing stressing true sense of promoting and clarity in pricing in accordance with Islamic teaching. Fifth, *customer centric* describes as an approach towards the customers, this mainly deals with the concept of marketing in general. It seems that Islamic marketing extends the general marketing concept with these interesting factors, which are dictated by the Islamic principles. Islamic marketing emphasizes developing marketing strategy in terms of marketing mix in accordance with the principles by offering permissible products, design, pricing, distribution and even promotions to meet customers' needs.

The principles prohibit public displays of affection, physical contact between members of the opposite sex, gambling, breaking fast in daylight during the holy month of Ramadan, consumption of pork and alcoholic drinks and inappropriate dressing (Zamani-Farahani and Henderson 2010). An Islamic destination can differentiate itself from other by placing a greater emphasis on Islamic attributes in their promotional campaigns. For example: a clear message on the availability of Halal foods and the availability of Islamic entertainments may represent a high priority for tourists in destination selection (Battour et al. 2014). Also, Saeed et al. (2001) explained that promotional techniques may not rely on sexual appeal, emotional appeal, fear appeal, pseudo research appeal, false testimonies or create the dullness of the mind and encourage extravagance. Then, it is not permissible for stereotyping of woman, excessive use of fantasy, the use of suggestive language and behaviour. This means that it strongly prohibited the use of women as promotional objects to attract customers/visitors' attention and influence their behaviour. In Islamic marketing, an advertisement must draw the true picture of destinations since Islam prohibits the practice of over promising and exaggeration. Promotional campaigns must follow the Islamic principles and must not contain false statements or promises or over praise the offering (Saeed et al. 2001; Abdullah et al. 2013). Therefore, making a false impression by giving incorrect and misleading information to promote a destination is strictly prohibited.

There is a common point of view between the Islamic tourism and Islamic marketing concepts that economic, social and religious benefits can be attained when enterprises adopt the Islamic principles. Those concepts highlighted that a destination must confirm Islamic principles to attract both Muslim and non-Muslim. This study is aimed to make preliminary investigation of the potentials to adopt Islamic tourism in Lombok, Indonesia. In turn now, we portray the methodology of the study.

# Methods

## *Research Context*

Lombok is located in Indonesia and is one of the two main islands of West Nusa Tenggara province. The island of Lombok has been popularly viewed as a thousand-mosque island where the majority of the population are practicing Islam (Saufi et al. 2014). In the beginning of tourism development, Lombok and its surroundings were promoted as '*beyond Bali'* tourist destination, to create an image of Indonesia's second Bali and attract visitors to enjoy and experience the beauty of Lombok (Dahles and Bras 1999).

To differentiate Lombok from Bali, Lombok tourism agencies created promotional campaign, i.e. '*you can see Bali in Lombok, but you can't see Lombok in Bali'*. Such *moto* was intended to create tourists' perception that Lombok is 'different' from Bali, while expecting the success of Bali influencing Lombok as a tourist destination. Nevertheless, such *moto* did not result in a clear and distinctive differentiation from Bali, and did not satisfy the majority of local people in Lombok (Saufi et al. 2014). Furthermore, Lombok as a tourist destination is often associated with Bali until recently. Consequently, it is reasonable to develop a new positioning for Lombok by adopting the Islamic principles. Knowing that Lombok's people are majority Muslim, such positioning may result in a clearer differentiation with Bali, and increase Lombok's tourism potentials.

## *Data Collection*

We collected data from electronic and non-electronic promotional materials used by Tourism Agencies and private tourism enterprises (travel agents, hotels and restaurants) in promoting Lombok. The promotional materials include websites, magazines (Lombok Magazine and Lombok Guide), brochures and other non-electronic promotional materials. The materials were analyzed in a highly iterative process and put into the unit of analysis. We focused our interpretation to the effect of the message from the texts and figures. We did not exclude figures in the analysis since they were main parts of marketing communication and contained important messages through which we identified the potentials of the destinations.

## *Analysis Process*

This study combines the frameworks built by Mayring (2000) and Thomas (2006) as the strategy for data analysis. Once the research methodology was established, it dictated the rest of the journey: it determined how the study engaged with data and theory. Identifying variations in terms of the messages was the first step of the analysis. Through constant comparison among the materials, it was possible to identify unique features/themes based on which categories were established. During the process, we also identified the frequencies of themes occurred in the promotional materials. The comparison also facilitated the identification of shared features across two categories to find patterns across the materials. The materials that had similarities were placed into the same group and when one or more materials showed differences, this led to the establishment of a new group/category of the potentials of Lombok and its surroundings' tourism in implementing Islamic principles.

The existence of various sects among Muslims leads to different interpretations of halalness and other religious strictures. Our interpretation is influenced by our social, cultural and educational backgrounds. In the process of analysis, our subjective interpretation on the materials played a great role. For example, materials which we considered as 'potential' or 'moderate' were based on our subjective judgments. Therefore, the categories were inductively and subjectively identified.

## Findings and Discussion

The aim of the study is to investigate the potentials of Lombok and its surroundings to be developed as a *halal*-friendly tourist destination in Indonesia. Using content analysis approach, the analysis focused on the promotional materials based on which we identified the potential of Islamic principles adoption. For the purpose of the analysis, we defined Islamic principles adoption as implementation of Islamic principles on all promotional materials, which are developed to influence visitors' behaviour and decision. The materials have to be clear about Islamic destination attributes, contain attractive messages without false contents/impression, use of woman as promotional objects, no display of affection and offering alcoholic drinks or non-halal meals. We then come up with two main categories including 'the *potential halal*-friendly tourist destination' and '*the moderate halal-friendly tourist destination*' as portrayed in Fig. 8.1.

### *Category I: Potential Halal-Friendly Tourist Destination*

The category reflects that materials confirm the Islamic principles containing attractive messages/pictures without relying on sexual appeal, affection, false statements but emphasize on the availability of mosques and halal foods. The promotional materials of this category do not contain false impressions and over praise the destinations. All promotional materials in this category were not only developed by tourism department but also private tourism enterprises. For example, a promotional material wrote: 'Lombok Island has beautiful tourism objects, unique

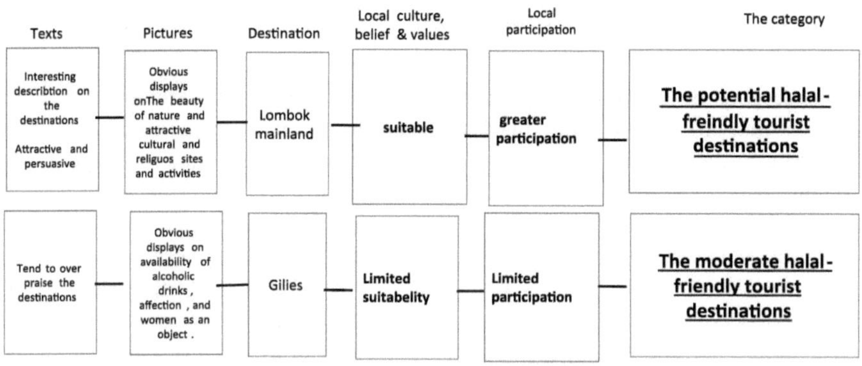


**Fig. 8.1** The emergent categories on the potentials of Islamic principles adoption

culture and traditions…' (Lombok Magazine 2015). Other promotional materials also contain objective information, for example: 'Lombok is a truly unique and enchanting holiday destination, pristine beaches with calm clear waters, abundant coral reefs, mystical ancient traditions, volcanic treks and spectacular sunsets make this the perfect destination' (Sheraton's leaflet). Such information is in line with Islamic principles as it is ethical and stresses on true sense (Abdullah et al. 2013).

The materials in this category mainly promote various destinations in the mainland of Lombok including religious sites/heritages to encourage Muslims to visit for *ziarah*. Visiting the sites may provide knowledge of how Islam developed and influenced local people's way of live. This may attract non-Muslim visitors. Accordingly, the figures displayed on the materials do not contain a false message and no sexual appeals. The materials that we collected regarding the mainland also provide, although still limited, information on availability of Halal food. The materials provide information on the availability of mosques which are easily found in every part of the island. This facilitates Muslim visitors to pray during their journey. This may create differentiations on Lombok's tourism, which may offer the beauty of nature, culture and availability of praying facilities.

The uniqueness and distinctive messages in the promotional materials seem to offer *Syariah* permissible destinations, i.e. nature and cultural attractions. They also contain information on local mosques and Islamic-cultural events. They deliver the differentiation and thus positioning of the destinations as Islamic tourism destinations. The destinations as illustrated in the materials may meet the visitors' needs and wants and thus they are beneficial to the visitors. The benefits may include supporting Muslim visitors to be good Muslims since the destinations are displayed as very beautiful, charming, enchanting and attractive God's creations. These may induce consciousness within individuals on the God's greatness and oneness. Visiting the destinations may be perceived as the act of worship (*ibadah*). Islamic principles, as suggested by the finding, are adopted by private tourism enterprises in Lombok by offering the beautiful island as *halal-friendly* destinations.

## *Category 2: Moderate Halal-Friendly Tourist Destination*

This category is characterized by the adoption of one or two main principles of Islamic tourism and Islamic marketing in the tourism promotional materials. Nevertheless, we did not identify any of promotional materials that totally reject the Islamic principles. The category is therefore entitled the *moderate halal-friendly* tourist destinations. For example: 'Gili Kondo, the quiet island like paradise' (Lombok Magazine, December 2014); 'This is once in a lifetime occasion to own your paradise…' (Gili Gede island in the Lombok Guide); 'The beautiful Gili islands of Lombok are paradise, with white sand beaches, few people and some great snorkeling and diving spots' (The Three gilis in the web of A&T Holidays)

The word 'paradise' is often used in the promotional materials to describe the beauty of the gilis and create an illusion of perfection in consumers' mind. This may be the way of developing Lombok's positioning in tourism: manipulating consumers' mind primarily through a communication exercise (Ries and Trout 1982). Nevertheless, the use of 'paradise' seems to over praise a destination as a paradise. Islamic marketing emphasizes the delivery of a true message and prohibit giving false impression and overpraising a destination. In Islamic marketing perspective, a differentiation can be developed through improving products and communicating them to the market by giving a true picture.

Indeed, the gilis are free from gambling areas since gambling is prohibited by the nation law. However, hotels and restaurants in the destinations clearly promote alcoholic drinks. Women in swimsuits are obviously displayed in the promotion. Tourism in the gilis seems to be closely positioning them close to Bali but they actually pose a different character. Tourism enterprises in the gilis rely their offering on beautiful beaches which are more peaceful, quieter and calmer compare to that of Bali's.

The gilis' promotions seem to overlook other attractions such as: reef, birds and turtle conservations. In northwest, among the three gilis, Gili Trawangan is the main tourist destination that has been popular as the party island yet the local culture is missing. Meanwhile, Gili Air has the largest local population and thus visitors can observe local culture (A&T Holidays 2015). In the southwest off Lombok, some gilis such as Gili Nanggu, Gili Tangkong, Gili Sudat, Gili Gede, Gili Ringgit, Gili Layar and Gili Asahan are largely uninhabited (The Lombok Guide Magazine 2015). However, as the Gilis are located near many villages close to villages in Lombok, tourists still have opportunity to interact with locals. Thus, the promotional materials on gilis should include information of places of prayer for Muslim. The lack of such information may discourage Muslim visitors to come, and inhibit the harmony between local people's culture and religious beliefs and tourism.

In Lombok, a wide variety of attractions is offered, through which visitors can see local culture and belief by visiting traditional villages, old mosques, Islamic festivals and traditions. Calendar event (2015) identified a wide variety of events in which indigenous people perform unique traditions such as '*baunyale*', '*peresean'* (stick fighting), '*begasingan*' game, *ruahsegare*, '*belanjakan'*, '*topat'* war, '*gendangbelek'* festival. The island is also rich with Islamic festivals/traditions for example: Celebration on the birth of the Prophet, '*berzanji'*, '*zikirzaman'* and '*burdah'*, '*Qasidah'* competition, and '*Takbiran'* night parade. All of the events need native people participation and contribution to perform and sustain the traditions. Social and local sustainability are two main dimensions in sustainable tourism. Tourism increase people's control over their lives and local identity without damaging indigenous culture by increasing local involvement and benefits to the local community (Brokaj 2014).

Islamic tourism attributes refers to the availability of Islamic norms and practices at the destinations (Battour et al. 2014). The attributes may improve visitors' religiosity as visiting the destination creates spiritual experiences in terms of the feeling of getting closer to the God, happiness, peace and secure. Muslim visitors may choose a destination, which offers Islamic atmosphere where they can hear call to prayer and find halal meals, hotel rooms with markers indicating the direction to Mecca and information of availability of praying facilities. Business practitioners who understand and implement the Islamic principles can be viewed as good hosts since they facilitate their visitors' *ibadah* and their service can be viewed as part of their worship of the God. The Qur'an emphasizes Muslim to be good hosts through indigenization of product and human capital in the tourism industry (Stephenson et al. 2010). A destination should not be in isolation from local community life. Indeed, they should play a great role and the main actors of tourism development. In Lombok's context, adopting Islamic principles in tourism leads to sustainable tourism in the destination.

The materials in the *potential halal-friendly* category generate a differentiation of the tourism destinations by uniquely displaying and illustrating the destinations as beautiful and peaceful places. In Lombok, mosques are easily found in almost every part of the island and the people cannot live apart from mosques. According to recent data, 95.33 % of local people in Lombok are Muslims who have 3970 mosques and 1244 Islamic schools (BPS NTB 2014). For the native people, a mosque is not only a place of worship but also a people's pride. Muslim tourists can hear '*azan'* (the call to prayer) from their hotel room or restaurant and easily access a local mosque.

The potential of adopting the Islamic principles is dominantly found in Lombok where tourism enterprises respect local culture or belief and highly regard the local community participation and marketing of the destination will result in a better welfare of wider society. The participation is argued to be a main component of sustainable tourism development as it may reduce conflict and foster local community support (Stephenson et al. 2010; Zamani-Farahani and Henderson 2010). In the Lombok context, according to Saufi et al. (2014), indigenous people's participation in the tourism industry is strongly determined by local culture and religion. Developing tourism industry on the island by adopting Islamic principles is actually in line with local culture and belief and thus generates sense of belonging and participation among indigenous people as indicated in the findings of this study.

Ghadami (2012) argued that Islam influences the mode of tourism development and marketing of destinations. They have to be ordered in conformity with the principles of Islamic law. This means that Islamic tourism needs Islamic marketing where Islamic principles are credible tools to apply and marketing activities can be viewed as *ibadah* or 'worship of God' (Stephenson et al. 2010). Similarly, destination marketing is an essential tourism management tool to inform and influence visitors' behaviour and decision (Zamani-Farahani and Henderson 2010). Therefore, marketing the destinations should emphasize positioning strategy involving four main elements, namely: tourists' emotional attachment to the destinations, tourists' prior expectation, the element of learning and satisfaction from travelling to the destinations (Kavaratzis and Ashworth 2008).

Learning God's creations in terms of natural beauty and local culture may lead to better knowledge and faith, which engender satisfaction and emotional attachment to the destination. Moreover, an Islamic destination produces 'meaningful experiences' by involving the local Muslim community and integrating the economic and religious motives in the destination (Stephenson et al. 2010). Community participation is now getting greater attention due to its crucial roles in sustainable tourism development as they have useful knowledge and may reduce conflicts (Zamani-Farahani and Henderson 2010).

## Theoretical Implications

The study found the potential for the adoption of Islamic principles in Lombok Island where the majority of the native people are Muslims and the religion plays a central role in their daily life. The adoption stimulates the people's participation and contribution since it is suitable to local culture, belief and values. Islamic marketing emphasizes value maximation for the well-being of a wider society by confirming Islamic principles in the marketing strategies and tactics. Accordingly, Islamic tourism stresses the adoption of Islamic principles into tourism industry. This study further provides understanding that the adoption will indeed ensure the sustainability of the industry and therefore, delivering the right differentiation of a destination in a more sustainable way by adopting Islamic tourism strategy.

## Managerial Implications

The study suggests religious values created and obtained by being good hosts in the tourism industry. This demands the adoption of Islamic principles by individual tourism enterprises on the gilis. Therefore, business practitioners of hotels, restaurants and other services need to understand and adopt Islamic principles of business conduct. The adoption generates individual enterprises' positioning in marketing their services. The promotional materials should display Islamic attributes of their service. This is particularly important to attract Muslim visitors who may still want to be in Islamic atmosphere and carry out their obligation to worship of God. Descriptive information on the availability of halal meals and praying facilities as well as Islamic hotels and entertainments should be clearly demonstrated.

### Limitation and Future Direction

This study used a content analysis approach to investigate the potential of Lombok and its surroundings to adopt Islamic principles. However, the study focused on promotional materials that may limit the findings on differentiation and promotion of the destinations. Marketing of Islamic destinations should be supported by marketing mix strategy, however in this study, two elements of the strategy, i.e. price and distribution might not be fully explored. Further research may uncover these issues by utilizing a more complex qualitative approach.

The study suggests that the adoption of Islamic principles may lead to sustainable tourism. Furthermore, studies are needed to reveal if this would also be true in other destination characterized by different situations.

## References

Abdullah JB, Hamali JH, Abdullah F (2013) Islamic marketing mix: the critical factors. In: Proceeding of 4th international conference on business and economic research, Bandung, Indonesia

A&T Holidays (2015) Lombok and beyond. www.lombokandbeyond.com. Retrieved on 17 Mar 2015

Battour M, Battor M, Bhatti MA (2014) Islamic attributes of destinations: construct development and measurement validation and their impact on tourist satisfaction. Int J Tourism Res 16:556–564

BPS NTB (2014) Nusa Tenggara Barat in Figures 2014. Mataram

Brokaj R (2014) Local Government's role in the sustainable tourism development of a destination. Eur Sci J 10(31):103–117

Dahles H, Bras K (1999) Entrepreneurs in romance tourism in Indonesia. Ann Tourism Res 26 (2):267–293

Dogan HZ (1989) Forms of adjustment sociocultural impacts of tourism. Ann Tourism Res 16:216–236

Elsamadicy AM, Negm EM (2013) Investigating the impact of the political change regarding the dominance of the Muslim brotherhood on tourists' intentions to visit Egypt before the 2013 Regime alteration. Am Acad Sch Res J 5(6):34–45

Ghadami M (2012) The role of Islam in the tourism industry. Elixir Mgmt, Arts 52:11204–11209

Haq F, Wong HY (2010) Is spiritual tourism a new strategy for marketing Islam? J Islamic Mark 1 (2):136–148

Henderson JC (2003) Managing tourism and Islam in Peninsular Malaysia. Tour Manag 24:447–456

Hussnain SA (2011) What is Islamic marketing. Global J Manage Bus Res 11(11):100–103

Jafari J, Scott N (2014) Muslim world and its tourism. Ann Tourism Res 44:1–19

Kavaratzis M, Ashworth G (2008) Place marketing: how did we get here and where are we going? J Plac Manag Dev 1 (2):150–165

Lombok.magazine (2015) Pink Beach. Lombok magazine No. 15 Feb–15 Mar, pp 16–17
Mayring P (2000) Qualitative content analysis, forum: qualitative social research, vol 1(2). http://www.qualitative-research.net/fqs/
Ries A, Trout J (1982) Positioning: the battle for your mind. Warner, New York
Sandikci O (2011) Researching Islamic marketing: past and future perspectives. J Islamic Mark 2 (3):246–258
Saeed M, Azmi IBAG (2014) Religion and brand switching behaviour of Muslim consumers. Middle East J Sci Res 21(9):1611–1617
Saeed M, Ahmed ZU, Mukhtar SA (2001) International marketing ethics from an Islamic perspective: a value maximization approach. J Bus Ethics 32:127–142
Saufi A, O'Brien D, Wilkins H (2014) Inhibitors to host community participation in sustainable tourism development in developing countries. J Sustain Tourism 22(5):801–820
Stephenson ML, Russell KA, Edgar D (2010) Islamic hospitality in the UAE: indigenization of products and human capital. J Islamic Mark 1(1):9–24
The Lombok Guide (2015) Gili Islands fast facts, p 79, Issue 189, 23 Mar–6 Apr 2015
Thomas DR (2006) A general inductive approach for analysing qualitative evaluation data. Amer J Eva 27:237–246
Wilson JAJ (2012) The new wave of transformational Islamic marketing, reflection and definitions. J Islamic Mark 3(1):5–11
Zamani-Farahani H, Henderson JC (2010) Islamic tourism and managing tourism development in Islamic societies: the case of Iran and Saudi Arabia. Int J Tourism Res 12:79–89

# Chapter 9
# Exploring the Rationales of Souvenir Purchase Behaviour and Roles of Behavioural Belief and Perceived Authenticity Among International Tourists in Malaysia

**Faiz Izwan Anuar, Astri Yulia, Hayatul Nadzirah Binti Hamden, Kamilia Fasiha Binti Abdul Aziz and Sharifah Maziah Bte. Wan Hamid Edruce**

**Abstract** Tourist attractions are associated with souvenirs; most tourists purchase souvenirs when travelling. In this study, we investigated the reasons for tourists purchasing souvenirs. In addition, we explored the factors that contribute to purchasing souvenirs. The two-way ANOVA results indicated that there was no significant difference in souvenir purchase between male and female tourists, and no difference was found between the tourists' origins (North America, South Africa, Europe, or Asia-Pacific). The multiple regression results showed that purchasing souvenirs as gifts significantly contributed to purchase frequency, while purchasing souvenirs as evidence and memories were not significant predictors of purchase frequency. Further, we evaluated the roles of behavioural belief and tourists' perceived authenticity about the souvenirs in the three purchasing behaviours: as gifts, evidence, and memories. Our findings extended knowledge of the importance of

F.I. Anuar (✉)
Department of Tourism Management, Universiti Teknologi MARA (UiTM), Shah Alam, Malaysia
e-mail: faizwanuar@salam.uitm.edu.my

A. Yulia
Faculty of Business, Universiti Selangor (UNISEL), Shah Alam, Malaysia
e-mail: dr.astri@unisel.edu.my

H.N.B. Hamden · K.F.B. Abdul Aziz · S.M.Bte.Wan Hamid Edruce
Faculty of Hotel & Tourism Management, Universiti Teknologi MARA (UiTM), Shah Alam, Malaysia
e-mail: hayatulnadzirah@yahoo.com

K.F.B. Abdul Aziz
e-mail: a.kamiliafasiha@yahoo.com

S.M.Bte.Wan Hamid Edruce
e-mail: maziah_edruce@yahoo.com



A. Saufi et al. (eds.), *Balancing Development and Sustainability in Tourism Destinations*, DOI 10.1007/978-981-10-1718-6_9

souvenirs for tourists, especially in Malaysia, and we concluded that tourists mostly purchase souvenirs for gifts compared to other purposes and that tourists' behavioural beliefs are a more significant factor than the souvenirs' perceived authenticity. The results of this research contribute to the development of the souvenir industry specifically and the tourism industry in general.

**Keywords** Souvenir purchase · Perceived authenticity · Behavioural beliefs · Tourism

## Introduction

Souvenirs are essential for international tourists visiting another country. They have become a central part of the tourism experience. Tourists want to take home souvenirs to remind them that they have been to a place. Some tourists may purchase souvenirs as gifts as part of their social responsibility and as evidence of visit. Souvenirs are a tangible product from the tourism industry that tourists can take and use wherever they go.

According to Fairhurst et al. (2007), expenditure on shopping comprises almost a third of the total travel spend. Tourism Malaysia Statistic (2013) reported that the number of tourist arrivals in 2013 was 25,720,000, and the total expenditure was RM 65.44 billion. It has been shown clearly that Malaysia has benefited a great deal from tourist dollars and could experience enormous economic gain from the souvenir business. In the first quarter of 2014, Malaysia received 7.09 million tourists in conjunction with Visit Malaysia Year (2014). Therefore, the increasing amount of tourist expenditure leads to a demand for souvenirs.

The souvenir purchasing process is influenced by internal factors, such as demographic profile, and external factors, which are the 5 Ps of marketing mix (Damrongpipat 2009). The 5 Ps are: personnel, price, product, promotion, and place. The quality and authenticity of souvenirs may also influence the decision-making process of an individual tourist. Tourists usually purchase souvenirs as gifts. Decorative items and handicrafts, creative design, uniqueness, durability, packaging, a theme of decoration, and the hospitality of the salesperson are found to be important when producing souvenirs (Damrongpipat 2009).

Moreover, tourists tend to seek the opportunity to explore more superficial aspects of travel, including the authentic life of the host community or region (MacCannel 1976) as cited by Shen (2011). Authenticity derives from "the property of connectedness of the individual to the perceived, to the everyday world environment and the process created it" (Hall 2007, p. 1140). The definition of authenticity varies with perception. The perception of authenticity is usually integrated into social processes, interests, and power relationships when the purchased souvenir describes the interaction between tourists and the host tourists (Xie et al. 2012). The authentic craft or souvenirs, which are usually associated with cultural beliefs and a way of life of the host community, contribute to providing momentous tourist experiences. Tourists intend to buy local souvenirs because this portrays and provides a glimpse into the cultural beliefs of the host community. Thus, authentic attributes might be the most significant factor for tourists buying local souvenirs (Wicks et al. 2004). The most important criteria for a souvenir to represent authenticity are its originality and uniqueness. Therefore, the purpose of this study was to determine the perception of tourists of the authenticity of local souvenirs in Kuala Lumpur. This study has been conducted in Kuala Lumpur with responses from international tourists.

A souvenir can be categorised as an essential element of the tourist experience, which provides evidence of the visit tourists use to recall memories. It is essential that authentic crafts and souvenirs represent the destination in order to provide a memorable experience for tourists. In addition, among all types of souvenir attributes, authenticity may be the most important factor that affects tourists' buying intentions (Wicks et al. 2004).

At the present time, there have been issues regarding the mass production of local souvenirs across Malaysia. According to Peoples and Bailey (2006), all goods and services have to compete for market share. Therefore, local traditional handmade products have to compete with mass-produced imported products. For example, mass-produced goods made of aluminium, ceramic, and tin compete with locally made pottery, handicrafts, and wooden baskets. The main competitive factors are the cost and quality of goods. Normally, local products are more labour-intensive than mass-produced goods, which reduces the cost of operations. As a result, the mass production of goods automatically replaces traditional handmade products, and the growing demand for souvenirs could be a profitable market for foreign traders.

Another issue that has been arising regarding the authenticity of local souvenirs is the localisation of identity among tourism souvenirs at a destination. According to TyMiranda (2014), most of the cheapest souvenirs do not originate in the respective city. Sometimes, the souvenirs that are available in other places are also available at the destination visited. Moreover, some souvenir shops are not locally owned by natives. There are several tips for the tourist when purchasing local souvenirs at a destination, such as stay local and cultural, visit shops that have a history of local products, and purchase the souvenir based on need. The tourist will be satisfied if the souvenir is made in the visiting country, thus proving that it is something unique.

In addition, according to Women's Web (n.d.), there was one common issue for travel souvenirs purchased by tourists, which is that they are all made in China. For example, the fridge magnets, keychains, and textiles from India are all made in China. Although it is difficult for tourists to find local souvenirs made specifically in the destination, it is advisable for them to do some research on the destination. Purchasing locally handmade souvenirs creates the best experience and satisfaction for the tourist, in addition to the money spent by the tourist benefitting the local natives and economies. Traditionally, the quality and design of souvenirs are an important segment for the tourist to interpret the authenticity of the product.

For the above rationales, this study formulated the following objectives: (1) to investigate souvenir purchase intention among tourists in Kuala Lumpur; (2) to determine tourists' perceptions of perceived authenticity of souvenirs in Kuala Lumpur, and (3) to analyze tourists' satisfaction with souvenir purchases in Kuala Lumpur. Specifically, the questions are: (1) What are the factors that influence tourists purchasing souvenirs in Kuala Lumpur? (2) What are tourists' perceptions of authenticity of souvenirs in Kuala Lumpur?, and (3) How satisfied are tourists with their souvenir purchases in Kuala Lumpur?

## Theoretical Framework

For the above objective, in the study framework, purchase intention has three components: gift, evidence, and memory. Purchase intention influences the behaviour belief as well as the satisfaction of purchasing the souvenir in Malaysia. For example, tourists feel satisfied when they buy souvenirs as gifts. Therefore, tourists' perception of purchase intention and authenticity of local souvenirs results in behaviour beliefs and satisfaction. To summarize, purchase intention is a factor that generates tourist-purchasing souvenirs. Authenticity influences the satisfaction of people purchasing the souvenir. For example, tourists feel satisfied when purchasing souvenirs that are authentic.

## Method

The purpose of this study was to investigate the factors influencing tourists' souvenir purchases. The data for this study were collected from international tourists visiting Kuala Lumpur, Malaysia from November to December 2014. A total of 110 tourists agreed to participate in this research. To collect the data, the researchers developed a set of questionnaires, which included questions about souvenir purchase frequency, potential receiver of souvenirs, souvenir preferences, perception of local souvenirs, behavioural beliefs about local souvenirs, perception of authenticity, and satisfaction with souvenir purchases. The study was piloted on 53 random respondents to test the reliability of the questionnaire. Cronbach's alpha was greater than 0.60. According to Nunnally and Bernstein (1994) as cited by Eid (2012), a Cronbach's alpha below 0.7 and above 0.6 is acceptable.

## Results

Table 9.1 presents the profiles of the respondents. The gender distribution of the survey respondents was 60 % female and 40 % male. The greatest number of respondents was in the age group 18–29 years old at 74 %, followed by 14.6 %

**Table 9.1** Demographic profile of respondents

| Demographic | Frequency | Percent |
|---|---|---|
| *Gender* | | |
| Male | 44 | 40 |
| Female | 66 | 60 |
| *Age* | | |
| 18–29 | 74 | 67.3 |
| 30–39 | 16 | 14.6 |
| 40–49 | 11 | 9.6 |
| 50 above | 9 | 8.5 |
| *Employment* | | |
| Student | 42 | 44.7 |
| Employed | 37 | 39.4 |
| Self-employed (own business) | 8 | 8.5 |
| Housewife | 1 | 1.1 |
| Retired | 3 | 3.2 |
| Others | 3 | 3.2 |
| *Origin* | | |
| Asia-Pacific | 62 | 56 |
| North America | 11 | 10 |
| South America | 5 | 4.2 |
| Europe | 33 | 29.8 |
| *Purpose of visit* | | |
| Holiday | 62 | 56.4 |
| Education | 26 | 23.6 |
| Leisure | 16 | 14.6 |
| Others | 6 | 5.4 |
| *Frequency of visit* | | |
| First time | 35 | 31.8 |
| Second time | 29 | 26.4 |
| Third time | 7 | 6.4 |
| More than 3 times | 39 | 35.4 |

aged 30–39 years old, 9.6 % aged 40–49 years old, and 8.5 % aged 50 years old and above. Regarding employment, most of the respondents were students at 44.7, 39.4 % were employed, 8.5 % were self-employed (own business), 3.2 % were retired and others, and 1.1 % were housewives.

Because the study focused on international tourists, the highest number of respondents in the survey originated in Asia-Pacific at 56 %, followed by 29.8 % from Europe, 10 % from North America, and 4.2 % from South America. The purpose of the visit to Kuala Lumpur for most of the respondents was a holiday at 56.4, 23.6 % for education, 14.6 % for pleasure-seeking, and 5.4 % for others, including a mid-stay camp for exchange student, business, and shopping. In addition, for the frequency of visit, 31.8 % of the respondents were visiting Kuala Lumpur for the first time, 26.4 % for the second time, 6.4 % for the third time, and about 35.4 % had visited more than three times.

Table 9.2 presents the types of souvenirs that respondents preferred. The highest numbers of respondents preferred the uniqueness of the souvenir at 69 responses. Wearable and local products were equal with 60 responses, usable souvenirs received 55 responses, souvenirs with the name of the destination received 50 responses, handmade souvenirs received 47 responses, low price received 37 responses, small (portable) received 21 responses, eco-friendly received 12 responses, and durable received 11 responds. High price was not a major tourist's preference when purchasing souvenirs. This means that the uniqueness of souvenirs was imperative for tourists when purchasing souvenirs.

Tourists often purchased different types of souvenirs when travelling. Table 9.3 presents the types of souvenirs tourists often purchased during travel. A total of 57 % of respondents purchased keychains and fridge magnets, 45 % purchased

**Table 9.2** Types of souvenir preferred during travel

| Types of souvenir | *N* |
|---|---|
| Unique | 69 |
| Wearable | 60 |
| Local products | 60 |
| Usable | 55 |
| Destination souvenirs | 50 |
| Low price | 37 |
| Small (portable) | 21 |
| Eco-friendly | 12 |
| Durable | 11 |
| Handmade | 11 |
| High priced | 0 |

**Table 9.3** Types of souvenir purchased during travel

| Types of souvenir | Percentage |
|---|---|
| Keychain | 57 |
| Fridge magnet | 57 |
| T-Shirt | 45 |
| Accessories | 44 |
| Traditional crafts | 42 |
| Postcard | 28 |
| Batik | 23 |
| Beads | 15 |
| Local handicraft (Mengkuang bags, Mat) | 7 |
| Mugs | 4 |
| Others | 4 |

T-shirts, 44 % purchased accessories, 42 % purchased traditional crafts, 28 % purchased postcards, 23 % purchased batik, 15 % purchased beads, and 7 % purchased local handicrafts, such as *Mengkuang* bags and mate. Other respondents stated that they preferred to buy mugs and other items, such as clothes, agarwood, carpets, chocolate, makeup, and other travel collectibles that are often purchased because of hobbies. Thus, most respondents bought keychains as their most preferred souvenirs in a previous or current trip to Kuala Lumpur. The study also asked the respondents for whom they bought the souvenirs: 85 % of respondents purchased souvenirs for their families, 75 % for their friends, and 70 % for themselves.

## Factor Analysis

Factor analysis was used to explore the rationales for souvenir purchase behaviour. All item loadings for each factor were larger than 0.06; hence, no items were removed. To analyse the data collected, we used SPSS for descriptive statistics purposes, ANOVA, and multiple regression for model evaluation. First, two-way ANOVA was performed on souvenir purchase data to evaluate the differences between gender and tourists' origins. The results indicated that there was no significant difference ($F = 0.494$, $df = 1$, $p = 0.484$). In addition, no difference was found ($F = 1.351$, $df = 3$, $p = 0.263$) between the tourists' origins (North America, South America, Europe, or Asia-Pacific).

## Multiple Regression

Multiple regression was used to examine the factors that influence souvenir purchase among tourists visiting Malaysia. The results showed that purchasing souvenirs as gifts significantly contributed to purchase frequency ($R^2 = 20.7$ %, $b = 1.304$ [$\beta = 0.216$], $p < 0.05$), while purchasing souvenirs as evidence ($b = 0.074$ [$\beta = 0.012$], $p = 0.936$) and memories ($b = 0.363$ [$\beta = 0.060$], $p = 0.699$) were not significant predictors for purchase frequency. In addition, we evaluated the roles of behavioural belief and tourists' perceived authenticity about the souvenirs in the three purchasing behaviours: as gifts, evidence, and memories. We found that behavioural beliefs had a significant effect on purchasing souvenirs as gifts ($R^2 = 15.8$ %, $b = 0.284$ [$\beta = 0.284$], $p < 0.05$); perceived authenticity significantly contributed to purchasing souvenirs as evidence ($R^2 = 9.3$ %, $b = 0.278$ [$\beta = 0.278$], $p < 0.05$); however, both factors did not have a significant effect on purchasing gifts as memories.

## Discussion and Conclusion

Souvenirs represent a precious moment of the travel experience, for instance, they portray images and recall unique memories of vacations. The current study investigated the reasons for tourists purchasing souvenirs. In addition, the study explored the factors that contribute to purchasing souvenirs. Interestingly, the results indicated that there was no significant difference in souvenir purchases between male and female tourists. Moreover, there was no significant difference found between the tourists' origins (North America, South America, Europe, and Asia-Pacific) in terms of souvenir purchases.

The multiple regression results showed that purchasing souvenirs as gifts significantly contributed to purchase frequency, while purchasing souvenirs as evidence and memories were not significant predictors of purchase frequency. Furthermore, the study evaluated the roles of behavioural belief and tourist's perceived authenticity about the souvenirs in the three purchasing behaviours: as gifts, evidence, and memories. Interestingly, the study found that behavioural belief had a significant effect on purchasing souvenirs as gifts; perceived authenticity significantly contributed to purchasing souvenirs as evidence; however, both factors did not have a significant impact on purchasing gifts as memories. Moreover, the study findings extended knowledge of the importance of souvenirs for tourists, especially in Malaysia: we were enlightened that tourists mostly purchased souvenirs for gifts compared to other purposes and that the tourists' behavioural beliefs were a more significant factor than the souvenirs perceived authenticity. The results of this research can contribute to the development of the souvenir industry specifically and tourism industry in general.

## References

Damrongpipat N (2009) Determinants of souvenirs purchasing behavior among international phuket visitors (Business administration in hospitality and tourism management). Prince of Songkla University

Fairhurst A, Costello C, Holmes AF (2007) An examination of shopping behavior of visitors to Tennessee according to tourist typologies. J Vacation Mark 12(3):311–320

Hall CM (2007) Response to Yeoman et al.: the fakery of "the authentic tourist". Tour Manag 28:1139–1140

MacCannell D (1976) The tourist: a new theory of the leisure class. Schocken Books, New York

Nunnally JC, Bernstein IH (1994). Psychometric theory (3rd ed). New York: McGraw-Hill

Peoples J, Bailey G (2006) Humanity: an introduction to anthropology. In: Cengage Learning, 8th edn. Wadesworth, Belmont, CA

Shen MJ (2011) The effects of globalized authenticity on souvenir. Int J Innovative Manag Inf Prod 2(1)

TyMiranda (2014) Buying the right souvenirs. Retrieved from http://www.travelerstoday.com/articles/13728/20141105/souvenirs-travel-tips.htm

Visit Malaysia Year Kuala Lumpur 15 million Tourists (2014) The star. Retrieved from http://www.thestar.com.my
Wicks B, Do K, Hsieh P-C, Komorowski A, Martin K, Qiu X, Yu G (2004) Direct marketing of crafts and souvenirs to Vladimir visitors. University of Illinois
Xie PF, Wu T-CE, Hsieh H-W (2012b) Tourists' perception of authenticity in indigenous souvenirs in Taiwan. J Travel Tourism Mark 29(5)

# Chapter 10
# Bidayuh Community of Malaysia: Presenting Culture and Nature as a Package for Sustainable Development of Indigenous Tourism

**Shazali Johari, Ahmad Shuib, Sridar Ramachandran, Syamsul Herman and Puvaneswaran Kunasekaran**

**Abstract** This paper aims to scrutinise the ethnic characteristics of the Bidayuh community in Sarawak, Malaysia, which could be conserved and capitalised on for the development of indigenous tourism. The community, also called 'people of the land', is characterised by its cultural identities such as 'O gong' (a gong troupe), massive drums, and 'pratuakng' (a percussion instrument made out of bamboo). Furthermore, this study brings to light the willingness of the indigenous community to partake in activities related to tourism. The study investigated the hindrances encountered by them to foster tourism in their region. The community's ethnic distinctiveness is acknowledged by foreign travellers as well as other Malaysian societies. However, questions have been raised over the actual act of developing tourism so as to reveal the customs and traditions of the community to the rest of the world. Although, according to initial analyses, the community is eager to partake in tourism activities by sharing their culture with foreigners, no effectual developmental initiatives have been undertaken by the government. Thus, this study holds significance from the point of view of generating awareness among the outsiders regarding the unexploited indigenous resources so as to promote sustainable cultural commercialisation.

S. Johari (✉) · S. Ramachandran · S. Herman
Faculty of Forestry, Universiti Putra Malaysia, Serdang, Selangor, Malaysia
e-mail: ahseng69@yahoo.com

S. Ramachandran
e-mail: sridarupm@gmail.com

S. Herman
e-mail: syamsulherman@gmail.com

A. Shuib · P. Kunasekaran
Institute of Agricultural and Food Policy Studies, Universiti Putra Malaysia, Serdang, Selangor, Malaysia
e-mail: mad.shuib@gmail.com

P. Kunasekaran
e-mail: puvaneswarankunasekaran@gmail.com

 
A. Saufi et al. (eds.), *Balancing Development and Sustainability in Tourism Destinations*, DOI 10.1007/978-981-10-1718-6_10

**Keywords** Cultural tourism · Initial inquiry · Community development · Sustainable tourism · Rural destination · Sarawak

## Introduction

The meaning of Bidayuh, among the many indigenous communities based in south Sarawak on the Borneo Island, is 'inhabitants of land'. The group hails from the western part of the Borneo Island, and is collectively known as 'Land Dayak', which was first used during the reign of Rajah James Brooke, recognised as the White Rajah of Sarawak. The community, which represents one of the key indigenous civilisations in Sarawak, resides in villages and towns around the Samarahan and Kuching regions of this state. The community constitutes around 8.1 percent of the state's populace and less than 1 percent of the entire populace of the country. According to Geddes (1954), the Bidayuhs are the "people of the interior", renowned for their longhouses and distinct traditions and customs. As a key respondent describes, the group is quite close to the nature.

> Bidayuhs are mainly Pagans or animist and they believe in ancestral worship and they also believe in the ancient spirits or nature. Due to that, they would have big celebrations like the Gawai (June 1st) which is a celebration to please the padi's spirit for good harvest. They would also hold small praise and worship sessions occasionally to the River and Jungle spirits to cure illness and ward of bad luck. However, at least 50 % of the Bidayuh community peoples have converted to Christianity since the arrival of missionaries into Sarawak (personal communication, 11 August 2012, Dihoi Anak Nyawen, Head of Village, Krokong-Tringgus).

Furthermore, Dihoi emphasised on the distinctiveness of the culture, which could be capitalised for business and tourism activities. The community's typical attributes include their ethnic dances, music, and food.

> Being the people of colourful culture, the Bidayuhs have their own cultural dress code (in the modern days, are only worn during festivities and celebrations). They play their own cultural music and did dance which comprises of a set of Gongs. The Bidayuhs are generally very soft-spoken people. While the men are proud and strong, the women are more gentle, less outspoken and would usually shy away from strangers and outsiders. The Bidayuh's staple food is rice but in a lot of places, wild-sago became their staple food" (personal communication, 11 August 2012, Dihoi Anak Nyawen, Head of Village, Krokong-Tringgus).

The community is generally categorised into six key subgroups: Biatah-Penrissen, Bau-Bidayuh, Padawan-Sembaan, Bukar-Sadung, Rara, and Salako. These subgroups were together known as Land Dayak; in 2002, the government officially named the entire community as Bidayuh.

## Literature Review

In rural areas, cultural resources generally are the driving factors of the entire tourism business, thereby contributing towards favourable socio-economic growth (Butler and Hinch 2007). The term cultural resources has been well defined by Pedford (1996). The author states that cultural resources refer to any living history aspects such as folklore, rural customs, family as well as local customs, ethics, views, and shared heritage. Culture is perhaps the key resource which indigenous communities possess. Their distinct rituals and traditions are promoted as tourist attractions in rural areas (Cohen 1988; Liu 2008; Brown and Cave 2010). Thus, cultural resource is essential for any community as a means of showcasing their identity and distinctiveness to others.

Culture can be perceived as the key product for indigenous tourism among native communities. From the cultural tourism context, commodification can help in boosting the domestic economic growth and development of customs by means of value expansion (Kunasekaran et al. 2013). For the indigenous communities, culture is the most renowned aspect emphasising on their authentic way of life. These groups often portray distinct customs, languages, practices, heritage, and food, which can be capitalised on to depict their distinctiveness to the rest of the world. Butler (1980) states that tourism grows this firmly dependent on their culture, legacies, and festivals. According to Butler and Hinch (2007), culture has evolved as a major attraction not just for tourists but also for businesspersons, government institutions, and academics.

Several academics see culture as something that has proved to be effectual in promoting growth of tourism (Cohen 1988; King 1993; Liu 2008; and Brown and Cave 2010). According to Besermenji et al. (2011), tourism can evolve as a helpful way of conserving the traditions and culture of Ontario Indian tribes. However, this is possible only with the involvement of the right people, and if tourism does not turn into a money-making source for some people. Tourism could aid in promoting and depicting several Indian traditions and cultures. Furthermore, it can aid in introducing their culture to the outsiders, and thus help in its conservation.

There are various kinds of tourism activities. Of these, travellers fare finder of adventure tourism, nature-based tourism, as well as arts and cultures that depict the distinctiveness of a group or community. Furthermore, indigenous groups might strive for new prospects as agencies which would be emboldened as a representative of cultural education. These benefits could be availed by an indigenous group that can be transformed into an emerging civic society through economic developments. Thus, the commodification of native culture would make the most of community economics (Mann 2012).

According to O'Gorman and Thompson (2007), commodification is the sole way of selling the indigenous culture to foreign travellers. From the context of cultural tourism, commodification can boost domestic economic growth and conserve traditions by decreasing poverty levels and raising the intrinsic value (UNWTO 2011). However, there are many academics that have criticised cultural commodification.

According to Cohen (1988), if indigenous culture is presented as a product of mass tourism, it can eventually lead to loss of intrinsic cultural appeal, authenticity, and significance. China's Naxi native community embraced cultural commodification and showcased its matchlessness to the rest of the world with the backing of the government authorities, as it believes that its culture has been conserved well and has not diminished (Brown and Luo 2012).

The cultural aspect plays a crucial part in cementing relationships for a sustainable growth of tourism by preserving the authenticity of a native community. Such communities often face extreme poverty because of their low income levels, triggered by poor living conditions. Tourism, along with a robust local culture, could strongly bolster efforts for local development (Rourke 2012). This development, in turn, would facilitate community capacity building by means of cultural commercialisation so as to reap economic as well as non-economic gains.

In northern Europe, particularly in Finland and Sweden, indigenous tourism is quite widespread, known for its cultural development (Petterson and Viken 2007). In both nations, the municipal affairs are regulated by the Sami Constitution System. The Sami is a renowned indigenous faction in the northern Europe, where the language and unique traits are determined by other surrounding ethnicities. As a dominant group with culture-driven native product of northern Europe, Sami maintains a dynamic presence in cultural tourism through various attractions, including cultural programmes, sites, and mementos, and the Sami Museum.

Tourism has resulted in several positive impacts on the indigenous societies of South East Asia. However, there are some groups that have faced adverse cultural impacts. The promotion of tourism has resulted in an enriching evolution at several places, causing the commodification of arts and culture (Menzies 2007). Indonesia's Tana Toraja community in particular has altered its traditions and rituals to cater to the inclinations of travellers. As per Adams (1990), the community is so desperate of exhibiting its 'staged authenticity' to the tourists that it has almost lost its own actual distinctiveness.

Ramachandran et al. (2006) too made some similar observations about the Semai community in Malaysia's Taman Negara. The researchers believe this community has been exposed to adverse externalities, and as a result, the youth has turned to drugs and alcohol, influenced by reckless foreign travellers.

Conversely, the conservation and marketing of cultural assets might enhance networking prospects for the native indigenous community (Bennett 2012) and, consequently, travellers might be able to see indigenous tourism as a distinctive, popular, pragmatic, and knowledge-offering attraction. The native communities in Eastern Australia are offered proper education on indigenous cultural tourism. This trend could bear likeness with cross-cultural exchange, wherein the retrieval and conservation of long-established aboriginal cultures could obtain appreciation and authenticity (Whyte 2010).

Indigenous tourism is characterised by a blend of rural tourism and native culture of far-flung areas that has the ability to draw outsiders to the specific region. A well-regulated tourism development by curbing the number of travellers to the indigenous rural areas enjoys the benefit of conserving the ethnicity of the indigenous community. Thus, the management of the aboriginal landscape could be harmonised with development of indigenous cultural tourism. Such tourism aims to offer a chance of association between travellers and indigenous communities as they are the two key parties of indigenous tourism.

## Methods

The study deploys an inductive qualitative methodology to comprehend the distinctiveness of the community in nurturing tourism. The objective is to collect data with no presumption of the main researcher (Creswell 2003). This approach facilitated elimination of any research biasness that could be triggered. A thorough interview was conducted to gather key data backed by observation technique. The gathered data was then scrutinised using the thematic analysis approach.

Table 10.1 depicts the respondents who appeared for the interview. Notably, all the interview conversations of the respondents are not publicised; only selected conversations which could be classified as related to natural and cultural resources were interpreted.

**Table 10.1** Key respondents of the study

| No. | Name of respondent | Position | Location | Gender | Age |
|---|---|---|---|---|---|
| 1. | Gabriel Tonga Noweg | Professor UNIMAS | UNIMAS | Male | 54 |
| 2. | Dihoi Anak Nyawen | Penghulu | Kpg. Krokong/Tringgus | Male | 58 |
| 3. | Nani Anak La jou | Woman Club Chairman | Kpg. Peros | Female | 47 |
| 4. | Lipik Anak Salau | Counsellor | Kpg. Bijulin | Male | 51 |
| 5. | Jorim Anak Ayup | Cooperation Member | Kpg. BatuSepit | Male | 53 |
| 6. | Risik Anak Giyan | Head of Village | Kpg. Belimbing | Male | 48 |
| 7. | Pisiew Anak Salam | Head of Village | Kpg. Kaman | Male | 60 |
| 8. | Takim Anak Sitam | Head of Village | Kpg. Peros | Male | 58 |
| 9. | Abam Anak Asoi | Head of Village | Kpg. PengkalanTebang | Male | 48 |
| 10. | Joseph Jap Anak Sayon | Homestay Chairman | Kpg. Krokong | Male | 45 |
| 11. | Jesen Anak Laun | Head of Village | Kpg. PuakKrokong | Male | 51 |
| 12. | Rajulin | Head of Village | Tringgus | Male | 48 |

## Natural Resources

For the community, their environs are their greatest strength in terms of packaging tourism slickly. The villages are situated closer to the forests and caves, which ensure travellers arrive from far and wide.

> However, the key attraction of this village is a beautiful river, which has crystal-clear water. Furthermore, our area is surrounded by several other beautiful natural resources which we can access without any limitations. The Pedi River is the largest (Respondent No. 9).

> There is a lovely and eye-catching place on top of the mountain which is indeed wide, and from that place, one can see the panorama of entire Kuching. People rarely go over there. It is journey of almost 4 to 5 h from Kampung Tringgus. The Bobot waterfalls are also an attraction which can be reached in around 4 h by climbing the staircase. Some tourists who are nature lovers go there (Respondent No. 12).

Few of the resources are managed by the Sarawak Cooperative Society with backing from the Sarawak Ministry of Tourism. This highlights involvement of the government in rural tourism development.

> There are separate packages that comprise exploring the forest, caves, and rivers in Bau, with an extremely low fee for foreign travellers. We have packages that comprise activities in beautiful places at Fairy Cave, Sarawak Kanan River, and exploration of the old forest. In 2013, three groups of students arrived here from Singapore and stayed for two nights to climb the Fairy Cave and explore the jungle (Respondent No. 10).

> A waterfall called Brobot exists below the Fairy Cave at Tringgus. It can be accessed by walking. On the high ground, there is a small river, two pools, and an orchard. It takes around six hours to reach this place (Respondent No. 4).

Few of the respondents said that there are certain hold-ups in the local natural assets that entail consideration from the local authorities from the context of tourism development.

> The issues which we encounter are primarily caused by Larian 3H, which has steadily ruined the crops of villagers and the forests. The majority of the issues which cropped up from the participants' safety aspects, such asif they fell down or get lost in the forests, are grave concerns (Respondent No. 8).

The majority of the respondents asserted that there are adequate natural resources in their village which could be capitalised on to draw tourists. However, actual participation is required if the tourism business has to emerge as successful (Arnstein 1969)

## Cultural Attraction

The community primarily banks on its culture to draw tourists. The dance, long-houses, and multi-coloured costumes are regarded as their key asset of cultural tourism. The community always seeks to exhibit its customs and traditions to outsiders.

> We have a distinctive culture which other people do not have. They may have money, but we have our culture, which can be transformed into money. A Baruk is a round-shaped structure typically located at the centre of the village. It is used as a meeting place for performing traditional ceremonies, dances as well as festivals (Gawais). The Bidayuhs observe many annual Gawais – for example, for the rice to spring out, for the rice to become healthy and strong, and when all the rice is harvested. Few of these traditional gawais are still observed; however, they are held on a particular date each year (Respondent No. 2).

> Annually, on 1st June, the Bidayuh community celebrates its Gawai, which lasts for a week. During Gawai, the majority of the Bidayuhs staying or working outside the villages return to meet and socialise with family and friends. It is not uncommon to merge Gawai with a wedding function as the entire family is already together. The Bidayuh is quite a peaceful community with respectable hospitality. Furthermore, we deliver group cultural performances at night and traditional food delicacies for one and all. Moreover, we conduct traditional sport activities for tourists as well as locals (Respondent No. 5).The Bidayuh community is renowned for its warrior mettle. In times of war, a Bidayuh man's status is adjudged by the number of skulls hanging on the ceiling just before entering his house. The more skulls, the higher is the warrior's rank (Respondent No. 7).

Besides, the area is bestowed with sufficient natural resources. Unfortunately, several of these resources have not been detected and nurtured as tourism products. Therefore, a robust infrastructure is required to draw tourists with the cooperation of the authorities. Financial constraints are often experienced with regards to tourism activities.

> There are many resources available in our region. However, most of them have not been detected and have been left unexplored, just like the Budu Cave. These can be turned into tourism attractions. Therefore, I suggest that cooperation is required from the government authorities to draw travellers to the area. I believe there are financial constraints for tourism projects (Respondent 8).

The community typically considers that it is opulent with resources, particularly cultural resource. They perceive their culture as distinctive—something that other people would not see anywhere else. Besides, they are vast volume of natural resources. However, these need to be explored and fostered as tourism products. Therefore, there is a need for cooperation to draw tourists to their region.

## Conclusion

The Bidayuh community possesses natural and cultural resources which can be commercialised for the purpose of tourism. The community thinks tourism will offer financial support as well as aid in safeguarding and sustaining its ethnicity. However, the government has so far not shown any genuine participation in any tourism expansion programmes. Even though they are part of the planning process, the ultimate decision is taken only by the governmental authorities. Therefore, the community is hoping that the government sees them as equal partners before arriving at any decision to foster tourism in its region.

## References

Adams R (1990) Self-help, social work and empowerment. BASW Macmillian, Basingstoke
Arnstein SR (1969) A ladder of citizen participation. J American Inst Planners 35(4):216–224
Bennett N (2012) A capital assets framework for appraising and building capacity for tourism development in aboriginal protected area gateway communities. Tour Manag 33(4):752–766
Besermenji S, Milić N, Mulec I (2011) Indians culture in the tourism of Ontario. Zbornikradova Geografskoginstituta Jovan Cvijić SANU 61(3):119–136
Brown KG, Cave J (2010) Island tourism: marketing culture and heritage–editorial introduction to the special issue. Int J Cult Tourism Hospitality Res 4(2):87–95
Brown K, Luo T (2012) Authenticity versus commoditization: the Chinese experience in the UNESCO heritage site of Lijiang. Int Cult Tourism Conf: New Possibilities 13(1):123–141
Butler R, Hinch T (2007) Tourism and indigenous people: issues and implications, 2nd edn. Butterworth-Heinemann, Michigan (first published: 1996)
Butler R (1980) The conception of a tourist area cycle of evolution: implications for management of resources. Can Geogr 24(1):5–12
Cohen E (1988) Authenticity and commoditization in tourism. Ann Tourism Res 15(3):371–386
Creswell JW (2003) Research designs: Qualitative, quantitative and mixed method approaches, 2nd edn. California: Sage Publication Inc
Geddes WR (1954) The land Dayaks of Sarawak: a report on a social economic survey of the land Dayaks of Sarawak presented to the colonial social science research council. HM Stationery Office for the Colonial Office, London
King VT (1993) Tourism and culture in Malaysia. In: Hitchcock M, King VT, Parnwell MJG (eds) Tourism in Southeast Asia. Routledge, London, pp 96–116
Kunasekaran P, Gill SS, Talib AT, Redzuan MR (2013) Culture as an indigenous tourism product of Mah Meri Community in Malaysia. Life Sci J 10(3)
Liu OP (2008) Packaging myths for tourism: the Rungus of Kudat. Universiti Kebangsaan Malaysia Press, Bangi
Mann B (2012) The protection of the bat community in the Dupnisa cave system, Turkey, following opening for tourism. Oryx 46(01):130–136
Menzies CR (2007) Traditional ecological knowledge and indigenous tourism. Tourism Indigenous Peoples Issues Implications 2:15–27
O'Gorman KD, Thompson K (2007) Tourism and culture in Mongolia: the case of Ulaanbaatar Naadam. Tourism and indigenous peoples: Issues and implications, 161–175
Pedford J (1996) Seeing is believing: The role of living history in marketing local heritage. In: Brewer T (ed) The marketing of tradition. Hisarlink Press, Enfield Lock, pp 13–20
Pettersson R, Viken A (2007) Sami perspectives on indigenous tourism in northern Europe: commerce or cultural development. Tourism Indigenous Peoples Issues Implications 177–187
Ramachandran S, Shuib A, Yacob MR, Mat Som AP (2006) Determining responsible tourism indicators: a case of Taman Negara, Malaysia. In: Proceeding of the international conference of sustainable tourism, Barbados
Rourke V (2012) Sustainable tourism and its use as a development strategy in Cambodia: a systematic literature review. J Sustain Tourism 23(5):797–818
UNWTO (2011) Sustainable tourism development. Retrieved http://sdt.unwto.org/en. Accessed 20 Sept 2012
Whyte KP (2010) An environmental justice framework for indigenous tourism. J Environ Philos 7 (2):75–92

# Chapter 11
# Movie Touring: The Influence of Film on Viewers' Travel Intention

**Mohd Noor Ismawi Ismail, Norzuwana Sumarjan, Nur Farihin Abd. Hadi Khan and Mohd Hafiz Hanafiah**

**Abstract** Screen tourism has played a significant role in expanding the number of tourists hoping to visit destination(s) portrayed in films and TV series. Therefore, identifying the drivers of screen tourism and understanding their effects on the tourism industry is crucial for destination managers. However, studies on-screen tourism, particularly in Malaysia, are limited and scarce. This study thus attends to a gap within screen tourism to identify potential factors that encourage viewers' visitation to a film location(s). In order to achieve this, the relationship between screen tourism and travel intention has been investigated. Using a quantitative research design with a structured questionnaire, the researchers distributed 300 online questionnaires, acquiring 176 usable responses. The respondents consist of Malaysian viewers who watched films that feature a particular destination within a 6-month period. Factor analysis and multiple regression analysis were used to confirm the proposed hypotheses. Place, performance, and personality were identified as the pull factors of screen tourism. This study confirmed past empirical evidence supporting the effects of the 3P Pull Factor theory on viewers' travel intentions. Additionally, it was found that the respondents were aware of the linkage between tourism and the film industry, which further supports the significant role that films play in influencing viewers' travel intentions.

M.N.I. Ismail (✉) · N. Sumarjan · M.H. Hanafiah
Faculty of Hotel and Tourism Management, Universiti Teknologi MARA, Puncak Alam Campus, 42300 Bandar Puncak Alam, Selangor, Malaysia
e-mail: ismawi@salam.uitm.edu.my

N. Sumarjan
e-mail: norzu161@salam.uitm.edu.my

M.H. Hanafiah
e-mail: hafizhanafiah@salam.uitm.edu.my

N.F.Abd.Hadi Khan
Faculty of Entrepreneurship and Business, Universiti Malaysia Kelantan, Pengkalan Chepa, 16100 Kota Bharu, Kelantan, Malaysia
e-mail: farihin.ah@umk.edu.my

 
A. Saufi et al. (eds.), *Balancing Development and Sustainability in Tourism Destinations*, DOI 10.1007/978-981-10-1718-6_11

**Keywords** Screen tourism · Travel intention · Push factor · Motivation, · And film

## Introduction

Film tourism, also known as film-induced tourism, movie-induced tourism, or screen tourism, is understood as a subset within the tourism discipline that is linked to a location(s) of a film or TV series, during or after production. In sum, film tourism presupposes that tourists become motivated to visit a location based on its on-screen reputation (Hudson and Ritchie 2006; Connell 2012). The area of research on cinematic tourism is relatively limited, yet increasingly researchers are acknowledging that film tourism is an interesting and worthwhile subject to attend to (Kraaijenzank 2009). Since the early 1990s, academics have studied the linkage between film television and tourism and it was during this time when screen tourism gained recognition within the tourism discipline (Kim and Wang 2012).

Preliminary research on-screen tourism has argued that there is little evidence to support the idea that films could attract tourists to visit actual film locations (Riley et al. 1998). Following suit, Rewtrakunphaiboon (2009) has encouraged future research to heed special attention to the successes of film tourism as well as the psychological aspects of tourists' behavior as it pertains to visiting film locations. Despite the fact that a number of studies have investigated the screen tourism phenomenon in several countries (Connell 2012), either for certain television shows (Wearing et al. 2011; Connell and Meyer 2009; Kim and Wang 2012; Kraaijenzank 2009) or across diverse subject matters and themes (Croy 2011; Macionis and O'Connor 2011; Butler 2011; Kim et al. 2006), more empirical research is needed in order to fulfill the calls to research and understand the dynamisms related to film-induced tourism.

In particular, there seems to be a lack of cross-disciplinary research between tourism and media studies in the Malaysian context. Beeton (2005) further confirmed that much of the recent academic literature that either reproduced preliminary studies of screen tourism or highlighted the role of films in promoting tourism have not contributed to this body of research. For example, because Malaysian films are often locally produced and thus targeted mainly for local consumption (Rosnan et al. 2010), they may not be particularly well-known for their film products on an international level. Furthermore, Malay films are often categorized as 'substandard' compared to imported films, which may also contribute to the downfall of the local film industry in Malaysia (Ahmad 2005). For that reason, the potential for Malaysia to exploit their local film industry in promoting themselves as a tourist destination remains relatively low. In response, an empirical investigation was undertaken, which examines the effects of the 3P's Pull Factor theory (Gjorgievski and Trpkova 2012) from the perspectives of Malaysian viewers to understand their travel intentions following media consumption. From a practical standpoint, this also benefits the Malaysian film industry in helping them to realize the drivers and subsequent benefits of screen tourism.

Such research is warranted, particularly concerning the local screen tourism industry, which might otherwise continue to be undervalued (Kim 2012). Without an adequate understanding of local screen tourism, many opportunities are lost for many destinations to build up their enchanting pull factors. Additionally, to the researchers' best knowledge, no research has conducted an empirical investigation that systematically examines the effects of the 3P's Pull Factors (place, performance, and personality) in relation to media products and their impact on viewers' intentions to visit film destinations. Consequently, the findings of this study can be considered a pioneering effort in the direction toward understanding Malaysia's film industry and the drivers of screen tourism.

## Screen Tourism

When research on film tourism commenced, the majority of studies were conducted either in the United States or in the United Kingdom. Subsequent research has been increasingly conducted from Australasian, Asian, and European perspectives (Connell 2012). The fundamentals of filming are all about drama and motion of pictures, which are very different from plastic arts, such as photography and painting (Vagionis and Loumioti 2011). The phenomenon of screen tourism is complex, due to its interdisciplinary underpinnings in fields such as sociology and psychology, as well as industry-based sectors such as the filmmaking industry, destination marketing, community development, and strategic planning of a destination (Beeton 2005). Hence, the study of screen tourism remains largely untapped within the broader field of tourism research.

Furthermore, Saltik et al. (2011) observed that much research has been conducted on film-induced tourism that focuses on how international movies have influenced the flow of tourism demand globally. For instance, Tooke and Baker (1996) agreed that motion picture films and television films have had a major influence on individuals' tastes and ideas. The notion of screen tourism was further elaborated by Busby and Klug (2001), who developed a diagram (Fig. 11.1) indicating the range of impacts that can occur as a result of movie-induced tourism.

## Literature Review

According to Crompton (1979), the "Push and Pull" model is a popular typology for understanding travel motivation. Lam and Hsu (2006) stated that most push factors originate from intangible or intrinsic desires of human beings, while pull factors attract individuals to a specific destination(s) once the decision to travel has been made. Additionally, Beeton (2005) contended that the television screen is

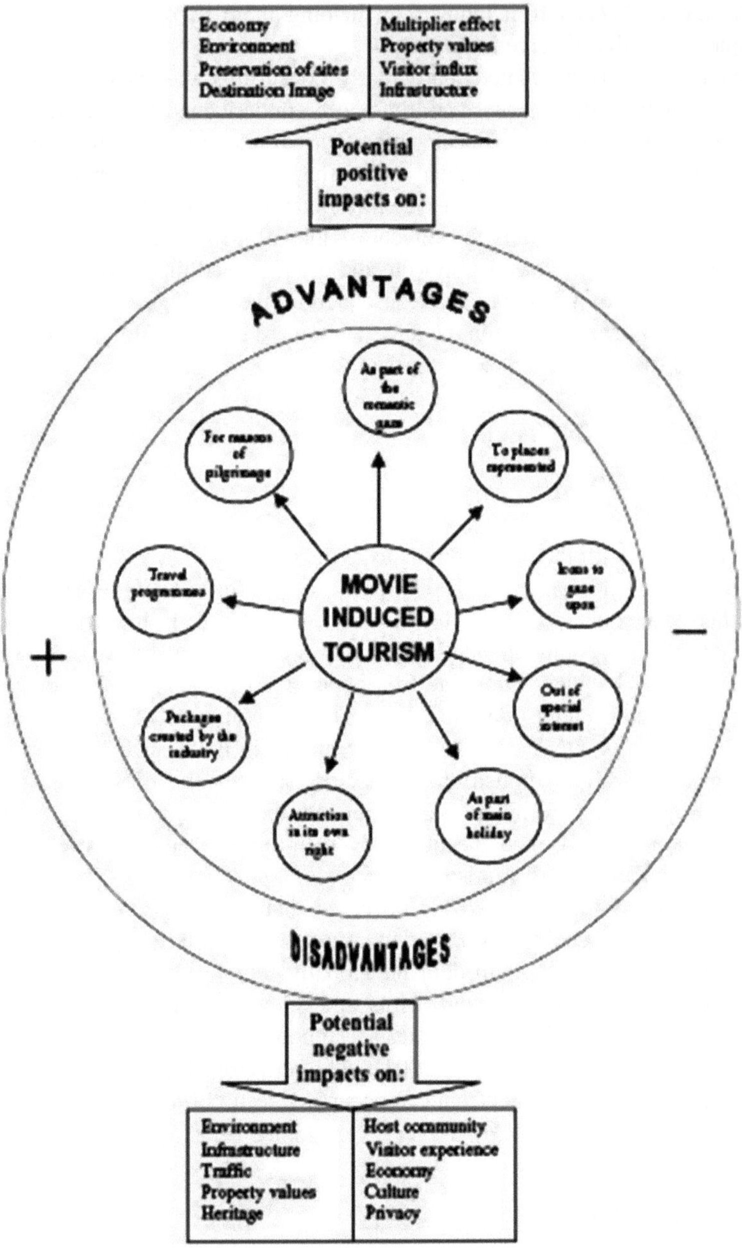


**Fig. 11.1** Overview of movie-induced tourism

another medium through which the tourists may experience a destination attraction, consequently serving as a strong motivation to visit the destination(s) featured.

In a similar vein, Rewtrakunphaiboon (2009) highlighted that films can induce viewers to travel to a filming location(s) based on the sceneries and landscapes, specific themes, plots, events, and actors that are central to a film; these work to influence an audience's thoughts, emotions, and attitudes toward certain destinations. More to the point, Croy (2011) argued that some characteristics, such as the picturesque locations featured in films (place), the engagement and tendency for an audience to personalize a story (performance), and adorations toward celebrities who took part in the film (personality), highlight the pre-visit role audiences to the film sites. Each of these factors are explored more in depth in the subsequent sections.

## *Place*

Preliminary studies on film-induced tourism considered a film location's geographic environment to be the main attraction and motivational pull factor driving audiences to visit a particular film destination (Tooke and Baker 1996; Riley et al. 1998). Much of this research asserted that an attractive landscape in and of itself provided viewers with enough motivation to overcome any negative aspects that may be embedded in a film's storyline (Beeton 2005). However, it is now understood that film induced tourism does not merely occur on the basis of a film's location, but also involves a film's off-site locations (Beeton 2005). Kim et al. (2006), for example, discovered that the attractive wild environments featured in *Crocodile Dundee* and *Lord of the Rings* are not the only places that attract visitors, but so too do the production studios and sets. Beeton (2005) also confirmed that off-site locations should not be disregarded with respect to their ability to foster screen tourism experiences.

## *Performance*

Tooke and Baker (1996) suggested that tourists might want to experience part of the narrative conveyed in a film rather than merely gaze at its location(s). This occurs when a film's plot or storyline engages the audience in such a way that they feel involved and connected (Kim et al. 2006). Riley et al. (1998) further expanded on the importance of elements such as storyline themes, exciting progression, and human relationships conveyed in a film; the authors argued that films have the potential to create out-of-the-ordinary worlds that can be recreated by visiting a film location. For instance, it was found that the high production values, beautiful scenery, and impressive storylines were among the most important elements, which have made Korean soap operas so popular among local and international viewers (Potipan and Worrawutteerakul n.d).

## *Personality*

Fascination with the famous (and infamous) is not new in the tourism industry. Since the late twentieth and early twenty-first centuries, celebrities have been renowned in many cultures, whereby connecting with celebrities has the ability to make the everyday person feel more important (Beeton 2005). While it has yet to be confirmed whether or not films featuring more attractive celebrities contribute to the success of a movie or product, Kim et al. (2006) have argued that it is not unreasonable to believe that celebrities do have significant pulling power in attracting audiences. In this way, visiting a film's destination(s) allows tourists the opportunity to get in touch with places closely associated with admired persons, and possibly even engage with the artifacts and memorabilia of the film on a more tangible level (Busby and Klug 2001). Su et al. (2011) also found that the ways in which characters interact with a film's location can influence individuals' attitudes toward locations. Similarly, Woodward (2000) found that Visit Britain, the tourist board of Great Britian, makes an effort to get endorsements from Bollywood actors to promote destinations in Britian, inasmuch as these actors have the power to add "kudos" to a destination, encouraging Indian and Asian tourists.

## *Visit Intention of Film Audiences*

Lam and Hsu (2006) asserted that it is beneficial to understand the reasons why people travel and to identify the factors that influence the behavioral intentions of selecting a specific travel destination, particualrly from a tourism or marketing standpoint. However, these authors have also noted that, currently, there is limited research on how push and pull factors guide individuals' attitudes and consequently drive their behavioral intentions with regard to selecting a specific travel destination. As familiarization of a destination increases among the tourists, so too does their awareness about holiday opportunities (Kim and Richardson 2003). Thus, it is believed that tourists' familiarity with a particular destination may also influence their intention to visit (Gibson et al. 2008). Croy (2011) pointed out that films play a major role in attracting and motivating one's desire for an experience, leading to an increase in film-induced tourism. Furthermore, film locations are purported to have attractions and sights that are easily reached, thus drawing the attention of many visitors across all seasons (Vagionis and Loumioti 2011).

## Method

This study applied a quantitative research design with a structured online questionnaire. Kim and O'Connor (2011) argued that the majority of the previous studies on-screen tourism applied qualitative methods, resulting in a dearth of appropriate instruments from which to measure tourists' travel intentions as they relate to destinations portrayed in films and TV series. Thus, a structured questionnaire was used as a tool to collect responses from a sample involving Malaysian audiences. With regard to the sample size, the researchers referred to Salkind (2003), who argued that the rule of thumb for an unknown population or sample size should be more than 30 but less than 500. Additionally, the researchers considered the requirements of multiple regression analysis, outlined by Hair et al. (1998), who claimed that is best to generate a sample size that is at least five times greater than the items to be analyzed.

The researchers distributed 300 online questionnaires and managed to generate 176 usable responses. A descriptive research design and correlational approach were also used. A self-administered questionnaire was used. The survey instrument was developed and modified following a comprehensive review of the literature and previous studies, which, as was previously mentioned, were primarily of a qualitative nature. The items were analyzed using an Exploratory Factor Analysis (EFA), so as to validate the developed items (Pallant 2005). The coefficient alpha for the place, performance, personality, and visit Intention dimensions was significant at value $>0 = 0.80$. An alpha value above 0.60 is considered acceptable and more than appropriate to proceed with further testing. The data were coded and analyzed using descriptive statistics and regression analysis.

## Findings

### *Respondent Profiles*

Based on the descriptive analysis, the sample size consisted of 44 males (25 %) and 132 females (75 %), accumulating a total of 176 respondents. Most of the respondents for this study were categorized as highly educated: 41.5 % reported to have a postgraduate qualification ($n = 73$), followed by 39.5 % ($n = 70$) undergraduate students. Based on the results of current employment, 62 respondents categorized themselves as (unemployed) students (35.8 %). This is not a surprise because, as was mentioned earlier, the majority of respondents were aged between 21 and 29 years old and as either graduated or postgraduate students.

The researchers included two additional questions about respondents' levels of exposure to films or television shows within a 6-month period, in order to determine their indication of recurrence in watching the same films or television shows. The findings showed that the majority of respondents watched more than one film every month (34.7 %, n = 61). Moreover, 131 respondents (74.4 %) indicated that they have watched the same films or TV shows repeatedly.

## *Factor Analysis*

Factor analysis was conducted to confirm the 3P's Pull Factor theory, by analyzing those items; Personality, Place, and Performance that constitute the theory. Factor analysis can ascertain whether a common factor(s) is present in the responses to the items. Therefore, 13 items on the 3P's Pull Factor scale (place, performance, and personality) were subject to Principal Components Analysis (PCA) using Statistical Program for Social Analysis (SPSS) Version 20. Prior to conducting PCA, the suitability of the data for the factor analysis was assessed. An assessment of the correlation matrix revealed the presence of many coefficients of 0.3 and above, as was recommended by Pallant (2005). The Kaiser-Meyer-Oklin (KMO) value was 0.769, exceeding the recommended value of 0.6 (Kaiser 1970, 1974) and the Barlett's Test of Sphericity (Bartlett 1954) also reached statistical significance, supporting the factorability of the correlation matrix.

To aid the interpretation of the three components, a Varimax rotation was performed. The rotated solution revealed the presence of a simple structure with the components revealing some considerable loadings. Most of the variables loaded onto one of the three components, as is illustrated in Table 11.1. The three-component solution explained a total of 56.46 % of the variance, with component 1 contributing to 20.07 %, component 2 contributing to 18.62 % and component 3 contributing to 17.77 %.

**Table 11.1** The rotated component matrix

| | Component | | |
|---|---|---|---|
| | 1 | 2 | 3 |
| Personality 2 | 0.812 | | |
| Personality 5 | 0.789 | | |
| Personality 4 | 0.667 | | |
| Performance 8 | 0.454 | | |
| Place 2 | 0.313 | | |
| Place 4 | | 0.890 | |
| Place 5 | | 0.878 | |
| Place 3 | | 0.722 | |
| Place 1 | | 0.504 | |
| Performance 1 | | | 0.843 |
| Performance 2 | | | 0.809 |
| Performance 5 | | | 0.623 |
| Performance 4 | | | 0.513 |

The interpretation of the three components were, to some extent, consistent with previous research on the 3P's Pull Factor theory, with personality pull factor items loading strongly on component 1, place pull factor items loading on component 2 and performance pull factor items loading on component 3, respectively. However, because 'place 2' generated the lowest score and loaded onto an unrelated component, the researchers removed the unnecessary item. In general, the results of this analysis support the use of the theory, as suggested by the authors of the 3P's Pull Factor theory (Gjorgievski and Trpkova 2012).

## *Regression Analysis*

A single-step regression analysis was conducted to analyze how strong the relationship is between the cultural, historical/heritage, and knowledge values and tour operators' behavior. The research dimensions were collapsed and entered into the equation as one. Table 11.2 displays the summary of the results.

From the table above, it is evident that personalities were able to explain 17 % ($R^2 = 17$, F-change = 23.289, $p < 0.01$**) of the examined variance of visit intention. The value of $\beta = 0.40$, $p < 0.01$** demonstrates that personality attributes have a positive contribution towards visit intention. This demonstrates that cultural values have a significant impact on the tour operators' behavior. Therefore, the first hypothesis ($H_1$) is supported.

The second hypothesis evaluates the relationship between the predictors comprising place attributes toward visit intention. The result show that place attributes were able to explain 24 % ($R^2 = 0.24$, F-change = 37.226, $p < 0.01$**) of the variance in visit intention. The results demonstrate that the place attribute ($\beta = 0.49$, $p < 0.01$**) positively influences visit intentions; therefore, hypothesis 2 ($H_2$) is supported.

For the third hypothesis, the relationship between the predictor variables of performance attributes and visit intention is considered. The results of the single-step regressions are exhibited in Table 11.2 above, which indicated that

**Table 11.2** Results of single-step regressions

| | Model I | Model II | Model III |
|---|---|---|---|
| Independent variable | Personality | Place | Performance |
| Dependent variable<br>B | Visit intention<br>0.40** | Visit intention<br>0.49** | Visit intention<br>0.51** |
| $R^2$ | 0.17 | 0.24 | 0.26 |
| Adj. $R^2$ | 0.16 | 0.23 | 0.25 |
| $R^2$ change | 0.17 | 0.24 | 0.26 |
| F-change | 23.289*** | 37.266*** | 41.917*** |

*Note* *$p < 0.05$, **$p < 0.01$, ***$p < 0.001$

performance attributes were able to explain 26 % ($R^2 = 0.26$, F-change = 41.917, $p < 0.01^{**}$) of the variance in visit intention. Again, the results show that the performance attribute ($\beta = 0.51, p < 0.01^{**}$) positively correlates to visit intention. Therefore, hypothesis 3 ($H_3$) is partially supported.

## Conclusion

The findings of the study validate the study framework developed based on the 3P's Pull Factor theory by Gjorgievski and Trpkova (2012). Based on the data set of this study, it was found that respondents consisting of Malaysian viewers were aware of the linkage between tourism and the film industry, thus acknowledging the significant roles of films and TV shows in inducing viewers' travel intention. Meanwhile, the effects of the 3P's Pull Factors on viewers' travel intentions were somewhat supported based on the statistical analysis conducted. Last but not least, all the pull factors involved (place, performance, and personality) were found to be significant predictors of viewers' travel intentions.

This study contributes to the literature on-screen tourism and the 3P's Pull Factor theory. The findings expanded the application of the 3P's Pull Factor theory to films and TV shows in Malaysia in order to understand the drivers behind why audience members visit film locations. Consequently, this study contributes to the literature on-screen tourism by analyzing Malaysian viewers' perspectives, to which future research could replicate in other contexts.

**Acknowledgments** This study was funded by the Ministry of Education Malaysia, the Universiti Teknologi MARA, Malaysia (UiTM), under the Research Acculturation Grant Scheme (RAGS).

## References

Ahmad M (2005) Colonial Policies, the Marketplace and the Malaysian Film Industry. Retrieved 27 Oct 2012. http://www.asef.org/images/stories/publications/ebooks/050602_pub_film-industries_03-chapter0.pdf

Bartlett MS (1954). A note on the multiplying factors for various chi square approximations. J Roy Stat Soc 16 (Series B):296–298

Beeton S (2005) Film-induced tourism. Cromwell Press, Great Britain

Busby G, Klug J (2001) Movie-induced tourism: the challenge of measurement and other issues. J Vacation Mark 7(4):316–332

Butler R (2011) It's only make believe: the implications of fictional and authentic locations in films. Worldwide Hospitality Tourism Themes 3(2):91–101

Connell J, Meyer D (2009) 'Balamory revisited': an evaluation of the screen tourism destination-tourist Nexus. Tour Manag 30(2):194–207

Connell J (2012) Progress in tourism management: film tourism—evolution, progress and prospects. Tour Manag 33:1007–1029

Crompton J (1979) Motivation for pleasure vacation. Ann Tourism Res 6(4):408–424

Croy WJ (2011) Film tourism: sustained economic contributions to destinations. Worldwide Hospitality Tourism Theme 3(2):159–164

Gibson HJ, Qi CX, Zhang JJ (2008) Destination image and intent to visit China and the 2008 Beijing Olympic Games. J Sport Manag 22:427–450

Gjorgievski M, Trpkova SM (2012) Movie induced tourism: a new tourism phenomenon. UTMS J Econ 3(1):97–104

Hair JF, Anderson RE, Tatham RL, Black WC (1998) Multivariate data analysis, 5th edn. Prentice-Hall, Upper Saddle River, N.J

Hudson S, Ritchie JRB (2006) Promoting destinations via film tourism: an empirical identification of supporting marketing initiatives. J Travel Res 44:387–396

Kaiser H (1970) A second generation little jiffy. Psychometrika 35:401–415

Kaiser H (1974) An index of factorial simplicity. Psychometrika 39:31–36

Kim H, Richardson SL (2003) Motion picture impacts on destination images. Ann Tourism Res 30 (1):216–237

Kim S, O'Connor N (2011) A cross-cultural study of screen-tourists' profiles. Worldwide Hospitality Tourism Themes 3(2):141–158

Kim S, Wang H (2012) From television to the film set: Korean drama daejanggeum drives Chinese, Taiwanese, Japanese and Thai audiences to screen-tourism. Int Commun Gazette 74(5)

Kim S (2012) Audience involvement and film tourism experiences: emotional places, emotional experiences. Tour Manag 33:387–396

Kim S, Robinson M, Long P (2006) Understanding popular media production and potential tourist consumption: a methodological agenda. Paper presented at the international tourism and media conference. Retrieved http://dspace.flinders.edu.au/xmlui/bitstream/handle/2328/26062/Kim%20Understanding.pdf?sequence=1

Kraaijenzank M (2009) Movie-induced tourism. Unpublished Master's Thesis, Aalborg University

Lam T, Hsu CHC (2006) Predicting behavioral intention of choosing a travel destination. Tour Manag 27:589–599

Lee SJ (n.d) The impact of soap opera on destination image: a multivariate repeated measures analysis. http://scholarworks.umass.edu/cgi/viewcontent.cgi?article=1325&context=gradc Retrieved 14 Oct 2012

Macionis N, O'Connor N (2011) How can the film-induced tourism phenomenon be sustainably managed? Worldwide Hospitality Tourism Themes 3(2):173–178

Pallant J (2005) SPSS survival manual: a step by step guide to data analysis using SPSS version 12. Allen & Unwin, Australia

Potipan P, Worrawutteerakul N (n.d) A study of the Korean wave in order to be a lesson to Thailand for establishing a Thai wave. Unpublished Master's Thesis, Malatdalen University, Sweden

Rewtrakunphaiboon W (2009) Film-induced Tourism: inventing a vacation to a location. http://www.bu.ac.th/knowledgecenter/epaper/jan_june2009/pdf/Walaiporn.pdf Retrieved 28 Oct 2012

Riley RW, Baker D, Van Doren CS (1998) Movie induced tourism. Ann Tourism Res 25(4):919–935

Rittichainuwat BN (n.d) Attractiveness of film-induced destinations. Paper presented at the 7th symposium of consumer psychology of tourism, hospitality and leisure. Retrieved http://hotelmag.siam.edu/images/stories/research/attractiveness%20of%20film-induced%20destinations%203.pdf

Rosnan H, Ismail MN, Mohd Daud N (2010) The globalization of film business and its effect on the Malaysian film industry. Bus Strategy Ser 11(5):325–332

Salkind NJ (2003) Exploring research. Pearson Education, New Jersey

Saltik IA, Cosar Y, Kozak M (2011) Film-induced tourism: benefits and challenges for destination marketing. Eur J Tourism Res 4(1):44–54

Su HJ, Huang Y-A, Brodowsky G, Kim HJ (2011) The impact of product placement on TV-induced tourism: Korean TV dramas and Taiwanese viewers. Tour Manag 32:805–814

Tooke N, Baker M (1996) Seeing is believing: the effect of film on visitor numbers to screened locations. Tour Manag 17(2):87–94

Vagionis N, Loumioti M (2011) Movies as a tool of modern tourist marketing. Tourismos Int Multidiscip J Tourism 6(2):353–362

Wearing S, Buchmann A, Jobberns C (2011) Free willy: the whale-watching legacy. Worldwide Hospitality Tourism Themes 3(2):127–140

Woodward I (2000) Why should the UK's tourism industry be interested in 'bollywood'films?. Insights 12

# Chapter 12
# Celebrity Chefs and the Enhancement of Culinary Tourism Experiences in Singapore

**Adam Abidin, Jenny H. Panchal and K. Thirumaran**

**Abstract** In recent decades, food has emerged as an important lifestyle trend in tourism. This paper explores the role of celebrity chefs in enhancing culinary experiences in Singapore. More specifically, our work identifies thematic determinants of how world-renowned chefs and their signature restaurants in Singapore contribute to the country's culinary scene. This study analyses 12 restaurant websites showcasing celebrity chefs. A range of themes such as the type of restaurant, cuisine, location, menu, length of time in business and promotional activities were examined to deduce the role and impact of celebrity chefs as an integral part of Singapore's desire to be a food capital of Southeast Asia. Celebrity status works in tandem with a mix of local and global cuisines, physical location and ambiance to carve out a niche for particular eating establishments in the landscape of worldwide culinary tourism, leading to the conclusion that celebrity chefs play a significant role in tourism promotion.

**Keywords** Celebrity chefs · Tourism promotion · Singapore

## Introduction

Singapore is a multifaceted tropical destination in SEA. Its tourism agenda strives to provide something for everyone. One strong appeal of the island city-state is the diversity of its cuisines. Less than a decade ago, street food and small restaurants serving local or international food were the main culinary attractions of the

A. Abidin (✉) · J.H. Panchal · K. Thirumaran
James Cook University Singapore, 149 Sims Drive,
Singapore 387380, Singapore
e-mail: adam.abidin@my.jcu.edu.au

J.H. Panchal
e-mail: jenny.panchal@jcu.edu.au

K. Thirumaran
e-mail: k.thirumaran@jcu.edu.au



A. Saufi et al. (eds.), *Balancing Development and Sustainability in Tourism Destinations*, DOI 10.1007/978-981-10-1718-6_12

city-state. Only a few five-star hotels catered to fine dining. However, in the past 5 years, a number of fine dining establishments have emerged in Singapore with additional boost from celebrity chefs as a big draw for both locals and internationals. Similar to Hong Kong, Singapore's particular blend of local culture and global cosmopolitanism has set the stage for its arrival on the scene of global culinary tourism (Tsai and Horng 2011).

This paper aims to explore the role of celebrity chefs in enhancing tourist experiences in Singapore. The specific objectives of this study are: (1) to identify the key variables that indicate how renowned chefs and their signature dishes and restaurants contribute to Singapore's culinary scene, and (2) to identify the extent to which restaurants websites succeed in promoting Singapore as a tourism destination. The paper unravels culinary tourism in the context of celebrity chefs by first surveying the literature on tropical destinations that have implemented food as a key attraction. Thereafter, through web content analysis method, findings are discussed in reference to Singapore's celebrity chefs. A broader perspective on culinary tourism and celebrity chefs in the context of destination marketing is discussed in our conclusion.

## Literature Review

Food is undoubtedly a crucial element in tourists' travel decisions. As Quan and Wang (2004) suggest, the increased interest in food tourism has resulted to its consumption as a key activity as well as an attraction. The consumption of food in the context of tourism can optimize a tourist's experience of a destination, hence, the significance of food's role in attracting visitors (Quan and Wang 2004; Hall and Sharples 2003; Ritchie and Zins 1978). The recognition of the food and service industry as well as food-related events have been identified as income-generating activities in tourism as early as the 1990s (Getz 1993). Similarly, Rao (2001) and Hjalager and Richards (2002) suggest that food-related activities and businesses also contribute to destination marketing and community development.

Mallon (1995) recognizes that the travelers search for unique gastronomical experiences has become a major consideration in the tourism industry, and is increasingly receiving attention from tourists, researchers and destination marketing organizations alike. Mallon argues that regional cuisine is an essential factor in tourism development and promotion of a destination. The author maintains that food as a driving force in drawing tourists to a destination should not be neglected.

Hall and Sharples (2003) and Ritchie and Zins (1978) define food tourism as a form of tourism in which a traveller visits a destination to seek gastronomical experiences. In order for culinary tourists to be attracted to the destination, two key conditions have to be met either independently or in combination. First, the destination has to have very popular and/or world-renowned dishes. Second, the destination marketing organization must be able to market food resources available in a palatable way, whether they engage celebrity chefs or through extensive marketing and advertisement programs. Taking into account these two key factors, we are able to state that Singapore employs both strategies to entice travelers to view Singapore as a culinary destination.

## Methodology

Web content analysis has been employed as a principle research method of this study. Krippendorf (1980) defines content analysis as a systematic, objective and quantitative method for studying communication messages and developing inferences concerning the relationship between messages and their environment. Based on Herring's (2001) and McMillan's (2000) separate works, content analysis is frequently used in the discipline of communications where it is also known as 'theme' or 'feature' analysis. It is important to note the terms "internet", "web" and "world wide web" are used interchangeably throughout this paper.

This study involves the analysis of 12 restaurants by celebrity chefs in Singapore. An initial search on Google using the key words "celebrity chef" "restaurant" and "Singapore" was undertaken. A list of celebrity chef-affiliated restaurants were provided by Singapore's official tourism website (www.yoursingapore.com), and the integrated resorts in Singapore, i.e. Marina Bay Sands (www.marinabaysands.com) and Resorts World Sentosa (www.rws.com). Twenty celebrity chef restaurants were identified.

Celebrity chef affiliated restaurants in this study refers to fine dining eateries that regularly or occasionally employ a famous chef to participate in the special event, cooking signature dishes or simply promoting the restaurant. A celebrity chef is usually a famous culinary expert and whose food and endorsements are noted by consumers and the media.

Several observations were made upon thoroughly studying the websites of these celebrity chefs and their respective restaurants located in Singapore. Prior to the opening of both the integrated resorts, there had been no attempts by these celebrity chefs to consider Singapore as a destination for establishing a foothold. Among those chefs identified for this study, the majority of them have their outlets based in Marina Bay Sands (MBS) and Resorts World Sentosa (RWS), while the rest are connected to independent restaurants located in the heart of Singapore. One important point to note is the type of menus that are offered. In the course of the research, the chefs' menu offerings were broken down into three different categories: local, localized and international. The definition of local would be the sale of products that are inherently unique to the country without any or much modification. Localized food refers to the fusion of local with international cuisines. The international menus refer to food originating from around the world.

## Findings and Discussion

The aim of this paper is to explore the extent of celebrity chefs' contributions in enhancing tourist experiences in Singapore. In this regard, it was essential to factor in variables in relation to the restaurants that are connected to the chefs. The chefs' background and restaurant variables such as menu offerings and restaurant locations were analyzed. These chefs are all individually talented, specializing in their own unique cuisines and dishes. Though they may have other brands overseas, they bring a new concept to Singapore that appeals to local palates. The reason for this is twofold. One is for the very simple fact of adhering to the strict import laws that Singapore has, which bars ingredients deemed illegal from entering the country. An example of this would be poppy seeds. They are commonly used for baking and cooking purposes. The Central Narcotics Bureau of Singapore deems this as a controlled substance and does not allow importation of the ingredient.

The second reason comes from the fact that some of the core ingredients to be used in the dishes are difficult to acquire. The common obstacle for these suppliers are due to the source of these items which have to be imported from far away countries, proving to be costly. As such, the suppliers offer alternative goods to the chefs, in an effort to aid in recreating what the original dish was meant to be. A chef from one of the celebrity restaurants in MBS, shared the same opinion about the challenges of importing some of the ingredients. Though there are situations where international ingredients are hard to come by in Singapore, celebrity chef affiliated European and American restaurants requisite local or Asian spices that are close substitutes.

Chefs from Tim Ho Wan and Waku Ghin, however, are not drastically impacted by such obstacles. As they are both Hong Kong and Japanese brands, the ingredients they need are easily acquired from within the region. As such, their menu offerings tend to stay authentic, thus offering to consumers the taste of their countries in the comfort of Singapore. Apart from the Asian restaurants, little or no mention was made of any Singaporean ingredients or cuisines that were listed in the menu.

The descriptions of the restaurant are another area where celebrity chefs have made a mark on the tourism landscape. The affiliated restaurants tend to promote their 'internationalness' to attract visitors. An example would be Gordon Ramsay and his new restaurant in Marina Bay Sands called Bread Street Kitchen.

> Bread Street Kitchen serves a British and European menu with fresh seasonal produce; whilst the bar combines well known classics with more innovative and fun cocktails, as well as a large collection of world wines. Bread Street Kitchen is open daily for lunch, dinner and drinks until late (Marinabaysands 2015).

Celebrity chefs, therefore, have an indirect influence in promoting Singapore as a destination for international cuisines. One of the most challenging obstacles these chefs have to face, is creating a menu that satisfies the local palate, or a generic international menu that caters to the mass tourists.

All of the chefs that have their restaurants in RWS and MBS do not have an independent website. Rather, their websites are a part of the resort's main website. In this scenario, the brand value of the celebrity chefs is seen as an offering by the resorts to attract international tourists. Tourists may visit the website of either MBS or RWS and might be motivated to visit the country due to the sheer number of famous chef names that the resorts have managed to acquire. It can be argued that these chefs see the resorts as a gold mine, and that being a part of it will greatly improve sales and revenue. The same can be said for the resorts as well, in that the celebrity chefs act as conduits for increased numbers of international tourists. It is a win-win situation for both parties from a financial standpoint, leveraging off each other to achieve corporate objectives. The by-product of this, however, is the deliberate promotion of Singapore through food tourism.

It is apparent that the chefs make minimal effort to incorporate Singaporean cuisine into their menu offerings. An argument for this might be the desire to stay true and authentic to the cuisine, bringing to locals and tourists alike new tastes from abroad. However, Susur Lee seems to be the only chef to have included Singaporean cuisines in his restaurant menus. Through partnering with the Tung Lok Group, the celebrity chef has introduced local cuisines as part of his creations. This is due to the fact that the Tung Lok group is a local restaurant conglomerate that prides itself upon sharing with the world the delicacies of Singapore and Asia as a whole. According to its 2008 annual report online, Tung Lok Group has paved the way for modern Chinese cuisine in Singapore by linking its brands and the talents of local chefs (e.g. Club Chinois, Chinois Susur Lee) among others (Tung Lok Restaurants 2015).

As Singapore strives to be the epicenter of Asian cuisines, the accessibility and ease of visiting Singapore allows for tourists to indulge in the myriad of cuisines that the country has to offer. This idea is supported by interviews with several locally renowned chefs, conducted by the Singapore Tourism Board (STB). The aim of the interviews was to establish why Singapore has seen so much success as one of the region's major food destination. As Singapore is a nucleus for trade and communication, it has the ability to attract people from across the globe. These people bring their own culture, sharing their influences through culinary techniques and ideas. The city-state offers a platform to celebrity chefs to present a wide array of culinary experiences. Clearly, the ability of the country to accept all forms of cuisines has transformed Singapore into an international gastronomic hub. It also involves a close working relationship with the private and public sectors. Through the STB's marketing efforts, the restaurant scene in Singapore has grown exponentially, showcasing sheer diversity of the vibrant food and beverage scene that the country has to offer.

## Conclusion

Food tourism worldwide has increasingly become a driving component of the tourism development. The fact that both local and international celebrity chefs have found a public space in the fine dining scene particularly amongst the middle and upper classes suggest that as a leading tropical Southeast Asian city, Singapore is becoming a competitor with Bangkok and Hong Kong in food tourism. As correctly observed by Van Esterik (2008), many Southeast Asians do patronize food stalls and more recently the affluent classes are partaking in fine dining that originally catered to tourists and expatriate communities. This confluence of tourists, expatriates and celebrity chefs in an emerging high society with a preference for fine dining indicates that the region is on the move in cuisine tourism.

The findings of this study reveal that websites of restaurants owned by celebrity chefs do not deliberately promote Singapore as a destination. Largely, each restaurant promotes itself as a food and beverage/dining establishment. It was observed that no explicit information was aimed directly at tourists. This is to say that these establishments are focused on serving both locals and tourists. Their menu offerings and restaurant locations bear no significant relationship to the extent to which the celebrity chefs promote Singapore as a destination through their respective restaurants. Further, it was found that the internationally acclaimed celebrity chefs' affiliated restaurants have not changed their menus to suit local taste, except for the local celebrity chef, Susur Lee. With respect to Singapore as a cuisine destination more effort should be made to define Singapore's place in the global food tourism landscape through its connection to celebrity chefs. Celebrity chefs are well placed to promote Singapore as a culinary tourism destination. Perhaps the epitome of a cuisine destination is one that people can equate food to a place. The issue of local and international food is another key finding that requires an understanding. As Cohen and Avieli (2004) suggest, not all tourists like the idea of consuming local dishes unless they are transformed to suit their own palates. In our findings, only the celebrity chef was able to successfully transform local food to suit international tastes while still maintaining its localness.

## References


Cohen E, Avieli N (2004) Food in tourism. Ann Tour Res 31(4):755–778

Getz D (1993) Festivals, special events. In: Khan MA, Olsen MD, Var T (eds) Encyclopedia of hospitality and tourism. Van Nostrand Reinhold, New York, pp 789–810

Hall CM, Sharples L (2003) The consumption of experiences or the experiences of consumption? An introduction to the tourism of taste. In: Hall CM, Sharples E, Mitchell R, Macionis N, Cambourne B (eds) Food tourism around the world: development, management and markets. Butterworth-Heinemann, Oxford, pp 1–24

Herring SC (2001) Web content analysis: expanding the paradigm. In: Hunsinger J, Allen M, Klastrup L (eds) The international handbook of internet research. Springer Verlag, London, pp 233–250

Hjalager AM, Richards G (2002) Tourism and gastronomy. Routledge, London

Krippendorf K (1980) Content analysis: an introduction to its methodology. SAGE, Newbury Park

Mallon P (1995) Gastronomy and tourism difficult but necessary union. Espaces 135: 21–23.

Marinabaysands.com (2015) Bread street kitchen by Gordon Ramsay—Award Winning Celebrity Chef Restaurant in Marina Bay Sands. Retrieved 23 Oct 2015, from http://www.marinabaysands.com/restaurants/celebrity-chefs/bread-street-kitchen.html

McMillan SJ (2000) The Microscope and the Moving Target: The Challenge of Applying Content Analysis to the World Wide Web. Journalism and Mass Communication Quarterly 77(1): 80–98.

Quan S, Wang N (2004) Towards a structural model of the tourist experience: an illustration food experience in tourism. Tour Manag 25:197–305

Rao V (2001) Celebrations as social investments: festivals expenditures, unit price variation and social status in rural India. J Dev Stud 38(1):71–97

Ritchie B, Zins M (1978) Culture as determinant of the attractiveness of a tourism region. Ann Tour Res 5:226–237

Tsai CS, Horng J-S (2011) Exploring marketing strategy of culinary tourism development in Hong Kong and Singapore: are source-based theory. Retrieved from http://www.ntnu.edu.tw/acad/rep/r98/a980001-1.pdf

Tung Lok Group (2015) Owner & Operator. Retrieved from http://www.tunglok.com/business/owner-operator

Van Esterik P (2008) Food culture in Southeast Asia. Greenwood Press, Santa Barbara

# Chapter 13
# Food Representation and Media: Experiencing Culinary Tourism Through Foodgasm and Foodporn

**Ikma Citra Ranteallo and Imanuella Romaputri Andilolo**

**Abstract** Some studies have found culinary tourism to support the development of cultural tourism. Through social media, the promotion of culinary tourism has influenced many people to consume food and drinks in variety of ways than they might have otherwise, and has drawn them to different tourist destinations to experience culinary attractions. Two of the most popular words in tourism social media jargon that have arisen with this phenomenon are #foodgasm and #foodporn. Neither of these has been widely studied by contemporary tourism researchers. Tourists using social media services, such as Instagram, Flickr, Tumblr, Youtube and Twitter, will use the hashtags (#)foodgasm and (#)foodporn on the photos that they share of particularly attractive food experiences, tastes and pleasures. Food representations via social media, especially in the form of photos, have created broader awareness of the diverse chain of global food production, distribution and consumption. Furthermore, pop culture involving aesthetics of food, as well as its social and cultural construction, has created food photography websites and food blogs. This research used content analysis to outline how tourists are using the two hashtags, all of which have implications on culinary tourism promotion and destination marketing.

**Keywords** Food representation · Social media · Culinary tourism · Foodgasm · Foodporn · Destination marketing · Food photography

I.C. Ranteallo (✉)
Department of Sociology, Faculty of Social and Political Sciences,
Udayana University, Denpasar, Indonesia
e-mail: ikmacitra@outlook.com

I.R. Andilolo
Department of Management, Faculty of Economics and Business,
University of Mataram, Mataram, Indonesia
e-mail: imanuelita@gmail.com

 
A. Saufi et al. (eds.), *Balancing Development and Sustainability in Tourism Destinations*, DOI 10.1007/978-981-10-1718-6_13

## Introduction

Food is an unforgettable element of travelling. The look, taste and colour coexist with the who, when, how and where it was consumed. Food memorabilia can also be brought home in the form of food specialty souvenirs (Lin and Mao 2015). Culinary and food tourism is a world travel trend because this type of travel offers not only the memory of food but also the rich storytelling traditions of indigenous peoples, i.e. the people who have always lived on the land as well as more recent newcomers who have made the land as their home (de la Barre and Brouder 2013). Apart from 'culinary tourism' and 'food tourism' terminology, 'culinary heritage' is a culturally constructed self-generating process, in which culinary artefacts are consumed by various clients, who attribute them with new meanings in different contexts. These dishes and their meanings are dynamic, responsive to social changes and involved with multi-directional processes of negotiation and interpretation (Avieli 2013).

Media is crucial in the construction of food representation for various purposes, among others is to promote culinary tourism. Some televised programs have been designed to incorporate culinary, tourism, taste and cooking into their concept. These programs involve viewers to actively or directly participate (Hartley 2012, p. 7–9) in the concept promoted, such as 'home cooking', 'do it yourself', 'do it with others', 'go green', 'organic product', 'slow food', 'raw food', etc. Some movies with culinary theme have also become travel references: *Fleeting Beauty* (2004), *Eating Sausage* (2004), *Coffee and Allah* (2007) researched by Fresno-Calleja (2013); *Like Water for Chocolate* (1992), *Big Night* (1996), *Food, Inc.* (2009), *Julie & Julia* (2009), *Ratatouille* (2007), *Jiro Dreams of Sushi* (2011) (Corliss 2014); *Tabula Rasa* (2014), and *Filosofi Kopi* (2015).

Tourism specifically in a tropical country, such as Indonesia, provides wider variety of culinary. Spices and produce in tropical countries may be hard to find in countries with different climate. The purpose of this article is to elaborate how social media and promotion of culinary tourism has influence people to consume food and drinks in variety of ways than they might have otherwise. Although internet is a popular social media in spreading information on tourism and culinary, however, not many studies have explore its opportunities and challenges with regards to contemporary issues in tourism.

In exploring contemporary issues in tourism by using social media to represent food in culinary tourism, we argue that two popular terms in this context are #foodgasm and #foodporn. Hashtag is a sign for trending topic worldwide. We begin with an overview of culinary tourism in popular culture. We then review the history and roots of culinary tourism related to tourism, food, image and taste. Next, we compare Google search results in searching for #foodporn, #foodgasm, #tourism and #culinarytourism shared in Instagram, Flickr, Tumblr, YouTube and Twitter. Finally, we argue that #foodgasm and #foodporn provide study opportunities on food representation and media in experiencing culinary tourism and taste.

Furthermore, the use of social media can increase promotion of culinary tourism and destination marketing as well as tourists' food knowledge.

## Methodology

We use content analysis to indicate the tourism activities of social media users in using both hashtags with implications on culinary tourism promotion and destination marketing. This qualitative study is an introduction for future studies to elaborate further the motivations and behaviours of tourists in culinary tourism with regards to food representation and media.

In essence, we use Google search to analyze and compare how many hits #foodgasm and #foodporn received related to #tourism and #culinarytourism by users of Instagram, Flickr, Tumblr, YouTube dan Twitter. Communication chain according to Harold Lasswell (1971, cited in Laughey 2007, p. 9) includes: who (control analysis); says what (content analysis); in which channel (media analysis); to whom (audience analysis); and with what effect (effect analysis).

## Culinary Tourism: Taste and Media

Culinary is a term used within the context of gastronomy, i.e. typical food of a country or region, food and food process, which are different from other country or region (Kivela and Crotts 2006). '…*Culinary tourism* was created by Long (2004, cited in Ottenbacher and Harrington 2013) to articulate the concept of experiencing local cultures through food and activities surrounding food'. Mannur (2010, p. 163) describes culinary tourism as: '… a cultural form that has become tooled in vastly different ways, thus producing seemingly dissonant narratives about the ethno-racial identity about the erstwhile foodie'. Culinary tourism can fulfil the objective of tourists on vision, touch, smell and taste (Franklin 2003). '…Smell is as important as the taste buds in the ability to taste, since flavours and aromas are perceived by the human brain with the olfactory bulb. However, taste remains profoundly subjective because the taste experience can never be physiologically shared' (Trubek 2008, p. 7).

Tourists travel to foreign places to experience the different geography, social and culture. Tourists must adapt to the different living condition, weather and food (Kuhn 2002, p. 118). They obtain the knowledge and detailed information on places they visit including its culinary. Some references may be obtained directly through friends and family, or provided by the social media such as food channels, whereby: '… celebrity chefs clearly plays a key role in their co-creation as new contemporary food experts, who, given their celebrity status and elevated voice within society, have access to many elite food actors to influence knowledge…' (Barnes 2014).

Ning Wang (2000, p. 20) states that taste and preference must be confronted with environmental technology within the scope of modernization. To this effect cultural phenomena is seen as part of romanticism attack on capitalism and industrialism as technology has separated humans from nature and natural amenities, which affects negatively on psychology and creates social problems. Romantic Movement came about towards the end of the eighteenth century to tackle the negative effect of technology and more so to be the foundation for tourism.

Romantic movement can be seen through gastronomic consumption in tourism that has become increasingly ruled by culturally embedded symbols, culinary products and traditions. This implies that food is also becoming a more important factor in influencing tourists' motivations for travelling to a particular destination (Gyimóthy and Mykletun 2009). Other natural tourism industry that came about is the leisure farming industry that offers food, accommodation, leisure farms, also engage in traditional agricultural activities and enjoy the tranquillity and closeness to natural resources (Huang 2006).

Food representation and media has an important role in culinary tourism. Media representations refer to *texts* or images that circulate in the media space and carry symbolic content, e.g. advertisements, blogs and Facebook pages. 'Representation refers to the *process* of re-presenting, the process by which members of a culture use systems of signs to produce meaning' (Orgad 2012, p. 47–48). Food images and #foodgasm and #foodporn texts may represent different meaning from the intended original meaning. Who and why these different meanings are produced?

Simply put, representation of food meaning is produced and reproduced with or without revision to increase product selling through advertisements. Tourism also uses advertisements for similar purpose. Agro-tourism, ecotourism, culinary rural tourism, gourmet tourism, cuisine tourism, local gastronomy, culinary heritage, gastronomic tourism, gastronomic heritage and culinary tourism reflect the romanticism concepts to give different food sensations to tourists. Food meaning is constructed to demonstrate social identities with collective memory and roots, which promote a sense of belonging to a territory (López-Guzmán et al. 2014).

Studies on representation are frequently elaborated in semiotic context. '… Books, films and so on are understood as *re-presentations*, that is, reflections or reproductions of the real world 'outside' them. [Stuart] Hall offers an alternative…: there *is* a real world outside representation but we can only make it signify and 'mean' through representation. Moreover, representations are not reflexive but *constitutive* and therefore have a real, material impact…' (Procter 2004, p. 125).

Contemporary studies on tourism are important to map food position as a popular theme in media. Food reviews can be found anywhere: food columns, cookbooks, gourmet magazines and celebrity chef television shows (Probyn 2000, p. 1–2). '…In postmodern society, making choices about who to be and how to live may be extremely complex, stressful and risky because information is fragmented… Once food becomes plentiful and varied, fashion takes over and the lure of novelty, the trendy become markers of identity, which are used to know and show who we are…' (Blichfeldt et al. 2012). Consumers are also involved in tackling issues on food risks raised by public concerns through informative campaigns. '…Consumers would like to have better quality foods but feel at the same time that public information about food risks disturbs their experience of cooking and eating. Consumers express trust in public control procedures regarding food production—but, at the end of the day, what they most enjoy buying are food products of which they have personal knowledge…' (Halkier 2001).

Psychologically media dwells in '…an emotional disposition, imaginative and cognitive activity, which has the potential to be converted into tourist activity. The tourist imagination as a concept is capable of capturing the mobility of relationships between tourism and the media' (Crouch et al. 2005, p. 1, 3). Tourist culinary activities are not only directed at consumption stage, but also from production as well as distribution stages. Ecotourism tour packages offer the pleasures of consuming food differently through Slow Food, '…a global social movement aimed at promoting food that is regionally, ethically, and sustainably produced, and convivially consumed. The movement uses culinary tourist events, such as food festivals and farmers' markets, to promote its philosophy and attract new members… affects foodways and highlights the contingent and challenging nature of practising ethical eating' (Williams et al. 2015).

Globalization as a cultural process involves the nutrition transition of diets, tastes and health profiles from region to region, and especially from the rich West (or 'North') to the poorer South (Lang 1999). One issue highlighted in culinary tourism is the Slow Food movement: '…a global social movement aimed at promoting food that is regionally, ethically, and sustainably produced, and convivially consumed. The movement uses culinary tourist events, such as food festivals and farmers' markets, to promote its philosophy and attract new members…' (Williams et al. 2015). In international relations, food is used as gastro diplomacy: '…another branch of Public Diplomacy where soft power rises as a tool of warfare…has obtained extra concentration and to symbolize food as the gravy to diplomacy's rice. Gastro Diplomacy is also an exceptionally powerful, nonverbal method of communication. Travelling back to a couple of previous millennia, culinary needs preceded diplomatic needs, opening up ancient trade routes and pathways that finally shaped today's global economic and political landscape…' (Nirwandy and Awang 2014).

Jukka Gronow maps tastes in social relations context. In the seventeenthand eighteenth century (Schümmer 1955, cited in Gronow 1997, p. 3), tastes were determined by certain social groups that legitimize other people's tastes based on what they perceived as the best tastes. By mid-eighteenth century a new interest in food, consumption, and health in England emerges at the same time as diet lifestyle (Turner 1982, Aronson 1984, Falk and Gronow 1985, cited in Gronow 1997, p. 6), as well as *English Malady* (sickness or problem in an English system—in this case food) that was rooted in greed and opulence. Modern society identifies modern gastronomy as a discipline and control of individual needs in order to maintain human desire on food and food consumption. Gastronomy has cultured modern consumer tastes by introducing differences and new classifications on food and drinks, as well as their nutrient content (Gronow 1997, p. 8).

According to Immanuel Kant in *Critique of Judgement* (1987, cited in Gronow 1997, p. 86), everyone has their own tastes, and nothing can dispute that. '…There are groups of food enthusiasts: not only does the traditional, exclusive gourmet survive, but there are also the so-called 'foodies', an object of British humour in the 1980s because of their excessive dedication to stylish and pernickety eating. But perhaps the most significant example of all is vegetarianism. Vegetarianism has become a social movement of considerable significance in the UK, where around 6 % of the population claim to be vegetarians, a much higher proportion than in other European countries…' (Warde 1997, p. 32). Collective taste according to Herbert Blumer (1969: 289, cited in Gronow 1997, p. 106) came about from imperfect fusion between uncertainty and dissatisfaction that stems from new experiences in fashion and the world at large. At this stage, collective taste is abstract, blurred, unbalanced and in need of direction. Fashion innovators and trend setters use this window of opportunity to determine its potential expression and shape. This opportunity especially in terms of food and food produce, is well used by food bloggers according to Alain Touraine (1974, p. 7, 50, 55, cited in Dant 2003, p. 13–14). They use digital technology to control, recommend, persuade, even elegantly force the readers to go and eat at a specified place because of the 'image' created in the food blog (Ranteallo and Sitowati 2014, p. 489; see also Mannur 2010, p. 171 about food writing).

Pierre Bourdieu (1984, p. 190, 196) stressed that taste in food is determined by social classes on their idea of body, effects of food to the body, health and beauty, as well as function. For example, the working class gives more emphasis on (male) bodily strength rather than the shape thus tend to consume cheaper and filling food. In comparison to the professional class, they tend to opt for tastier, healthier, brighter and non-fattening food. Bourdieu concludes that food taste of social classes becomes the foundation that shapes categories of human body. Food presentation and consumption as well as food settings are strictly determined on arrangements that are pleasing to the eyes. Food presentation of edible shape and colour (like an artwork); ethics that determine attitude and body movement; ways to serve oneself and others during meal and the use of different utensils; seating arrangements based on social hierarchy; awareness and controlled on body movements are social rules that need attention in certain food culture. '…Taste is the difference between food as a mere form of sustenance and food as part of life's rich pageant, a part of sociality, spirituality, aesthetics, and more…' (Trubek 2008, p. 6).

Bell and Hollows (2006, p. 11; see also Trubek 2008, p. 21–22) state that Bourdieu created a category of character called new bourgeoisie that acts as 'cultural intermediaries' or taste-makers that legitimizes life style and combines the art of consumption, spent and enjoyment. '…food fads, fashions and sects as evidence of atomization and the disintegration of social regulation. However, what Fischler (1980, cited in Warde 1997, p. 32) disparages as cultish may be seen by someone else as commitment to a set of values or mode of life, not pathological but admirable practice. These phenomena might instead be evidence of stylization, niche consumption and neo-tribalism'.

## Using #foodgasm, #foodporn, #tourism and #culinary Tourism to Experience Taste

Social media becomes the bridge for tourists and tourism agents to enjoy a culinary adventure. Instagram, Flickr, Tumblr, YouTube and Twitter are popular social media used to share free information (specifically texts and images) on food. Instagram allows for photo and video upload and to edit the uploaded file. Instagram is also connected to and can be shared directly on Facebook, Twitter, Tumblr and many others (https://instagram.com/). 'YouTube allows billions of people to discover, watch and share originally-created videos. YouTube provides a forum for people to connect, inform, and inspire others across the globe and acts as a distribution platform for original content creators and advertisers large and small' (https://www.youtube.com/yt/about/). Flickr can present photo, video and blog of memorable pictures taken with phone camera, or to showcase the best picture or video around the world to become a web celebrity. Flickr user can allow friends, family and other contacts to comment on the creation as well as adding notes and tags (https://www.flickr.com/about). Tumblr 'lets you effortlessly share…text, photos, quotes, links, music, and videos from your browser, phone, desktop, email or wherever you happen to be. You can customize everything, from colours to your theme's HTML' (https://www.tumblr.com/about). Twitter is an information network of messages with 140 characters called Tweet to find news with interesting topics and updated factual information (https://support.twitter.com/). Apple App Store even produce an application called InstaFood Pro for smartphone with the tagline 'if you want to show everybody what you are eating and do it with style!'.

Advertisements and producers have created a food fetish and co-modification that surpass its own value. The need for food as is for sex, involves attraction, sensuality, pleasure and enjoyment. At certain stage food has been treated similar to pornography with exploitation and co-modification as such to please the human senses. At this stage, food is said to be in the context of foodporn. Whereas foodgasm—taken from orgasm; culmination point of sexual activity—refers to the pleasure after consumption and enjoyment of food and drink. However, Stierand and Lynch (2008, 338–339) view food porn more as culinary innovation.

Tourists who are users of Instagram, Flickr, Tumblr, YouTube and Twitter often put #foodgasm and #foodporn on pictures of food to share experience, taste and enjoyment of consuming particular culinary at a specific place. Food representation by the media has increase people's attention to chain of production, distribution and consumption. Furthermore, pop culture incorporates aesthetics and social and cultural construction of food to produce food photography and food blogs (Ranteallo and Sitowati 2014, p. 487–489).

Instagram, Flickr, Tumblr, YouTube and Twitter expect to fulfil the need of its users. We want to identify the actual level of public interest on tourism and culinary tourism on these applications. Based on the findings in Table 13.1, search results for culinary tourism is far less than tourism. This should be taken as an important sign and challenge on studies of contemporary tourism. We opted for Google as our

**Table 13.1** Comparison on Google results (hits) (16 June 2015)

| | Instagram | Tumblr | Flickr | YouTube | Twitter |
|---|---|---|---|---|---|
| #foodgasm #foodporn #tourism | 39,400 | 2,920 | 1,840 | 4,290 | 4,760 |
| Seconds | 0.47 | 0.40 | 0.43 | 0.21 | 0.44 |
| #foodgasm #foodporn #culinarytourism | 64 | No results | 9 | 7 | 5 |
| Seconds | 0.52 | – | 0.79 | 0.41 | 0.40 |
| Related hashtags | #foodie #foodphotography #foodculture #foodtravel #foodblogger #taste #gastronomy | #realfoodtravel #eating #tasty #yummy #delicious | #foodadventure #vacation #holiday #travelgram #instatraveling #foodpic #cuisine | #traveler #instatravel #culinary #tourism | #foody #travel #travelling #foodoftheday |

search engine instead of Google Scholar to identify the level of public interest on culinary tourism. Nevertheless, further researches into similar topics are needed to make this identification.

Motives behind the use of #foodgasm, #foodporn, #tourism and #culinarytourism in the context of contemporary tourism are controlled by the social media users to convey ideas related to food. We used random search hence we are not able to elaborate on the identities of these users. However, communication chain who—says what—in which channel—to whom—with what effect can be seen in food TV channel Asian Food Channel (AFC) that broadcast Food Hero Asia by displaying #FoodHeroAsia in its broadcast. #FoodHeroAsia is shared on Twitter, Pinterest and Instagram (asianfoodchannel.com).

This day and age, tourists cannot be described by a single definition as mere users and recipients that tend to be passive. Present day tourists come from different backgrounds and with different intentions that serve as active users. A general example, i.e. food traveller, food blogger, chef blogger, food photographer even extreme food traveller. The main benefit of using #foodgasm, #foodporn, #tourism and #culinarytourism is to make public ones activities and travels apart from reasons of narcissistic, selfie photographs and acknowledgement of social economic status by the public. In many social media, popularity is determined by the amount of followers. Popularity is important to support food knowledge in order for the user to inform somebody else what to do and how to think related to culinary tourism.

Culinary tourism at present is going back towards romanticism with the global spread of back to nature movement. This ideological movement is expected to help humanity to face the threats of climate change, continuous raise in population and extreme activities. Culinary tourists are also expected to become knowledgeable and aware of food issues such as food security, global rise in food prices and long-term food supply agreements or to acquire land for agriculture (see Obidzinski et al. 2013). Culinary tourism activities in the context of sociological functions of food consumption support the notion of intergenerational transmission of "good products" by emphasizing the nostalgic touch related to childhood, used by companies as strategic marketing (Alexandra and Pichon 2014).

The users of social media as active viewer are also affiliating and forming social movement related to culinary and food in Indonesia, among others: (1) Gerakan Pangan Aman dan Sehat Indonesia (Indonesia Food Safety and Healthy Movement) to increase critical awareness in the society on Indonesia's food implementation (*Facebook*: GPASIndonesia. *Twitter*: @KonsumsiPangan); (2) Komunitas Cinta Pangan Lokal Indonesia (Community of Local Food Lovers) is a social movement to introduce and raise interest on the variety of local foods in Indonesia (http://pangandindonesia.wordpress.com; *Facebook*: Cinta-Pangan-Lokal-Indonesia); (3) Aku Cinta Masakan Indonesia (I Love Indonesian Food) (@acmID); and (4) Underground Secret Dining, a culinary show for gastronomic adventures with most of the culinary are Indonesian (http://undergroundsecretdining.wordpress.com). These social movements have educated farmers, fishermen and breeders to produce commodity that can compete in the national and international market. Nonetheless, local ways must be maintained in order to preserve culinary taste (Ranteallo 2015). The impacts of increasing culinary tourism worldwide have given rise to innovation on food literacy; culinary arts; cultural identities in tourism; indigenous entrepreneurship in tourism; and balancing development and sustainability in tourism destination.

## Conclusion

Users of social media services, such as Instagram, Flickr, Tumblr, Youtube and Twitter, use the hashtags (#)foodgasm and (#)foodporn on the photos that they share of particularly attractive food experiences, tastes and pleasures. Furthermore, pop culture involving aesthetics of food, as well as its social and cultural construction, has created food photography websites and food blogs. Food representations via social media, especially in the form of photos, have created broader awareness of the diverse chain of global food production, distribution and consumption, as well as increased knowledge on local cuisines that give rise to the attraction of culinary tourism. The impacts of increasing culinary tourism worldwide stimulate innovation on food literacy; culinary arts; cultural identities in tourism; indigenous entrepreneurship in tourism; and balancing development and sustainability in tourism destination. Further researches on how tourists are using social media services especially the two hashtags, #foodgasm and #foodporn, are needed to assess their implications on culinary tourism promotion and destination marketing.

## References


Alexandra V, Pichon P-E (2014) A taste of nostalgia. Links between nostalgia and food consumption. Qual Mark Res Int J 17(3):225–238. doi:10.1108/QMR-06-2012-0027

Avieli N (2013) What is 'Local Food?' Dynamic culinary heritage in the world heritage site of Hoi An, Vietnam. J Herit Tour 8(2–3):120–132. doi:10.1080/1743873X.2013.767812

Barnes C (2014) Mediating good food and moments of possibility with Jamie Oliver: Problematising celebrity chefs as talking labels. Geoforum, http://dx.doi.org/10.1016/j.geoforum.2014.09.004

Bell D, Hollows J (2006) "Towards a history of lifestyle" in historicizing lifestyle. In: Bell D, Hollows J (eds) Mediating taste, consumption and identity from the 1900s to 1970s. Ashgate, Hampshire, pp 1–20

Blichfeldt BS, Mikkelsen M, Andersen LB (2012) Cooking up the self. Acad Minutes http://www.akademiskkvarter.hum.aau.dk/pdf/AM2012/Bodil_Marie_Lisa_Cooking_G.pdf. Accessed 8 June 2015

Bourdieu P (1984) A social critique of the judgement of taste (trans: Nice R). Massachusetts: Harvard University Press

Corliss R (2014) "Movie-able feast: the 8 most delicous films about food", Time. http://time.com/93672/best-food-movies/. Accessed 8 June 2015

Crouch D, Jackson H, Thompson F (2005) Introduction: the media and the tourist imagination. In: Crouch D, Jackson R, Thompson F (eds) The media and the tourist imagination: converging cultures. Routledge, London & New York, pp 1–13

Dant T (2003) Critical social theory: culture, society and critique. Sage, London & New Delhi

de la Barre S, Brouder P (2013) Consuming stories: placing food in the Arctic tourism experience. J Herit Tour 8(2–3):213–223. doi:10.1080/1743873X.2013.767811

Franklin A (2003) Tourism: an introduction. Sage, London & New Delhi

Fresno-Calleja P (2013) 'Food for thought': filmic recipes for New Zealand's multiculturalism. Continuum J Media Cult Stud 27(6):850–861. doi:10.1080/10304312.2013.794191

Gronow J (1997) The sociology of taste. Routledge, London & New York

Gyimóthy S, Mykletun RJ (2009) Scary food: commodifying culinary heritage as meal adventures in tourism. J Vacat Mark 15(3):259–273. doi:10.1177/1356766709104271

Halkier B (2001) Consuming ambivalences. Consumer handling of environmentally related risks in food. J Consum Cult 1(2):205–224

Hartley J (2012) Digital futures for cultural and media studies. Wiley-Blackwell, West Sussex

Huang L (2006) Rural tourism revitalization of the leisure farm industry by implementing an e-commerce strategy. J Vacat Mark 12(3):232–245. doi:10.1177/1356766706064620

Kivela J, Crotts JC (2006) Tourism and gastronomy: Gastronomy's influence on how tourists experience a destination. J Hosp Tour Res 30:354–377. doi:10.1177/1096348006286797

Kuhn L (2002). Trusting tourists: an investigation into tourism, trust and social order. In Dann GMS (Ed) The tourist as a metaphor of the social world. CAB International, Oxon & New York

Lang T (1999) Diet, health and globalization: five key questions. Proc Nutr Soc 58:335–343

Laughey D (2007) Key themes in media theory. McGraw-Hill, New York

Lin L, Mao P-C (2015) Food for memories and culture—a content analysis study of food specialties and souvenirs. J Hosp Tour Manag 22:19–29. doi:10.1016/j.jhtm.2014.12.001

López-Guzmán T, Di-Clemente E, Hernández-Mogollón JM (2014) Culinary tourists in the Spanish region of Extremadura, Spain. Wine Econ Policy, 3:10–18. 10.1016/j.wep.2014.02.002

Mannur A (2010) Culinary fictions: food in South Asian diasporic culture. Temple University Press, Philadelphia

Nirwandy N, Awang AA (2014) Conceptualizing public diplomacy social convention culinary: engaging gastro diplomacy warfare for economic branding. Procedia Soc Behav Sci 130:325–332. doi:10.1016/j.sbspro.2014.04.038

Obidzinski K, Takahashi I, Dermawan A, Komarudin H, Andrianto A (2013) Can large scale land acquisition for agro-development in Indonesia be managed sustainably? Land Use Policy 30:952–965. doi:10.1016/j.landusepol.2012.06.018

Orgad S (2012) Media representation and the global imagination. Polity Press, Cambridge & Malden

Ottenbacher MC, Harrington RJ (2013) A case study of a culinary tourism campaign in Germany: implications for strategy making and successful implementation. J Hosp Tour Res 37(1):3–28. doi:10.1177/1096348011413593

Probyn E (2000) Carnal appetites: foodsexidentities. Routledge, London

Procter J (2004) Stuart hall. Sage, London, New York

Ranteallo IC (2015) Media dan ketahanan pangan Indonesia: Studi sosiologi selera. Paper presented in Konferensi Nasional Sosiologi IV, Universitas Samratulangi Manado, 22 Mei 2015 (in press)

Ranteallo IC, Sitowati I (2014) Konsumsi, food blog dan digitalisasi makanan [Consumption, food blog and digitalization of food]. In: Nazaruddin M (ed) Membayangkan Indonesia baru. Program Studi Ilmu Komunikasi Universitas Islam Indonesia, Yogyakarta, pp 486–493

Stierand M, Lynch P (2008) The art of creating culinary innovations. Tour Hosp Res 8(4):337–350. doi:10.1057/thr.2008.28

Trubek AB (2008) The taste of place. A cultural journey into terroir. University Of California Press Berkeley, Los Angeles & London

Wang W (2000) Tourism and modernity. A sociological analysis. Pergamon, Oxford

Warde A (1997) Consumption, food and taste. Culinary antinomies and commodity culture. Sage, London & New Delhi

Williams LT, Germov J, Fuller S, Freij M (2015) A taste of ethical consumption at a slow food festival. Appetite 91:321–328. doi:10.1016/j.appet.2015.04.066

# Part II
# Balancing Sustainability in Community Tourism Development

**Imanuella Romaputri Andilolo**

University of Mataram, Mataram, Indonesia

Tourism has a significant impact on people's lives. We have all been connected in one way or another through tourism. Tourism can bring positive changes to community lives through the increase of economic capacity, improved infrastructure and education, the preservation of culture and heritage, and environmental protection. Needless to say, left untended, tourism can alter cultural values, destroy local heritage, degrade the environment, and essentially damage the very fabrics of communities.

The United Nations has defined sustainable development as "development that meets the needs of the present without compromising the ability of future generations to meet their own needs." It is imperative to note that the sustainability of individual parts does not necessarily make a sustainable community. Sustainable development entails a combination of factors that is essential to the makeup of a community, including the economy, education, socio-cultural traditions and values, the natural environment, and physical infrastructure. On the other hand, community development without a sense of community is a hollow development. Community participation and involvement is, therefore, a necessary antecedent to obtaining sustainability.

Jiang (Chap. 14) identifies the characteristics of urban residents' leisure activities and leisure skills in Hangzhou, China. She touches upon Chinese cultural behaviors that shape their leisure activities and the forms of tourism development in a place. The author finds that there are opportunities for those who manage civic life, as well as for commercial tour operators, to include more assisted leisure programs for Chinese tourists. Greater diversity and choice in new leisure activities, particularly those likely to appeal to the better educated, are immediate areas in need of innovation to improve the overall life satisfaction in increasingly sophisticated cities such as Hangzhou. Samsudin and Othman (Chap. 15) discuss issues and challenges of Islamic tourism development in Malaysia. The Muslim market is a rapidly growing throughout the world, but it is not without complications. Complications arise in the relationships and disagreement between tourism practices and Islamic religious practices.

Aminuddin et al. (Chap. 16) investigate the influence of age and gender on the motivation and satisfaction of special event volunteers in Malaysia. By identifying and understanding the factors that influence volunteer motivation and satisfaction, event managers can better develop strategies for the retention of volunteers. This is valuable to understand as special events are increasingly becoming tourist attraction worldwide. Related to keeping volunteers happy, Muntasib et al. (Chap. 17) discuss community social capital at the Gunung Padang tourist site in West Java, Indonesia. They recognize the positive attitudes of the community in terms of being involved in various social networks and developing links with outside stakeholders. The community has also taken proactive action in relation to experience sharing, participation in social organizations, and protection of the environment. However, the study also identifies a negative side to the community's social capital in that some individuals retain jealousy of others with regard to job opportunities. This could pose a threat to the success of tourism development within the community.

Rahayuningsih et al. (Chap. 20) explore how tourists create community through their behavior in photo taking at the Nglanggeran ancient volcano in Indonesia. Photos are mainly shared through online social media and the mostly widely used means of photo taking is on mobile phones or smart phones.

Naidoo et al. (Chap. 18) focuses on community wellbeing situated in Mauritius. The study is set within the context of tourism development in Small Island Developing States (SIDS) and examines the positive and negative outcomes of enclave tourism to community wellbeing. A stakeholder approach with inputs from the host community, amongst others, is highly desirable to further understand the role that enclave tourism can play to harness community wellbeing. Tourism enclaves tend to be dominated by corporate interests, and Triansari and Yuniarta (Chap. 19) provide a study of how corporate social responsibility (CSR) activities are perceived by hotels in Bali. It is interesting to note that majority of the hotel managers interviewed have the misunderstanding that CSR is a government regulation. In addition, they feel that adopting a CSR policy has a positive effect on the brand image of the hotel.

Government responsibility, or the lack thereof, is examined in post-war Sri Lanka by Ratnayake and Hapugoda (Chap. 22). The post-catastrophic tourism context in Sri Lanka has been influenced by immediate commercial motives and no attention has been paid to the political significance of the traumatic historical memory of the communities affected by the long lasting conflict. Post-war tourism situations in other countries have generated *dark tourism* sites in impacted areas, where interests in seeing the 'monuments' of the fierce battle have supported a local tourism economy.

Gani et al. (Chap. 21) review public participation practices in planning for tourism in natural settings in Malaysia. The results indicated an overall lack of standard methods, the extent of information presented, and opportunity for the public to share views and opinions in decision making. These findings call for more thoughtful consideration by authorities to improve the current practice of public participation. Lagarense and Walansendow (Chap. 23) also emphasize the need for public participation in their examination ofthe potential of a cluster-based model for waterfront urban tourism for the city of Manado, on the island of Sulawesi, Indonesia. This study asserts the need for a critical review on what has been done in terms of protecting and improving the environment to ensure a sustainable development approach. It also highlights the significance of public participation to ensure that all members of the community gain equal access to the waterfront area. Busaini (Chap. 24) concludes this part of the book with an outline of a possible sharia tourism development strategy for West Nusa Tenggara Province in Indonesia. He further explores the potentials for sharia tourism development and what is required from the government to achieve a successful strategic development of sharia tourism products.

# Chapter 14
# Leisure Activities and Leisure Skills of Urban Residents in Hangzhou, China

**Yan Jiang**

**Abstract** This study identifies the characteristics of urban residents' leisure activities and leisure skills from a 952 person sample in Hangzhou, China. The research documents, the frequency, and the categories of leisure activities in which the urban residents participate. The study also assesses their skills at those leisure activities. Four key findings emerged. First, residents generally had a low frequency of leisure activities, and these were mainly casual activities. Second, the respondents' demographic characteristics influenced their participation. Third, leisure skills were generally low. Fourth, demographic characteristics were related to their leisure skills. These results affirmed the following points. First, the forms of leisure activities undertaken by residents were mainly static, and this may be a reflection of Chinese traditional culture. Second, residents felt they had low leisure skills and that many of the leisure activities they undertook were monotonous or boring. Third, people are facing more life pressures, especially in urban cities, but the influx of other people from different cities does add to the variety of the leisure culture. There are opportunities for those who manage civic life as well as tour operators to include more assisted leisure programs for Chinese tourists. Diversity and choices of new leisure activities, particularly some likely to appeal to those with more education are immediate areas for innovation and development as a means to improve overall life satisfaction in increasingly sophisticated cities such as Hangzhou.

**Keywords** Leisure activities · Leisure skills · Characteristics · Urban residents

Y. Jiang (✉)
Department of Tourism, Zhejiang International Studies University, Hangzhou, Zhejiang Province, China
e-mail: yanjiang1978@foxmail.com

 
A. Saufi et al. (eds.), *Balancing Development and Sustainability in Tourism Destinations*, DOI 10.1007/978-981-10-1718-6_14

## Introduction

The newly developing leisure industry in China has shown strong potential in economic terms. More leisure products are being provided, and government departments at many levels are developing leisure policies. The landmark document is The National Tourism and Leisure Outline (2013–2020) promulgated by the State Council General Office in February 2013.[1] This document stresses that is important to improve not only the amount of leisure time but also the leisure quality of the residents. Improvements to the current situation can be maximized if there is a solid awareness of how the residents currently spend their free time. The present study speaks to this aim by surveying residents' leisure activities and the skills they believed they had achieved through leisure activities. This information could assist governments and domestic tour operators to understand Chinese leisure demands, and provide activities that may be of interest.

There has been a large amount of research on leisure activities. Leisure has been defined from the perspective of activities and time (Parker 1976). Kelly (1978) defined leisure as a freely chosen activity, while Dumazedier (1967) described leisure activities as all activities excluding obligatory work, family, and social duties. Some Chinese authors such as Xu (1980) emphasized the free characteristics of leisure activities while Liu (2003) emphasized the lack of pressure in leisure time which can bring such benefits as relaxation, pleasure, satisfaction, and pleasant memories.

According to the previous research, the core feature in defining leisure activities lies in the optional nature of the time spent. In a major addition to the definitional work Stebbins (1992, p. 3, 1997) proposed the concept of serious leisure and casual leisure. Serious leisure such as reading was defined as voluntary and highly involving requiring and developing special skills for the participant (Stebbins 1992, p. 3). Casual leisure, such as rest and some forms of television watching, was defined as the spontaneous use of time, with a low skill base, mildly pleasant, and enacted without any special training (Stebbins 1997).

Research on the definition of leisure activities has provided a basis for classifying activities. It is hard to provide a complete list of leisure activities. As Kraus (1990) noted, the boundary between leisure and work can be so subtle that it is very difficult to list all the leisure activities. DeGraaf et al. (1999, p. 121) also thought it was hard to classify leisure activities, since the activities might be infinite, and the classification approach might be arbitrary. There are studies both on the methods to classify leisure activities and the actual classification of these pastimes. Three ways have been used to classify leisure activities: the form, the meaning, and the level of leisure activities. For the form of leisure activities, the classification has used subjective appraisals, as well as statistical tools, such as factor analysis and multidimension scaling work. There have been many studies on the subjective analysis

[1]The central government official website, 18 Feb. http://www.gov.cn/gzdt/2013-02/18/content_2334106.htm.

of leisure activity classifications (Hills and Argyle 1998; Chen and Lee 2011), and many studies employing factor analysis (Ragheb 1980; Brown et al. 1991; Tinsley and Eldredge 1995). Factor analysis is more objective, but is hard to analyze the characteristics of leisure items. In this study, a subjective analysis approach was applied to classify leisure activities because it was perceived to be more subtle for the Chinese array of pursuits.

There has been only limited research on leisure skills, and the topic is usually discussed under the framework of serious leisure put forward by Stebbins (1992, p. 3). He thought leisure that involved more skills would bring more satisfaction. The concept of leisure skills is a promising one but requires more analysis and work. Potentially, the relationship between leisure skills and leisure activities is very close, and interrelated. This paper analyzes the leisure characteristics of Chinese people in Hangzhou using both the recording of leisure activities and the perceived skills associated with these pursuits.

## Method

### *Sample Characteristics*

The respondents were chosen in order to gain a wide range of demographic attributes. All the interviewees were citizens of Hangzhou city. Peasants, migrant workers, and students in the universities were excluded. These three groups of people were excluded because:

(i) There are generally large differences between peasants and urban residents in China, especially differences in income. Urban residents have superior living conditions facilitated by their higher incomes, better jobs, and better medical insurance. The study examines and is limited to the urban residents.
(ii) University students who were not local residents were also excluded, because of cultural differences. There are great differences between the regions in China, especially between the north and south. People have different culinary tastes, different interpersonal communication modes, and arguably different values toward leisure.
(iii) Migrant workers were excluded, because nearly all are peasants from developing areas, and so they have different characteristics from the local citizens.

Questionnaires were distributed as widely as possible to obtain results which fully reflected the characteristics of the city residents, but there are also differences between districts. The six main districts of Hangzhou city, Uptown district, Downtown district, Gongshu district, Jianggan district, Westlake district, and Riverside district were covered. However, Yuhang district and Xiaoshan district were excluded, because they are the peripheral districts where many peasants rather than city residents live.

The questionnaires were delivered in places frequented by many citizens, such as shopping malls, public squares, and public leisure areas in communities. One thousand and twenty five (1025) respondents were interviewed, of which 952 provided properly completed questionnaires (validity rate is 92.88 %). Table 14.1

**Table 14.1** The demographic characteristics of the respondents

| | Composition | Number of respondents | Valid percentage (%) | Accumulative percentage (%) |
|---|---|---|---|---|
| Age | 15–24 | 206 | 21.6 | 21.6 |
| | 25–34 | 405 | 42.5 | 64.2 |
| | 35–44 | 179 | 18.8 | 83.0 |
| | 45–60 | 127 | 13.3 | 96.3 |
| | Over 61 | 35 | 3.7 | 100.0 |
| Education | Junior high school or below | 59 | 6.2 | 6.2 |
| | Senior high school or technical secondary school | 150 | 15.8 | 22.0 |
| | Undergraduate college or junior college | 631 | 66.3 | 88.2 |
| | Postgraduate | 112 | 11.7 | 100.0 |
| Marital status | Single | 392 | 41.2 | 41.2 |
| | Married | 535 | 56.2 | 97.4 |
| | Other (eg. divorced/separated/widowed) | 25 | 2.6 | 100.0 |
| Vocation | Managers in companies or leaders in public institutions | 126 | 13.2 | 13.2 |
| | Ordinary staff in public institutions | 139 | 14.6 | 27.8 |
| | Government workers | 32 | 3.4 | 31.2 |
| | Business service personnel or salesman | 102 | 10.7 | 41.9 |
| | Ordinary staff in companies | 205 | 21.5 | 63.4 |
| | Self-employed people | 76 | 8.0 | 71.4 |
| | Workers | 28 | 2.9 | 74.4 |
| | Students | 69 | 7.2 | 81.6 |
| | Teachers | 63 | 6.6 | 88.2 |
| | The retired | 51 | 5.4 | 93.6 |
| | Laid-off people including housewives | 40 | 4.2 | 97.8 |
| | Community workers | 21 | 2.2 | 100.0 |
| Monthly income (USD) | 250 or below | 89 | 9.3 | 9.3 |
| | 251–500 | 246 | 25.8 | 35.2 |
| | 501–833 | 326 | 34.2 | 69.4 |
| | 834–1666 | 214 | 22.5 | 91.9 |
| | 1667 or above | 77 | 8.1 | 100.0 |

*Source* Author's classifications of demographic characteristics of the sample

describes the demographic characteristics of the respondents. Income units have been converted from RMB into US dollar (1 USD$ = 6 RMB).

## Questionnaire

The questionnaire sought to address specific questions:

1. How frequently and in what kind of leisure activities do residents participate? The researcher applied a subjective analysis approach to classify leisure activities. It was hard to avoid the problem of activity overlap, so the paper classified leisure activities from the aspect of leisure motivation to improve the reliability of the questionnaire. Much research has shown that leisure motivation affected leisure participation positively (Ragheb 1980; Ragheb and Tate 1993; Wu 2008). The paper applied the hierarchy of needs theory, founded by Abraham Maslow, which included physical need, safety need, social need, respect need, and self-realization need, to the classification of leisure activities. Thus leisure activities were classified into five categories, leisure activities for the purpose of relaxation (e.g., napping, watching TV, reading newspaper, etc.), enjoyment (e.g., Shopping, theater cinema, etc.), social contacts (e.g. Chatting and dining out at a restaurant with friends, playing cards, etc.), sports (physical activities, including Taichi, yoga, etc.), and self-improvement (e.g., hobbies, religious activities, etc.). In order to assist interviewees understand the leisure activities, many examples were listed. Many of these activities, such as surfing, walking, climbing, shopping, reading, etc., were identified from the survey results undertaken by the magazine *Insight China,* in 2011. The interviewees were asked to describe how frequently they pursued their leisure activities on a rating scale from 1 to 5 which equates to "never"(1), "seldom"(2), "sometimes"(3), "frequently"(4) and "nearly every day"(5).
2. How are residents' leisure skills?
   Leisure skills were assessed by interviewees' judging themselves across three elements, which included the status of leisure time, leisure activities, and leisure skills. The interviewees were asked to describe their status during free time by three ratings, "feeling empty, unsatisfied, bored," "being able to find some interesting things," and "feeling fulfilled." Opinions about their status when participating in leisure activities used the ratings "dull and limited," "diversified but limited," and "undertake all kinds of leisure activities" were assessed by three ratings. Assessing the capacity for leisure participation was measured by "very few skills commanded," "some skills commanded" and "many skills commanded."

## Results

### *Residents Had Generally a Low Frequency of Leisure Activities, Which Were Mainly Casual Leisure Activities*

Leisure activities for the purpose of relaxation, enjoyment, and social contacts belong to casual leisure, while leisure activities for the purpose of sports and self-improvement belong to serious leisure according to the definitions by Stebbins. The results of the survey indicated that most respondents generally had a low or moderate participation rates in leisure activities, and those activities they undertook were mainly static activities which belong to casual leisure. Very few people undertook serious leisure activities.

The frequency that people undertook activities for the purpose of relaxation had a mean value of 3.65, with over half of the respondents pursuing some relaxation activity daily, while 11.9 % of the respondents did not take any leisure time for rest and relaxation. The frequency of the other non-relaxation activities undertaken was quite low, with the mean rating values between 2.53 and 2.81. The frequency of enjoyment and social contacts was also low, and only half participated in these activities while the percentages of the respondents who rarely or never participated were 35.4 and 35.6 %, respectively. The frequency of participation in sports was very low, with more than half (54.2 %) rarely or never undertaking sporting activities. Self-improvement was also very low, and more than half (54.3 %) rarely or never pursued any activity relating to self-improvement. Table 14.2 describes the frequency of different kinds of leisure activity participation for the urban residents.

According to the mean rating value and standard deviation of the frequency which leisure activities are undertaken, the leisure activities can be categorized into three groups. Group 1, which has a mean rating value more than 3.5 and a standard deviation less than 1, can be called the common leisure activity, which means the urban residents preferred these common pursuits. Group 2, with a mean rating value less than 3.5 and a standard deviation more than 1, can be called the auxiliary leisure activity, which means that few residents undertook the activity and the

**Table 14.2** Percentage (%) of the population and the frequency of participation in different kinds of leisure activities

| | Relaxation | Enjoyment | Social contacts | Sports | Self-improvement |
|---|---|---|---|---|---|
| Never | 0.6 | 3.7 | 2.9 | 43.3 | 44.7 |
| Seldom | 10.7 | 31.7 | 32.7 | 27.6 | 32.0 |
| Sometimes | 33.8 | 46.8 | 47.2 | 13.6 | 10.5 |
| Frequently | 33.1 | 16.3 | 15.1 | 4.6 | 3.2 |
| Nearly every day | 21.7 | 1.5 | 2.1 | 10.9 | 9.6 |

*Source* Author's field observations

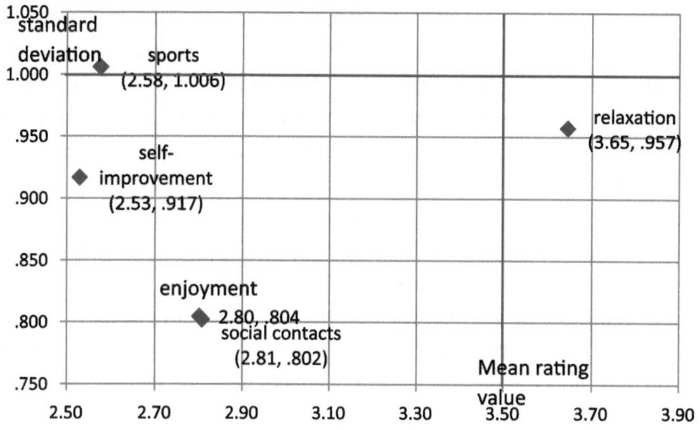


**Fig. 14.1** Mean rating value and standard deviation of leisure activities

frequency at which they undertook the activity was widely spread. Sporting leisure activities (Mean = 2.58, Standard Deviation = 1.006) belongs to Group 2. Group 3, with a mean rating value of less than 3.5 and standard deviation less than 1, can be called minor leisure activities, which means the residents had low preference for this kind of use of their time and few undertook this form of leisure activity. The other three leisure activities identified in the survey of social contacts (2.81, 0.802), enjoyment (2.80, 0.804), and self-improvement (2.53, 0.917) belong to Group 3.

Figure 14.1 shows the groupings of the leisure activities and the mean rating of frequency of undertaking the activities.

## *Different Demographic Characteristics Influenced the Participation of Leisure Activities*

There was a difference in the preference of leisure activities of enjoyment, social contacts, and sports between genders. The independent-sample test indicated there was no apparent difference between genders in the leisure activities of relaxation and self-improvement, but females preferred enjoyment ($\text{Mean}_{\text{female}} 2.89 > \text{Mean}_{\text{male}} 2.70$),[2] while males preferred social contacts ($M_{\text{male}} 2.86 > M_{\text{female}} 2.76$) and sports ($M_{\text{male}} 2.75 > M_{\text{female}} 2.42$).

The age of the respondents affected their participation in the leisure activities of relaxation, enjoyment, social contacts, and sports, but had no apparent influence in undertaking self-improvement activities. Generally, the age group between 25 and 44 participated in all leisure activities less frequently than any other age groups. This might be because this age group was the busiest group among respondents,

[2]M represents Mean rating value in the following paragraphs.

most probably preoccupied with developing careers and supporting families. According to the Scheffe*post hoc* comparison and mean analysis, the people over the age of 61 undertook relaxation more frequently (M = 4.23) than other age groups ($M_{max}$ = 3.86), especially compared to the age group under 44 ($M_{max}$ = 3.69). The age group between 35 and 44 participated in the leisure activity of enjoyment apparently less frequently than other age groups, and especially the younger people between 15 and 24. The age group between 25 and 44 also participated in the leisure activity of social contacts apparently less frequently than other age groups, and especially those between 45 and 60. The age group between 25 and 44 participated in the leisure activity of sports apparently less frequently than other age groups, and especially those above 61. Generally in China those aged over 61 are retired, and have enough time to participate in sports such as Taichi and other physical activities for health and wellbeing, while many younger people especially those between 25 and 44 have to look after their children and parents, and have great pressure from work, which lowers the frequency of their leisure activities.

Education levels influenced peoples' leisure activities. People with an education level of junior high school or below (M = 4.02) had higher frequency of leisure activity of relaxation than those with an education level of undergraduate college or junior college ($M_{max}$ = 3.62), while those with postgraduate qualifications (M = 2.79) had higher frequency of involvement in self-improvement than those with an education level below senior high school or technical secondary school ($M_{max}$ = 2.43). Education levels had significant correlation with age according to Pearson correlation test. Most people of the age below 44 had an education level of undergraduate college or junior college, while only about one-third of the age over 44 had the same education level. Typically older respondents had less education.

Vocations influenced leisure activities of relaxation and sports. The retired residents participated in relaxation more frequently (M = 4.31) than others, such as managers of companies or leaders in public institutions (M = 3.44), government workers (M = 3.34), business service personnel or salesman (M = 3.57), staff in companies (M = 3.58), and self-employed people (M = 3.47). The retired residents participated in the leisure activity of sports more frequently (M = 3.24) than business service personnel or salesman (M = 2.31) and ordinary staff in companies (M = 2.53). These data might be explained by retired residents having more free time.

There were no apparent differences between income levels. It might be because many public leisure facilities are free in Hangzhou, so income levels affect how they spend money, but do not affect whether they pursue activities. For example, nearly all the people can afford the pursuit of dining out; the difference is what kind of restaurants they go.

Marriage status influenced leisure activities. The single respondents participated more in the leisure activity of enjoyment than married respondents and the mean rating values were 2.91 and 2.73 respectively.

## *Leisure Skills Were Generally Low*

About 60 % of the respondents thought their leisure activities were dull and limited, and the respondents assessed their own leisure skills as low, and about one-third considered they had a few skills when participating in leisure activities. It is possible that these assessments underestimate their skill levels as it may be seen as immodest to assess oneself as highly competent in Chinese culture. The respondents had a higher evaluation of how they used their free time, and about two-thirds of the respondents said they could find interesting things to do.

Overall leisure skills were assessed by respondents themselves to be quite low, since the average mean rating value of leisure skill assessments was 1.751, and the standard deviation was 0.460. The result could also be related to the low frequency of participation in sports and self-improvement.

## *Different Demographic Characteristics Influenced Leisure Skills*

There were some differences in leisure skills between genders. The result of independent-sample test indicated that males had higher self-assessment of their status in leisure activities and skills they commanded than females, but both rankings were not high and the mean rating values were lower than 2. There were no apparent differences between genders in the self-assessment of status during their free time.

Different age groups had different levels of assessment of their leisure status during free time. According to the ANOVA correlation test and Scheffe*post hoc* comparison and mean analysis, the age group between 25 and 44 (M = 1.87) had apparently lower self-assessment than other age groups, and especially those between 35 and 60 ($M_{min}$ = 2.10), while the age group between 25 and 34 (M = 2.00) had apparently lower self-assessment than others, and especially those between 45 and 60 (M = 2.20).

Groups with different education levels had apparently different self-assessments of leisure status during free time and leisure skills they commanded. The respondents of the postgraduate education level (M = 2.21) were able to find more interesting things during free time than those with an education level of junior high school or below (M = 1.92) and those with an education level of undergraduate college or junior college (M = 1.99). Generally for the self-assessment of leisure skills they commanded, there was a direct and positive correlation between education level and self-assessment. Those with the education level of junior high school or below (M = 1.47) considered they commanded the lower leisure skills than those with an education level above undergraduate college or junior college (M = 1.78), and those with an education level of below senior high school or technical secondary school (M = 1.67) considered they commanded the lower skills than those of the postgraduate education level (M = 1.92). The mean rating value of 2 means the respondents command some leisure skills, while the Mean rating value 1.5 means the respondents only command a few leisure skills.

Marriage status affected the self-assessment of leisure skills. Single respondents had apparently the lower self-assessment of the status during free time than the married respondents, and the mean rating values were 1.89 and 2.12, respectively. There were no apparent differences of the self-assessment of leisure skills commanded between vocations.

There were differences in the self-assessment of leisure skills respondents felt they possessed and the status during free time between groups with different incomes. Those who had the monthly income between US $834 and 1666 believed they had more fulfilling status during free time (M = 2.12) than those who earned less than US $250 (M = 1.88), and those who earned more than US $1667 (M = 2.00) had higher self-assessment of leisure skills commanded than those whose monthly income was between US $501 and 833 ($M_{max}$ = 1.75). It means those who earned more money had more skills required during leisure activities and more abilities to enjoy their leisure time.

## Conclusions and Suggestions

Several key features about leisure participation and skills for the residents of Hangzhou, a leading leisure and tourism center in Eastern China, were identified in this study. First, the forms of leisure activities undertaken by the residents were mainly static, and this may be a reflection of Chinese traditional culture. Second, the residents considered they had low leisure skills and considered that many leisure activities they undertook were monotonous or boring, which may reflect that the new leisure culture considered by the western world is developing.

There were many leisure activities undertaken in ancient China, which were mainly religious activities or in teahouses where people could listen to traditional operas or meet socially with different people. Currently, however, there are few remaining temples and the traditional Chinese opera is not as popular with the younger generation. The traditional leisure activities have not transferred easily between generations and the modern leisure activities, including sports, have not yet been well developed or accepted by all Chinese people. In Hangzhou, one of the most popular leisure activities is the dining culture symbolized by teahouses. Tea houses remain important for enjoyment and social contacts. Overall though, it can be suggested that the traditional leisure culture is being abandoned while the modern leisure culture has not yet developed.

Third, people are facing more living pressure, and people from other cities bring different leisure cultures. In this context of taking a broader view of the results it can be suggested that there are great differences between Chinese and western tourists in terms of leisure activities and skills. Some Chinese tourists are looking forward to adopting aspects of a western leisure life.

Opportunities for tour operators as well as government and civic bodies may be to provide more assisted leisure programs for Chinese tourists, so that they can attend and experience them without worrying about having the skills. Some of these leisure and tourism linked training opportunities may even facilitate the participation by Chinese citizens when they travel abroad. For example, dive and Great Barrier Reef day trip companies could provide assisted snorkeling and diving programs, but also investigate the tourists' swimming skills in advance so that alternatives could be provided if swimming is not a skill. Many Chinese people cannot swim, especially those from the inland. If more knowledge is known about Chinese tourist leisure skills in advance, international tourist organizations and companies may be able to provide programs for Chinese locals that build their capabilities and provide necessary basic training for a more memorable experience. If the international tourist activities and style of leisure activities are not suitable for Chinese tourists, they will not participate, thus both parties lose opportunities to interact and benefit from the leisure options.

In developing the theme that international tourism can benefit from studies of within Chinese leisure it can be noted that the Chinese in Hangzhou had a preference for relatively undemanding activities. As many Chinese have a preference for static leisure activities it may be advantageous to provide some leisure facilities that Chinese tourists like and are familiar with in international resort settings. Skills can be improved with training, but it is hard to change people's preferences, and one way is to meet their interests. For example, despite enjoying a day touring many Chinese tourists may enjoy a game such as mahjong in the evening and these facilities could be provided to enhance the tourist experience. Some Chinese tourists might be happy to rent or borrow such facilities to pass time during their trips. Mahjong parlors could be provided for tourists if there is enough demand.

Particular opportunities exist for those Chinese who are better educated. They had the highest scores for the activity of self-improvement. Again both within China and abroad there are ways to enhance the leisure time of this group. Young and well-educated groups are more inclined to travel overseas, and these people might be the ones that want to learn new skills. We cannot expect all the people to try new-skill programs, but the younger and better-educated people might be more willing to try new things, and participate in training courses. Sometimes preferences and skills are related, such as swimming. It is hard to persuade a person who cannot swim to learn, but some may try if given the right circumstances. As China modernizes and develops a strong interest in the quality of life for all its citizens, the role of leisure grows as a key activity to manage and develop. Research in one city is a beginning to understand who is doing what in their free time in modern day China and points to some options for building new skills and behaviors to create a better life for the community at home and abroad.

## References

Brown BA, Frankel BG, Fennell M (1991) Happiness through leisure: the impact of type of leisure activity, age, gender and leisure satisfaction on psychological well-being. J Appl Recreat Res 4:368–392

Chen SH, Lee CK (2011) Relationships between leisure participation, leisure satisfaction and well-being for the member of senior college in Tainan. NPUE J Sports Sci 7:173–186

DeGraaf D, Jordan DJ, DeGraaf KH (1999) Programming for parks, recreation, and leisure services: a servant leadership approach. Venture Publishing Inc, State College, PA

Dumazedier J (1974) Sociology of leisure. Elsevier, Amsterdam

Hills P, Argyle M (1998) Positive moods derived from leisure and their relationship to happiness and personality. Personal Individ Differ 3:523–535

Kelly JR (1978) A revised paradigm of leisure choices. Leisure Sci 4:345–363

Kraus RG (1990) Recreation and leisure in modern society, 4th ed. HarperCollins, New York

Liu YL (2003) A study of leisure activities participating, work pressure, and work fulfillment of rank and file firemen. National Yunlin University of Science & Technology, Yunlin (unpublished master's thesis)

Parker S (1976) The sociology of leisure. George Allen and Unwin, London

Ragheb MG (1980) Interrelationships among leisure participation, leisure satisfaction and leisure attitudes. J Leisure Res 2:138–149

Ragheb MG, Tate RL (1993) A behavioural model of leisure participation, based on leisure attitude, motivation and satisfaction. Leisure Stud 1:61–70

Stebbins RA (1992) Amateurs, professionals, and serious leisure. McGill-Queen's University Press, Montreal and Kingston

Stebbins R (1997) Casual leisure: a conceptual statement. Leisure Stud 1:17–26

Tinsley HEA, Eldredge BD (1995) Psychological benefits of leisure participation: a taxonomy of leisure activities based on their need-gratifying properties. J Couns Psychol 2:123–132

Wu MC (2008) A study on causal relationship model of leisure attitude, leisure motivation, leisure satisfaction on leisure behavior of college students. Educ Res J 2:83–100

Xu YX (1980) The meaning, content and methods of leisure. Phys Educ J 2:27–40

# Chapter 15
# Attitudes, Lifestyle and Social Benefits of Local Communities Towards Islamic Tourism Development in Malaysia

**Sabrina Samsudin and Nor'Ain Othman**

**Abstract** Muslim market is one of the latest growing market segments in the world. Muslim countries, such as Malaysia, United Arab Emirates, Turkey and Indonesia have developed Islamic tourism in marketing the country as a holiday destination for Muslim travellers. Non-Muslim countries, such as Japan, Korea, Thailand and Europe also show great interest to attract this market. There are various issues and challenges faced by these countries. Malaysia, being a multiracial and multi-religious country with Islamic modernization and revivalism has provided an interesting research context for Islamic tourism development. The issues and challenges such as the sensitivities and sensibilities of the other religion towards the implementation of Islamic tourism in the country may exist. These issues are also supported based on the previous studies where problems such as complications of the relationship and disagreement between religious practices occurred when tourism and Islamic religion came in contact. Thus, this study aimed to (i) identify the most influential socio-cultural attributes towards the development of Islamic tourism in Malaysia, (ii) to investigate the relationship between socio-cultural attributes and the development of Islamic tourism in Malaysia, and (iii) to examine the government support as the moderating variable towards the development of Islamic tourism in Malaysia. The findings and results from this research may benefit the local communities as well as the government particularly the Ministry of Tourism and Culture and also Islamic Tourism Centre (ITC). The outcome of this study also could be used as an indicator in setting up the particular

S. Samsudin (✉) · N. Othman
Faculty of Hotel and Tourism Management, Universiti Teknologi MARA, Puncak Alam, Malaysia
e-mail: ninasabrina20@yahoo.com

N. Othman
e-mail: norain568@salam.uitm.edu.my



A. Saufi et al. (eds.), *Balancing Development and Sustainability in Tourism Destinations*, DOI 10.1007/978-981-10-1718-6_15

policy standard for Islamic tourism development. As for the local communities which consist of multiracial community, they surely will have a better understanding regarding the proper Islamic tourism perspective.

**Keywords** Islamic tourism · Local communities · Attitudes · Lifestyle · Social benefits

## Introduction

Tourism industry nowadays is constantly on the watch for ways to expand by incorporating new market sections. For instance, Islamic tourism is the latest market segment that is developing rapidly in the Muslim countries (Battour et al. 2014). The development of Islamic tourism has risen various business components within the tourism industry including tour operators, hotels, airlines, and restaurants align in restructuring their products and services with halal or sharia compliance in order to meet the demand of the growing market segment (Rahman 2014).

In regard to the relationship of local communities with the development of tourism, Farahani and Henderson (2010) mentioned that community reactions towards tourism may be different within countries. In some Islamic nations, tourism industry is considered to have the potential to violate Islamic cultural values and traditions, so it is not taken as a major development priority for them. Moreover, tourism may gradually alter the society's values, beliefs and cultural practices. By observing tourists' behaviour, local residents may change their lifestyle by modifying the way they dressed, eating habit, entertainment, recreational activities, etc. While these influences may be interpreted positively as an increment in the standard of living, it may as well be viewed negatively as an indication of socialization (Brunt and Courtney 1999).

Adding up to its socio-cultural impacts, Ap (1992) mentioned that tourism is perceived to change the value systems, individual behaviour, family relations, lifestyle and community organizations. These forms of social impacts may be positive or negative, which could act upon the changes in the social structure of the community. Consequently, this study will examine the situation in Malaysia, a country in which Muslims are dominant, and also assess the socio-cultural impacts of the Islamic tourism development as well as the effects on the local residents in Malaysia.

## Problem Statement

Farahdel (2011) has mentioned in her study that people with different religions have different cultures, social norms, beliefs and behaviours since the religion shapes them in high extend. In another study, Din (1989) also emphasized that the multi-religious population in Malaysia plus the tension between Islamic modernization and revivalism have provided an interesting research context for tourism development. For that reason, in line with the development of Islamic tourism, there would be challenges to enforce Islamic tourism in a multi-racial country. Issues such as the sensitivities and sensibilities of different practitioners towards the implementation of Islamic tourism in the country will exist. These issues are supported by Henderson (2003) in which problems such as complications of the relationship and the disagreement between religious practices occur when tourism and Islamic religion come in contact. Islam can rule the behaviours of its believers' in a significant way; however, it could result in some conducts and behaviours of non-Muslim residents that can be seen as particularly inconvenient and inappropriate in a Muslim context (Henderson 2003). As a result, in some destinations, the coexistence between Muslim and non-Muslim tourism practices can be problematic. Islamic tourism can be targeted on both Muslim and non-Muslim market and could be viewed as seeking Islamic knowledge. Therefore, this study aims to investigate whether there is a demand for Islamic tourism and what are the local communities' opinion and expectation of Islamic tourism.

## Literature Review

In the literature review, each explanation would relate the variables of the study which are attitudes, lifestyles, and social benefits of local communities to Islamic tourism development in Malaysia. Each review is from a range of the literature on the concepts and theories which are relevant to the topic. These ranges of sources which are collective might be perceived to constitute a suitable body of the literature.

### *Overview of Islamic Tourism Development in Malaysia*

Islamic tourism is the latest tourism destination in the world nowadays (Bhuiyan et al. 2011). A study by Kamarudin and Ismail (2012) noted that the emergence of Islamic tourism has begun since 2000. In addition, the 11 September 2001 tragedy had consequently caused restrictions for Muslim tourists to enter the western countries. Thus, Muslims particularly from the Middle East has changed their directions of travelling from Western countries to Muslim countries mainly in Asia. By seeing this as an opportunity, Malaysia is also concerned with the development of Islamic tourism. Tourism Malaysia has taken an excellent initiative by establishing an Islamic Tourism Centre (ITC) on February 2009. Since then, ITC has taken the responsibility to develop strategies in order to develop Islamic tourism product in Malaysia successfully.

## *Islamic Tourism*

Bhuiyan et al. (2011) have emphasized that Islamic tourism is considered as a significant fragment of religious tourism because it is mainly designed for Muslims who abide by sharia rules (Islamic law). As such, many Muslims scholars have introduced the concept of Islamic tourism which has its own characteristics which is flexible, rationale, simple and balanced (Laderlah et al. 2011). In addition, there are several opinions given on what is Islamic tourism is all about. Nevertheless, the specific definition Islamic tourism concept is still unclear and the implication of the Islamic tourism terms itself is not well-defined. Since there is no precise definition on Islamic Tourism, researchers have agreed the most with the explanation was given by Farahani and Henderson (2010). This is because based on their study, the perspectives of Islamic Tourism are focusing more on the residents and tourism development. They stated that Islamic Tourism is a form of tourism targeted to Muslim community who wishes to indulge the services, facilities, and activities compliant with Islamic principles. For this reason, their explanation on Islamic tourism is more suitable since researchers are also focussing on the same context.

## *Attitudes*

According to Wang et al. (2006), attitudes are built upon the perceptions and beliefs of reality, but are closely related to deeply held values and to personality. In another study by Lankford et al. (1994), they stated that attitudes are not fixed as they can change according to social influence or other factors. In terms of tourism, the support for the basic theory of attitude can be explained as the residents' attitudes that do only reflect the residents' view about tourism and its impacts, but also show the result of interaction between residents' perceptions and the factors that affect their attitudes. The influential factors of residents' attitudes are varied by which almost all the impact factors are the variables that could change at different stages of tourism development and with different experiences. Hence, understanding the residents' current perspectives on local tourism development is necessary since residents' perceptions of tourism can predict how locals will usually behave towards visitors, their values, and their requirements (Lin 2014). Most of the previous studies have stated that local residents' support for tourism development is critical because successful operation and sustainability depend heavily on their goodwill (Jurowski et al. 1997; Gursoy et al. 2002). Meanwhile, Andriotis (2005) has mentioned that the support for tourism can be measured by the perceptions of the local population which can dictate the extent of the host community's acceptability of tourism.

## *Lifestyle*

Based on the study by Farahani and Musa (2012), local community lifestyle is categorized under the socio-cultural impacts attribute. The impacts consist of the changes in social structures and relations which are induced by the development of tourism in a destination. Apart from the changes in the host culture (local residents), the impacts of tourism also show stern influences on the social structure of the local society (Chilembwe 2014). This is because tourism activities bring together visitors and residents from different societies and cultures. Therefore, even though they are not directly involved in tourism, residents who are living in popular tourist destinations are probably affected by the existence of visitors and tourism industry (Fagence 2003). According to a research by Zaei and Zaei (2013), tourism might have different effects on the social and cultural aspects of life in a specific region which depends on the cultural and religious strength of that region. The level of socio-cultural impacts or influences experienced by residents may depend on a number of factors, including the number and the type of tourists, the nature of tourism development in the area, and the rapidity of development (Brunt and Courtney 1999).

## *Social Benefits*

Tourism development can contribute both positive and negative benefits to the local residents (Farahani and Musa 2012). The examples for socio-cultural benefits are cultural development and exchange, improve the image of local residents, social and amenity improvements, education and also conservation. In addition, a study has identified that among the positive impacts of the socio-cultural are the enhancements of various local services, infrastructure development, and the participation of local in community-based tourism. From this, employment opportunities are created and traditional culture is preserved and revitalized. Moreover, the rapid growth of tourism might improve the legislative and local services, produce greater investment in education, and upgrade the healthcare system, social services, leisure facilities, as well as the variety of social entertainment and recreational activities (Pizam and Milman 1988; Brunt and Courtney 1999). While on the other hand, the negative benefits of socio-cultural include host cultural destruction, social instability, crimes and changes in traditional values (Pizam and Milman 1988). Some residents perceive tourism as having negative social and cultural impacts and some tend to see tourism as having positive economic, social and cultural impacts. This notion is then reinforced by Harisson (2001) on the role of personal benefit obtained by the local community from tourism development which may commute adverse and positive perception on tourism development.

### *Government Support*

In the tourism context, various forms of government support programme can be defined as financial and training, government funding policy, provision of basic infrastructure and protection against competition from big business (Yusuf 1995), education and training programmes (Robertson et al. 2003), and the provision of soft loans (Mahajar and Mohd Yunus 2006). The tourism industry is inefficient without its proponents of government support (Hughes 1984). According to Perce (1989), generally, local government is the most important authority in establishing tourism development policies. Therefore, the principle of tourism planning and development that should involve broader community is important (Ap 1990; Ap and Crompton 1989; Brayley et al. 1990). Moreover, the studies on local populations' perception of tourism impacts are useful in setting up the programmes to minimize the friction between tourists and residents, helping government to understand the social impacts (Sethna 1980), and in formulating plans to gain resident support for tourism ventures (Belisle and Hoy 1980). Currently, the number of studies on local community attitudes towards future tourism development is increasing due to the nexus between community support and government future development (Hanafiah et al. 2013). It is undeniable that local community support for tourism is necessary in order to ensure commercial, socio-cultural, physiological, political and economic sustainability of the industry. Therefore, their role in influencing tourism development activities through working together with the government is vital (Jamaludin et al. 2009).

## Methodology

The respondents for this study were the multi-racial residents living in Kuala Terengganu. This location was selected for this study since the states in East Malaysia is still exposed to the tourist development from Islamic perspectives which gives priority to the creation and redevelopment of mosques and other Islamic infrastructures (Cleary 1997). In addition, it has mentioned in a study that Terengganu (East Coast Economic Region) has huge opportunities to develop Islamic tourism because this location has a lot of natural beauties, cultural and archaeological heritage, historic civilizations and religious places which might be a potential tourism destination (Bhuiyan et al. 2011). The questionnaire was divided into six major sections in which each section contains questions addressing the variables to suit the research objectives. The items used in the questionnaire were derived from various related literature (McCool and Martin 1994; Hanafiah et al. 2013; Eraqi 2007). They were measured using five-point Likert scale from 'strongly agree' to 'strongly disagree'. Lastly, since the respondents were among the locals and have different educational background, the questionnaires were designed in bilingual, English as well as Bahasa Malaysia since both languages are widely spoken in this country.

Since some of the items were developed and modified for the purpose of the study, a pre-test of the measurement instrument is necessary to validate the items in the scales. For the pilot test, 30 survey questionnaires were distributed to several locals. The response from the pre-test were analyzed to test the reliability of the measurement items and the final version of the questionnaire was developed. The exact data distribution was conducted during a period of 2 months in the month of January and February 2015 at several Islamic tourism attraction areas in Kuala Terengganu such as Islamic Heritage Park (Taman Tamadun Islam) and Crystal Mosque (Masjid Kristal) because most of the residents here are more prone to the development of Islamic tourism. Based on the statistics obtained from Official Portal Kuala Terengganu City Council, the number of local residents in Kuala Terengganu in 2014 was 111,822 people. Based on Krejcie and Morgan's (1970) table for determining the sample size, for a given population of more than 100,000, the suitable sample size of 384 would be needed to represent a cross-section of the population. Hence, in order to achieve the sample size, a total of 384 questionnaires were distributed using non-probability sampling. Overall, 384 questionnaires were distributed and only 300 were collected, resulting in a response rate of 78 % which were used for analysis.

## Findings

### *Sample of Respondents*

Based on the result, gender distribution of the respondents was slightly unequal where the number of the male respondents exceeded the female respondents with a percentage of 56.7 % ($n$ = 170) and 43.3 % ($n$ = 130), respectively. The result also illustrated that the age of locals in the 21–30 age group accounted the highest respondents (32.3 %), followed by those aged from 31–40 (27.0 %), 31–40 age group (20.0 %), and respondents aged above 50 (12.7 %). On the other hand, respondents in the age below 20 represented the smallest number (8.0 %). As for the distribution of religion, 50.0 % of the respondents are Muslims, followed by the second largest group that consists of Buddhist (20.3 %), while the Hindus respondents were 15.0 %. Lastly, Christians accounted for 14.7 % respondents. Hence, the distribution of religion was equal between Muslims and non-Muslims.

Regarding the educational level, majority of the respondents could be considered well educated. More than half of the respondents have a Bachelor Degree (53.0 %) and a Master Degree (16.7 %). 26.7 % of the respondents have a Diploma. In addition, only small number of respondents have secondary education (3.7 %) while none of them were from primary and doctorate level. Lastly, in terms of occupation, 38.3 % of the respondents are private employees, followed by the second largest group that consist of students (22.7 %). The respondents who own a business and also a government servant were both 17.0 % each. Apart from that, only 5.0 % of the respondents are retirees.

## *What Is the Most Influential Socio-cultural Attribute Towards the Development of Islamic Tourism in Malaysia?*

Based on the descriptive analysis, it indicates that most influential socio-cultural attributes are the attitudes of local communities' towards the development of Islamic tourism in Malaysia. This is because most of the respondents showed positive attitudes and 'strongly agree' scale in each statement in the attributes. This proves that they are accepting the development of Islamic tourism in their area. This is also supported by a previous study by Jurowski and Gursoy (2004) who mentioned that many communities are encouraging the development of tourism as a mean to improve the quality of life for residents. In addition, most studies thus far have not found residents to be greatly concerned about the negative aspects of tourism on a general level (Andereck and Vogt 2000).

## *Is There Any Relationship Between Socio-cultural Attributes and the Development of Islamic Tourism in Malaysia?*

The results were analyzed by using standard multiple regression analysis. Subsequently, the results explained that all three variables have a significant relationship with Islamic tourism development. Firstly, both attitude and lifestyle attributes have stated 0.000 values for a significant relationship. Next, social benefits stated 0.016 values for a significant relationship. Hence, it showed that all the three independent variables have a significant relationship with Islamic tourism development since all of the p-value were smaller than 0.05.

## *What Is the Government Support as Moderating Variable Towards the Development of Islamic Tourism in Malaysia?*

Since this research has a moderating variable, the hierarchical multiple regression technique is appropriate for analyzing the result. From the result of the model summary, it showed that the value of $R^2$ changed from 0.206 to 0.362 after the moderating variable of government support was added. Hence, it means that a total of 15.6 % of the government attributes were added to the actual socio-cultural attribute influencing the dependent variable which is the Islamic tourism development.

## Conclusion

In conclusion, the research found that most of the respondents are aware of the Islamic tourism development but they do not clearly understand its fundamental concepts. Interestingly, most of the respondents which include non-Muslims residents show their support towards the development of Islamic tourism. This means that Islamic tourism will definitely have a good potential and could be sustain in the future if it is carefully planned. Hopefully, the existence of this study will make several important contributions to the existing literature of Islamic tourism. Previously, many literature have discussed Islamic tourism in general; however studies that focus on local residents in a particular area and their acceptance towards it are limited. Therefore, this study definitely extends the existing body of the literature in tourism research and Islamic tourism research in particular.

In addition, local authority and government bodies could use this study to determine how to educate the local communities of getting know the fundamental concept of Islamic tourism. Although Malaysia is widely known as an Islamic country, not every resident clearly understand the proper concepts of Islamic tourism. As for the possible further research, researchers may focus on identifying proper marketing plan for Islamic tourism development in Malaysia. According to Bhuiyan et al. (2011), tourism industry in dominated countries adopts modern and suitable marketing strategy for developing their tourism industry. They also focus their tourism places to the tourists with market-oriented ways. However, due to improper marketing strategy, Islamic tourism is not given greater attention from the tourists. Therefore, further research on the marketing problem could be developed.

## References

Andereck K, Vogt C (2000) The relationship between residents' attitudes toward tourism and tourism development options. J Travel Res 39:27–36

Andriotis K (2005) Community groups' perceptions of and preferences for tourism development: evidence from crete. J Hospitality Tour Res 29(1):67–90

Ap J (1990) Residents' perception research on the social impacts of tourism. Ann Tour Res 17, 610–616

Ap J (1992) Residents perception on tourism impacts. Ann Tour Res 19(4):665–690

Ap J, Crompton J (1989) Residents' strategies for responding to tourism impacts. J Travel Res 31:47–50

Battour M, Battor M, Bhatti MA (2014) Islamic attributes of destination: construct development and measurement validation, and their impact on tourist satisfaction. Int J Tour Res 16(6):556–564

Belisle FJ, Hoy D (1980) The perceived impact of tourism by residents: a case study in Santa Marta, Columbia. Ann Tour Res 7:83–101

Bhuiyan MA, Siwar C, Mohamad Ismail S, Islam R (2011) Potentials of Islamic tourism: a case study of Malaysia on east coast economic region. Aust J Basic Appl Sci 5(6):1333–1340

Brayley R, Var T, Sheldon P (1990) Perceived influence of tourism on social issues. Ann Tour Res 17:285–290

Brunt P, Courtney P (1999) Host perceptions of sociocultural impacts. Ann Tour Res 26(3):493–515

Chilembwe JM (2014) Examination of socio-cultural impacts of tourism in Chembe village in Mangochi District Malawi. Int J Bus Quant Econ Appl Manage Res
Cleary M (1997) Colonial and post—colonial urbanism in north-west Borneo. In: Shaw B, Jones R (eds) Contested urban heritage: voices from periphery. Ashgate, Aldershot, pp 11–30
Din KH (1989) Islam and tourism: patterns, issues and options. Ann Tour Res 16(4):542–563
Eraqi M (2007) Local communities' attitudes towards impacts of tourism development in Egypt. Tour Anal 12:191–200
Fagence M (2003) Tourism and local society and culture. In: Sigh S, Timothy DJ, Dowling RK (eds) Tourism in destination communities. CABI Publishing, Cambridge, pp 55–78
Farahani HZ, Henderson JC (2010) Islamic tourism and managing tourism development in Islamic societies: the cases of Iran and Saudi Arabia. Int J Tour Res 12(1):79–89
Farahani HZ, Musa G (2012) The relationship between Islamic religiosity and residents' perceptions of socio-cultural impacts of tourism in Iran: case studies of Sare'in and Masooleh. Tour Manage 33(4):802–814
Farahdel F (2011) Islamic attributes and it's impact on muslim tourists' satisfaction: a study of Iran
Gursoy D, Jurowski C, Uysal M (2002) Resident attitudes: a structural modeling approach. Ann Tour Res 29:79–105
Hanafiah M, Jamaluddin M, Zulkifly M (2013) Local community attitude and support towards tourism development in Tioman Island, Malaysia. Procedia Soc Behav Sci 105:792–800
Harisson D (2001) Tourism in small islands and microstates. Tour Recreat Res 26(3):3–8
Henderson JC (2003) Managing tourism and Islam in Peninsular Malaysia. Tour Manage 24 (4):447–456
Hughes HL (1984) Government support for tourism in the UK: a different perspective. Tour Manage 5(1):13–19
Jamaludin M, Othman N, Awang AR (2009) Community based, homestay program. Procedia Asian J Environ Stud (ajE-Bs) 3(9)
Jurowski C, Gursoy D (2004) Distance effects on residents' attitudes towards tourism. Ann Tour Res 31(2):296–312
Jurowski C, Uysal M, Williams R (1997) A theoretical analysis of host community resident reactions to tourism. J Travel Res 36(2):3–11
Kamarudin LM, Ismail HN (2012) Muslim tourists' typology in Malaysia: perspectives and challenges. In: Tourism and hospitality international conference
Krejcie R, Morgan D (1970) Determining sample size for research activities. Educ Psychol Meas 607–610
Laderlah SA, Rahman SA, Awang K, Man YC (2011) A study on Islamic tourism: a Malaysian experience. In: 2nd international conference on humanities, historical and social sciences. IACSIT Press, Singapore
Lankford S, Chen J, Chen W (1994) Tourism's impacts in the Penghu national scenic area, Taiwan. Tour Manage 15(3):222–227
Lin Y (2014) Residents' attitudes toward tourism impacts in Zhouzhuang Canal Town. Mid-Sweden University, Sweden
Mahajar A, Mohd Yunus J (2006) The effectiveness of government export assistance programs on Malaysia small and medium enterprises (SMEs). Probl Perspect Manage 1:58–71
McCool S, Martin S (1994) Community attachment and attitudes towards tourism development. J Travel Res 32(3):29–34
Perce P (1989) Tourism development, 2nd edn. Longman Scientifics, London
Pizam A, Milman A (1988) Social impacts of tourism on central Florida. Ann Tour Res 15:191–204
Rahman MK (2014) Motivating factors of Islamic tourist's destination loyalty: an empirical investigation in Malaysia. J Tour Hospitality Manage 2(1):63–77
Robertson M, Collins A, Medeira N, Slater J (2003) Barriers to start-up and their effect on aspirant entrepreneurs. Educ Training 308:316

Sethna RJ (1980) Social impact of tourism in selected caribbean countries. In: Hawkins ELDE (ed) Tourism planning and development issues. George Washington University, Washington DC, pp 239–249

Wang Y, Pfister R, Morais D (2006) Residents' attitudes toward tourism development: a case study of Washington. In: Proceedings of the 2006 northeastern recreation research symposium. CABI Publishing, Wallingford, pp 414–418

Yusuf A (1995) Critical success factors for small business: perceptions of South Pacific entrepreneurs. J Small Bus Manage 33(2):68–73

Zaei ME, Zaei ME (2013) The Impacts of tourism industry on host community. European J Tour Hospitality Res 1(2):12–21

# Chapter 16
# Do Age and Gender Influence Motivation and Satisfaction Among Special Event Volunteers?

**Norliza Aminudin, Hassnah Wee and Nur Aleeya Adlin Zahid**

**Abstract** People have various motives in participating as volunteers in special events. In fact, motivation theories of pull and push factors have been the common dimension in identifying motivation among volunteers toward satisfaction. Nevertheless, a limited number of studies have investigated the influence of demographics (age and gender) on motivation and satisfaction among special event volunteers. As special events have been expanding around the globe, identifying, and understanding the factors that influence volunteer motivation and satisfaction would enable event managers to develop better strategies for the retention of volunteers. Hence, this article specifically discussed if age and gender influence motivation and satisfaction among special event volunteers, with reference to Malaysian context.

**Keywords** Volunteer · Age · Gender · Special event · Motivation · Satisfaction

## Introduction

Volunteers are essential in helping reduce organizational capital in the form of reducing the cost of provision and the budget of full-time staff (Wong et al. 2010; Cemalcilar 2009). Additionally, having people to volunteer would allow event managers to plan their events well by adding and expanding the resources without exhausting the budget (Cnaan and Goldberg-Glen 1991, p. 270). Volunteers are involved in many fields including arts, education, health, welfare, environment

N. Aminudin · H. Wee (✉) · N.A.A. Zahid
Faculty of Hotel and Tourism Management, Universiti Teknologi MARA (UiTM), Shah Alam, Malaysia
e-mail: hassnah739@salam.uitm.edu.my

N. Aminudin
e-mail: norliza@salam.uitm.edu.my

N.A.A. Zahid
e-mail: aleeyazahid@gmail.com



A. Saufi et al. (eds.), *Balancing Development and Sustainability in Tourism Destinations*, DOI 10.1007/978-981-10-1718-6_16

protection, recreation, and youth service. Like the organizers, volunteers have become a priority in an event for it to be successful. Furthermore, Lockstone and Smith (2009) noted that volunteering has its own place and has served as a key element in an event. Based on Getz (1997), event volunteers usually participate in a special event that has been held once or is held annually, usually sponsored and organized by a company and has its own size and scale (Hall 1992). Volunteers are also considered as important stakeholders in event management and have been called the "third force" in event tourism as they are the movers to the success of an event (Olsson et al. 2013).

## The Important Role of Special Event Volunteers

Some previous researchers determined that volunteers had been employed as a dependent variable for an event because they provide a solution to the organizer (Costa et al. 2006; Goldblatt 2002; Meyer et al. 2007). However, the situation can be complex as it requires many people, considerable resources, as well as time to process in planning and coordinating with the management. Moreover, volunteers need commitment in planning and delivering the right message for an event. Plus, several events have been conducted by many volunteers, such as arts, music, cultural, and sports events, as well as festivals (Costa et al. 2006), which further prove that they are the movers for an event to happen.

Therefore, the role of volunteers in special events is very important. Without volunteers, events that are run on limited budget would cease to exit. Thus, there is a critical need to expand the understanding of volunteers' satisfaction and what motivates volunteers to reparticipate in future events. Hence, the better event organizers understand these sources of satisfaction or dissatisfaction with event operations, the more effective the management of volunteers as well as the overall efficiency of the event operations can be (Farrell et al. 1998).

For example, in Malaysia, special events have become an important tourist attraction. More than a hundred events are held annually, which include sports, cultural, entertainment, and business events. Hence, the importance of volunteers is emphasized not only because of their contributions to the success of an event but also because of their contribution to cultural diversity in Malaysia that consists of multiracial citizens. Therefore, organizations such as Rakan Muda (Youth Corp) and 1Malaysia For Youth (iM4U) have been established to provide such support for event organizers. The iM4U, for instance, is an organization that is supported by the government as the best platform to encourage volunteerism. This establishment involves not only volunteer jobs in charity work but also special event volunteers. In fact, based on the statistics of iM4U for 2013, they had more than 1 million members. In view of this, the Malaysian Ministry of Tourism and Culture had established "1Malaysia Voluntourism," which is an aggressive effort taken by the government to foster the spirit of volunteerism to help accelerate development in the tourism sector. This tourism volunteer force is tasked to kick-start five major programs under the initiative of 1Malaysia Voluntourism, including 1Malaysia Event Volunteers, which assists in organizing national tourism and cultural events. In addition, during the announcement of the 2015 budget, the government declared 2015 as the "Year of Festivals." This declaration proves that festivals are important to the country, especially to its tourism industry. Besides, with the increasing number of festivals (Raj 2003), the function of volunteers in events has increased, including in art and cultural events (Arts and Economic Prosperity III 2007).

In a big event, the existing staff or employees are not sufficient to manage and run event operations; hence, the organizer will have to look for volunteers for support (Downward and Ralston 2005; Formadi and Mayer 2009; Nassar and Talaat 2009). Not only are those volunteers "vital stakeholders," but they also need motivation and satisfaction while taking part in the event. Thus, it is important to retain the volunteers as they might be encouraged to participate again. Therefore, event organizers must find ways and means to retain these volunteers and provide them motivation so that they will reparticipate in upcoming events (Millette and Gagné 2008). This is essential because volunteers contribute their skills without asking for direct personal income, and they can make an event successful. Understanding the term related to volunteerism, motivation, and satisfaction would enhance the importance of volunteers' participation in a special event.

## Volunteerism

Volunteering is defined as "any activity in which time is given freely to benefit another person, group or cause" (Wilson 2000, p. 215). Volunteerism is also the willingness of people to help others without any expectation of payment or any other tangible gain. There are some volunteer activities that include monetary exchanges (i.e., they are paid for experience), while some are in the form of obligation (i.e., volunteer for school credit) (Holmes and Smith 2009). Besides, Kemp (2002) stated that volunteerism is where people spend their time with no reward, and it is called as their personal investment. According to Penner (2002), volunteerism is an activity or work done by volunteers without any intention of having monetary incentives and just to support the goal of an organization, a group, or an individual.

## Motivation

Motivation is both internally and externally constructed in detail. Internal motivations are linked to personal satisfaction, while external motivations are more related to rewards and tangible benefits (Munro 2001). It had been found that volunteers are slightly motivated by the joy of being in the social activity of the festival with like-minded people (Ryan and Deci 2000). Moreover, a research carried out at the Whistler Men's World Cup of Skiing displayed similar results, where promoting the town and helping to make the event a success were the highest motivators (Williams et al. 1995). In fact, many studies concerning volunteer motivation have also included the element of satisfaction (Cnaan and Goldberg-Glen 1991; Farrell et al. 1998; Love et al. 2011). According to Houle et al. (2005), different volunteering tasks are seen as satisfying different motives, and individuals prefer activities with characteristics and benefits that match their volunteer motives. The following are reviews of the three characteristics of motivation that influence volunteers' satisfaction in a study carried out in 2014 among 200 special event volunteers during Malaysia Event Volunteers' Day. In this study, it is seen that satisfaction is influenced by three motivation factors, namely, career orientation, interpersonal contact, and personal growth. According to Davis et al. (2003), there is a positive relation between motivation and satisfaction. Therefore, if the event organization can meet the volunteers' motivation, the result will be a positive impact on their satisfaction, which is similar to this research's findings. The result also reveals that to ensure satisfaction among the volunteers, the suitable motivation factors must be identified appropriately.

## Demographic Factors

Scholars in business literature have regarded gender as a key demographic variable that influences people's motives. In fact, two related theories have attempted to explain this phenomenon: (1) socialization theory by House (1981), who stated that permanent gender identity is developed through socialization processes in childhood, resulting in "care orientation" in the case of females and "justice orientation" in the case of males (Gilligan 1982), and (2) social structural theory by Eagly and Crowley (1986), who suggested that individuals behave according to the stereotypes associated with the social roles they occupy, with males behaving more aggressively and independently and being more results oriented as opposed to females who are more nurturing, caring, and relationship oriented (Franke et al. 1997; Roxas and Stoneback 2004). However, the motives for volunteer involvement are varied. Smith (1999) proposed that special attitudinal factors influence volunteer participation in addition to demographic factors, such as age, gender, education, and income level. These attitudinal factors include perceived benefits that individuals gain from volunteering in the organization (personal benefits). Furthermore, these factors may affect the level of satisfaction among volunteers as different individual characteristics play an important perspective.

On top of that, demographic statistics has demonstrated that majority of the volunteer workforce is composed of women and highly educated individuals (Rokach and Wanklyn 2009). Besides, in several studies concerning gender differences in volunteer motivation, researchers have found that women scored higher on most, if not all, functions than men (Chapman and Morley 1999; Fletcher and Major 2004). On the contrary, several researchers have found that men tend to favor instrumental motivators such as career functions, while women tend to favor concern-related motivators such as values (Prentice and Carlsmith 2000; Switzer et al. 1999).

With that, several theories have been put forth by researchers to explain age differences in volunteer motivation; the most prominent is the socioemotional selectivity theory. People of different ages are faced with various challenges and concerns. For individuals, volunteerism is a way to address some of these concerns (Omoto et al. 2000), but when people become older, these concerns change, and they come to ascribe different meanings to their roles as a volunteer. Whereas in the case of age, Briggs et al. (2010) noted that age difference influences the reasons individuals volunteer. For instance, young adults are more likely to be motivated by reasons centered on themselves (personal motivation), but older adults are more likely to be motivated by other reasons and values. Wei et al. (2011) also stated that adults are more willing to take part in volunteer activities because of their sense of responsibility toward the local community, because of certain worthy social causes, and because they can contribute to the well-being of other people. Thus, it is evident that demographic factors such as age and gender influence the motivation of volunteers to participate in events.

## Motivation and Satisfaction

Previous studies stated that volunteer motivation and satisfaction showed significant impact. According to Farrel et al. (1998), there is a relationship between volunteer satisfaction, motivation, and actual experience. The theoretical basis for this linkage has been drawn from consumer behavior literature that suggests that an individual's satisfaction results from a comparison between the rewards and the costs of an experience relative to expectation (Oliver 1980; Wu 2002). Besides, consumers tend to purchase a product or participate in a service based on their previous satisfying experiences (Farrell et al. 1998). That is, if volunteers are satisfied and their motivational needs are met, then they would likely come back to volunteer for future special events. Moreover, volunteer job satisfaction can be important to the development and maintenance of a volunteer's organizational commitment (Bennett and Barkensjo 2005; Costa et al. 2006; Davis et al. 2003). Therefore, it is important to identify the various factors that influence their satisfaction with experience. Job satisfaction is defined as "the pleasurable emotional state resulting from the perception of one's job as fulfilling or allowing the fulfillment of one's important job values, provided the values are compatible with one's needs" (Locke 1976, p. 1304).

## Motivation and Demographic Influence

McCullagh (2005) expressed motivation as the driver and the direction for volunteers. Furthermore, demographic factor could affect volunteers' motivation in participating in the event. Meanwhile, Downward et al. (2005) found that age could influence volunteers' motivation as well, whereas Tang (2008) stated that gender and race influenced the type of event volunteers chose to participate in.

Moreover, a few studies have indicated that there has been very limited analysis that had looked into age difference in motivation to volunteer (Black and Jirovic 1999; Okun and Schultz 2003). For example, some studies have shown similarities between younger and older volunteers (Black and Jirovic 1999; Clary and Snyder 1999; Marriot Seniors Living services, 1991), whereas certain studies have indicated that younger volunteers pay more attention to career motives (for example, Black and Kovacs 1999; Clary and Snyder 1999; Okun et al. 1998; Okun and Schultz 2003) compared with older volunteers who consider social motives as an important reason to be involved in volunteer activities (Greenslade and White 2005; Okun and Schultz 2003; Okun et al. 1998).

In reference to Malaysian context, a study was carried out in 2014 where 200 special event volunteers participated as respondents of the survey. In this study, gender and age had been selected as the moderating variables to investigate their influence on the motivation and satisfaction of volunteers and their intention to reparticipate. This study confirmed that satisfaction had been influenced by motivation factors, namely, career orientation, interpersonal contact, and personal growth. Therefore, if the event organization could meet the volunteers' motivation needs, the result would be a positive impact on their satisfaction, which is in line with the findings obtained in this study. The results also revealed that to ensure satisfaction among the volunteers, the right type of motivation must be identified appropriately. This outcome is consistent with that of Davis et al. (2003) as they identified a positive relationship between motivation and satisfaction.

This study also proved that volunteers were motivated by career functions that regarded volunteer service as a means to help their career and personal growth. Besides, Clary et al. (1998) measured an individual's career motivation with items such as "volunteering allows me to explore different career options." Therefore, special event volunteers would be motivated and satisfied if the event organizers could provide such career opportunities for their personal growth. In fact, Megginson et al. (1982) also stated that people experience job satisfaction when they feel good about their jobs, and this feeling is often related to doing their jobs well or becoming more proficient in their professions or being recognized for good performance. Volunteers are also motivated to explore their own strengths and expand their understanding of the cause, of other volunteers, and of the organization (Clary et al. 1998). Special event volunteers would feel more satisfied if they could establish good interpersonal contact with the event organizers. This means that having established interpersonal contact with the event organizers is important for special event volunteers to seek future volunteering opportunities.

With regard to the volunteers' demographic profiles, this study confirmed the moderating effect of age and gender on the link between motivation and satisfaction. However, this study revealed that only age influences the motivation toward satisfaction, which indicated that when the age becomes older, the relationship between interpersonal contact, career orientation, personal growth, and satisfaction would also become stronger. According to Omoto et al. (2000), the motives for volunteering may change predictably with changes in roles and agendas over a person's lifetime. For instance, younger adults engage in volunteer services to meet motivational agendas related to interpersonal relationship considerations, whereas older adults seek to meet motivational agendas related to service and community obligation concerns. Furthermore, Funes (1999, cited in Maria and Juan 2009) concluded that the life stage of a person affects the articulation of all such variables with respect to one's predisposition and decision to act. Hence, the accumulation of experiences with the passing of time and changes in perceived social conditions across life stages contribute to these age effects.

On the contrary, in a similar study, the results on gender moderating analysis confirmed that only interpersonal contact displayed a positive relationship with gender. In fact, a few researchers found that females had the tendency to participate because of interpersonal contacts during the event (Chapman and Morley 1999; Fletcher and Major 2004). Meanwhile, males failed to show any tendency to favor any variable other than their age that chose interpersonal contact. In addition, several researchers found that men tend to favor instrumental motivators such as career functions, while women tend to favor concern-related motivators such as values (Prentice and Carlsmith 2000; Switzer et al. 1999). The results appear to have put on the view that females are more likely to be sociable and to interact with people than males.

In conclusion, the first key finding of the study indicated that the motivational factor that had been regarded as the most important by volunteers was interpersonal contact. This finding was confirmed by the analytical results. The significant impact of interpersonal contact on motivation and its high factor loading had been congruent with a previous research conducted by Bang et al. (2009). Second, this finding also exhibited that females were largely involved in special event volunteering rather than males. This finding had been consistent with that of Zweingenhaft et al. (1996), which stated that women were more likely to engage in volunteering activities, so female voluntary work has greater impact than that performed by men. Furthermore, Malanska (1993) stated that social work is mainly performed by women, so volunteering is a largely female task. Meanwhile, Dolnicar and Randle (2004) also noted that the behavioral and sociodemographic characteristics of individuals are closely related with the profile of the typical volunteer. For instance, Dolnicar et al. (2007) discovered that male individuals were more predisposed to undertake volunteer activities in sports or recreational areas, while women were more predisposed to undertake social activities in the community.

The third key finding of the moderating effect of age displayed a positive relationship between the moderator and the interpersonal contact with satisfaction, which implied that if the age becomes older, the relationship becomes stronger. This means that adult volunteers are more willing and more satisfied in volunteering work as compared to youth or teenagers. This finding was supported by Wei et al. (2011), who stated that adults are more willing to take part in volunteer activities because of their sense of responsibility toward the local community, because of certain worthy social causes, and because they can contribute to the well-being of other people. In addition, Briggs et al. (2010) further supported that age difference influences the reason individuals volunteered. For example, young adults are more likely to be motivated by reasons centered on themselves (personal motivations), but older adults are more likely to be motivated by other reasons and values. Meanwhile, Marta and Pozzi (2008) stated that young adults are an important source of potential volunteers and that their commitment to volunteering can have its own specific meanings and characteristics for this particular period in their lives. Such individuals normally show a tendency to adopt prosocial attitudes and behavior (Marta et al. 2006). In fact, there have been similarities and differences found in previous studies in comparison to this study.

This study has practical implications to the event tourism industry that employs volunteers as part of their organization team, which is very significant to the success of the event. Event managers should make an effort to provide better motivation factors, such as long-term career opportunities, to enhance their personal growth and volunteering experiences with the event organization. Besides, establishing good interpersonal contact with volunteers contributes to better satisfaction among event volunteers. It is important also for the event managers to understand how to retain their volunteers, or in other words, to manage a proper volunteers' retention program. According to Silverberg (2001), the need to understand volunteer satisfaction in an effort to help managers avoid costly recruitment, retention, and volunteer management processes is indeed significant.

## Conclusion

The role of staff and volunteers in special events is very important. Without volunteers, many programs or events that are run with little to no budget would cease to exit. Thus, there is a critical need to expand the understanding of volunteers' satisfaction and what motivates volunteers to reparticipate in future events. The better event organizers understand these sources of satisfaction or dissatisfaction with event operations, the more effective the management of volunteers as well as the overall efficiency of the event operations can be (Farrell et al. 1998). By understanding the motivation and satisfaction of volunteers based on their demographic influence, event organizations can come up with the right strategy on how to retain them for future events. In fact, two demographic variables (age and gender) were used by Wymer (2003), and he found that gender and age are strong factors considered by volunteers when they intend to participate in events.

## References

Bang H, Alexandris K, Ross SD (2009) Validation of the revised volunteer motivations scale for international sporting events (VMS-ISE) at the Athens 2004 Olympic games. Event Manage 12 (3–4):119–131

Bennett R, Barkensjo A (2005) Internal marketing, negative experiences, and volunteers' commitment to providing high-quality services in a UK helping and caring charitable organization. Int J Voluntary Nonprofit Organ 16(3):251–274

Black B, Jirovic RL (1999) Age differences in volunteer participation. J Volunteer Adm 7:38–47

Black B, Kovacs PJ (1999) Age-related variation in roles performed by hospice volunteers. J Appl Gerontol 18(4):479–497

Briggs E, Peterson M, Gregory G (2010) Toward a better understanding of volunteering for nonprofit organizations: explaining volunteers; pro-social attitudes. J Macromarketing 30 (1):61–76

Cemalcilar Z (2009) Understanding individual characteristics of adolescents who volunteer. Pers Individ Differ 46:432–436

Chapman JG, Morley R (1999) Collegiate service learning: motives underlying volunteerism and satisfaction with volunteer service. J Prev Interv Community 18:19–33

Clary EG, Snyder M (1999) The motivations to volunteer: theoretical and practical considerations. Curr Dir Psychol Sci 8(5):156–159

Clary EG, Snyder M, Ridge RD, Copeland J, Stukas AA, Haugen J, et al. (1998) Understanding and assessing the motivations of volunteers: A functional approach. J Pers Soc Psychol 74:1516–1530

Cnaan Ram A, Goldberg-Glen Robin S (1991) Measuring motivation to volunteer in human services. J Appl Behav Sci 27(3):269–284

Costa AC, Chalip BL, Green BC, Simes C (2006) Reconsidering the role of training in event volunteers' satisfaction. Sport Manage Rev 9:165–182

Davis MH, Hall JA, Meyer M (2003) The first year: influence on the satisfaction, involvement, and persistence of new community volunteers. Pers Soc Psychol Bull 29(2):248–260

Dolnicar S, Randle M (2004) What moves which volunteers to donate their time? An investigation of psychographic heterogeneity among volunteers in Australia. Voluntas 18:135–155

Dolnicar S, Gruny B, Randle M (2007) Segmenting the volunteer market: learnings from an Australian study. Faculty of Commerce—Paper, University of Wollongong, Wollongong

Downward P, Lumsdon L, Ralston R (2005) Gender differences in sports event volunteering: insights form Crew 2002 at the XVII Commonwealth Games. Managing Leisure 10:219–236

Eagly AH, Crowley M (1986) Gender and helping behavior: a meta-analytic review of the social psychological literature. Psychol Bull 100:283–308

Farrell Jocelyn M, Johnston Margaret E, David Twynam G (1998) Volunteer motivation, satisfaction, and mangement at an elite sporting competition. J Sport Manage 12(4):288–300

Fletcher T, Major D (2004) 2004 medical students' motivations to volunteer: an examination of the nature of gender differences. Sex Roles 51(12):109–114

Formadi K, Mayer P (2009) Employees value anticipation in event management. J Tour Challenges Trends 2(2):115–127

Franke George R, Crown Deborah F, Spake Deborah F (1997) Gender differences in ethical perceptions of business practices: a social role theory perspective. J Appl Psychol 82(6):920–934

Funes MJ (1999) Jóvenes y acción voluntaria: la edad como factor condicionante en la acción participativa. Estudios de juventud, 45(99):87–92

Getz D (1997) Event management and event tourism. Cognizant Communication Corporation, Elmsford

Gilligan C (1982) In a different voice: psychological theory and women's development. Cambridge, MA: Harvard University Press

Goldblatt J (2002) Special events: twenty-first century global event management. Wiley, New York

Greenslade JH, White KM (2005) The prediction of above-average participation in volunteerism: a test of the theory of planned behaviour and the volunteers functions inventory in older Australian adults. J Soc Psychol 145(2):155–172

Hall CM (1992) Hallmark Tourist Events: Impacts, Management, and Planning. John Wiley, London

Holmes K, Smith K (2009) Managing volunteers in tourism: attractions, destinations and events. Elsevier, Oxford, pp 4, 7–8, 12, 29–32, 83, 88

Houle BJ, Sagarin BJ, Kaplan MF (2005) A functional approach to volunteerism: do volunteer motives predict task preference? Basic Appl Soc Psychol 27(4):337–344

House JS (1981) Work stress and social support. Addison-Wesley, Reading

Kemp S (2002) The hidden workforce: volunteers' learning in the olympics. J Eur Ind Training 26(4):109–116

Locke EA (1976) The nature and causes of job satisfaction. Dunnette MD (ed) Handbook of industrial and organizational behavior. Rand McNally, Chicago, pp 1297–1349

Lockstone L, Smith K (2009) Episodic experiences: volunteering flexibility in the events sector: people and work in events and conventions: a research perspective. CABI, Wallingford, pp 108–124

Love A, Hardin RL, Koo GY, Morse A (2011) Effects of motives on satisfaction and behavioral intentions of volunteers at a PGA tour event. Int J Sport Manage (12)1:86–101

Malanska H (1993) Women volunteers at GMHC. Women and AIDS, London

María Celeste Dávila and Juan Francisco Díaz-Morales (2009) Age and motives for volunteering: Further evidence. Eur J Psychol 2:82–95

Marriot Seniors Living services (1991) Marriott seniors volunteerism study. Washington, DC

Marta E, Pozzi M (2008) Young people and volunteerism: a model of sustained volunteerism during the transition to adulthood. J Adult Dev 15:35–46

Marta E, Guglielmetti C, Pozzi M (2006) Volunteerism during young adulthood: an Italian investigation into motivational patterns, Voluntas Int J Voluntary Nonprofit Organ 17:221–232

McCullagh P (2005) Sport Exerc Psychol Lect. Cal State University East Bay, 10/27

Megginson LC, Mosley DC, Pietri PH (1982) Management concepts and applications (4th ed). Harper Collins, New York

Meyer P, Edwards D, McDonnell I, Grabowski S, March R (2007) The future of volunteer managed festivals: where do we go from here? (eds) In: Proceedings of the 17th annual CAUTHE conference: tourism: past achievements, future challenges: 11–14 February 2007 Manly, Sydney, Australia

Millette V, Gagné M (2008) Designing volunteers' tasks to maximize motivation, satisfaction and performance: the impact of job characteristics on volunteer engagement. Motiv Emot 32:11–22

Munro J (2001) Motivations and enduring involvement of lieuser program volunteers: a study of the Waterloo region Track 3 ski school. In: Leisure Research Symnposium: A Leisure Odyssey

Nassar N, Talaat N (2009) Motivations of young volunteers in special events. Tourismos. Int Multi J Tour 4(1):145–152

Okun MA, Schultz A (2003) Age and motives for volunteering: testing hypotheses derived from socioemotional selectivity theory. Psychol Aging 18(2):231–239

Okun MA, Barr A, Herzog AR (1998) Motivation to volunteer by older adults: a test of competing measurement models. Psychol Aging 13(4):608–621

Oliver RL (1980) A cognitive model of the antecedents and consequences of satisfaction decisions. J Mark Res 17(4):460–469

Olsson A, Therkelsen A, Mossberg L (2013) Making an effort for free–volunteers' roles in destination-based storytelling. Current Issues in Tourism. 1–21

Omoto A, Snyder M, Martino S (2000) Volunteerism and the life course: investigating age-related agendas for action. Basic Appl Soc Psychol 22(3):181–197

Penner LA (2002) Dispositional and organizational influences on sustained volunteerism: an interactionist perspective. J Soc Issues 58:447–467

Prentice DA, Carlsmith KM (2000) Opinions and personality: on the psychological functions of attitudes and other valued possessions. In: Olson J, Maio GR (eds) Why we evaluate: functions of attitudes. Erlbaum, Mahwah, pp 223–248

Raj R (2003) The impact of festivals on cultural tourism. In: Paper presented at the 2nd DeHann tourism management conference, Nottingham, England. Retrieved 27 Aug 2008, from http://www.nottingham.ac.uk/ttri/pdf/conference/raj%20razaq.pdf

IM4U. Retrieved May 12, 2013, from 1M4U: http://www.1m4youth.com.my/

Rokach A, Wanklyn S (2009) Motivation to volunteer: helping empower sick children. Psychol Educ Interdisc J 46(1):7–25

Roxas ML, Stoneback JY (2004) The importance of gender across cultures in ethical decision-making. J Bus Ethics 50:149–165

Ryan RM, Deci EL (2000) Intrinsic and extrinsic motivations: classic definitions and new directions. Contemp Educ Psychol 25(1):54–67

Silverberg ME (2001) Measuring job satisfaction of volunteers in public parks and recreation. J Park Recreation Adm 19:79–92

Smith S (1999) The concept of the good quality of life. The Greeks and the good life, pp 17–32

Switzer CL, Switzer GE, Stukas AA, Baker CE (1999) Medical student motivations to volunteer: gender differences and comparisons to other volunteers. J Prev Interv Community 18(1–2):53–64

Tang F (2008) Socioeconomic disparities in voluntary organization involvement among older adults. Nonprofit Voluntary Sect Quart 37:57–75

Wei Y, Donthu N, Bernhardt KL (2011) Volunteerism of older adults in the United States. Int Rev Public Nonprofit Market 9

Williams PW, Dossa KA, Tompkins L (1995) Volunteerism and special event management: a case study of Whitler's men's world cup of skiing. Festival Manage Event Tour 3(2):83–95

Wilson J (2000) Volunteering. Ann Rev Sociol 26:215–40

Wong LP, Chui WH, Kwok YY (2010) The volunteer satisfaction index: a validation study in the chinese cultural context. Soc Indicators Res 1. doi:10.1007/s11205-010-9715-3

Wu HC (2002) Exploring the relationships between motivation and job satisfaction of volunteer interpreters: a case study from Taiwan's national part. In: Paper presented at IUCN/WCPA-EA-4 Taipei Conference. Taipei, Taiwan

Wymer WWW (2003) Differentiating literacy volunteers: a segmentation analysis for target marketing. Int J Nonprofit Voluntary Sect Mark 8(3):267–285

Zweingenhaft RL, Amstrong J, Quintis F, Riddick A (1996) The motivational and effectiveness of hospital volunteers. J Soc Psychol 136(1):25–34

Prentice DA, Carlsmith KM (2000) Opinions and personality: on the psychological functions of attitudes and other valued possessions. In: Olson J, Maio G (eds) Why we evaluate: functions of attitudes. Erlbaum, Mahwah, pp 22[illegible]–24[illegible]
Ram Y (2007) The impact of [illegible]. Paper presented at the 2nd [illegible] tourism management conference, Nottingham, England. Retrieved 22 Aug 2008, from [illegible]
[illegible]
[illegible]
[illegible]
[illegible]
[illegible]
[illegible]

# Chapter 17
# Community Social Capital in the Ecotourism Development of Gunung Padang Site, Cianjur, West Java

**E.K.S. Harini Muntasib, Arzyana Sunkar and Resti Meilani**

**Abstract** The Gunung Padang Site (GPS) is located in Karyamukti Village, Campaka subdistrict, Cianjur District, and is believed by locals to be a legacy of Prabu/King Siliwangi. The location is recognized by the tourism industry as a tourist destination but there has been limited research on its tourism development, particularly in respect of the community's social capital (which could play an important role in the success of its tourism and ecotourism). This research was designed to provide data on social capital in the community around GPS, and employed interviews based on a modified 'Social Capital Assessment Tools' questionnaire. The findings indicate that the community of Karyamukti Village has positive social capital that could be exploited for the ecotourism development of GPS. Community members trust the GPS management (with the exception of the Kompepar). They also trust their informal leader. There are small number of people who still practice customary law, which could be beneficial as a tourist attraction. The community tends to be open towards visitors and nongovernmental organizations (NGOs). In terms of networking, the people are involved in various social networks, and have started to develop links with other stakeholders. They have taken proactive action in relation to experience sharing, participation in social organizations, and protection of the environment. However, there was also a negative side to their social capital: they still retain some jealousy of each other with regard to job opportunities. This could pose a threat to the success of any tourism development in that there would likely be unfair competition by individuals or subgroups, which in turn could affect tourism operations in the location.

E.K.S.H. Muntasib (✉) · A. Sunkar · R. Meilani
Department of Forest Resources Conservation and Ecotourism, Faculty of Forestry, Bogor Agricultural University, Bogor, West Java, Indonesia
e-mail: harinimuntasib@yahoo.com

A. Sunkar
e-mail: arzyanas@gmail.com

R. Meilani
e-mail: meilani_resti@yahoo.com



A. Saufi et al. (eds.), *Balancing Development and Sustainability in Tourism Destinations*, DOI 10.1007/978-981-10-1718-6_17

**Keywords** Community · Ecotourism development · Gunung padang site · Social capital · Indonesia

## Introduction

The Gunung Padang Site (GPS) is located in Karyamukti Village, Campaka Subdistrict, Cianjur District. The GPS covers an area of 3 ha, including 900 $m^2$ of ancient structures. Imprints of Kujang (a Sundanese traditional weapon) and tiger footprint can be found at the site. The surrounding community believes that the site is a legacy of Prabu/King Siliwangi. The location is recognized by the tourism industry as a tourist destination. Various researches, focused on historical, archaeological, and geological aspects, have been conducted at the site. However, there has been limited research on the development of its tourism potential, particularly in respect of the community's social capital, which it is believed, can play an important role in the success of tourism and ecotourism development.

Social capital is the resources that exist because of, and arise out of, relationships between members of distinct social groups, and between any such groups, and is defined as "relationships among actors (individuals, groups, and/or organizations) that create a capacity to act for mutual benefit or a common purpose" (Spellerberg 2001, p. 9). Social capital puts emphasis on group potential and the pattern of relationships between individuals in a group and between groups, with a focus on social networks, norms, values, and mutual trust, all of which arise between group members, and become the group norms (Harpham et al. 2002 as cited in Jones 2005; Hasbullah 2006). These regulations, norms, and sanctions, play important roles in placing group interest over individual interest. Formal and informal regulations, norms, and sanctions are also important in the generation of positive attitudes and behavior, crucial for biodiversity and ecosystem preservation, and the efficient management and use of public goods. Jones (2005) even suggested that social capital is the 'missing link' in development, and should be the focus of policy, practices, and research.

In a human development context, social capital has had great influence, since several development dimensions are greatly affected by social capital; these include the ability to solve problems collectively in order to promote rapid changes in the community, to develop collective awareness of ways of improving quality of life, and to identify chances to increase welfare. Strong social bonds should lead to welfare improvement.

The components of social capital include trust, cohesiveness, proactive action, and internal–external relationships that help develop social network (Pretty and Ward 2001; Pretty 2003; Pretty and Smith 2003). Hasbullah (2006) suggested that network participation, reciprocity, trust, social norms, values, and proactive action are the main elements of social capital. Pretty and Smith (2003) pointed out that many studies have shown an increase in natural resources conservation activity within a given community living inside and around a protected area, when good individual, group and network interrelations tended to exist, and when local knowledge was used in the planning and implementation process of any conservation activity and development. Pongponrat and Chantradoan (2012) considered social capital to be a crucial mechanism for the successful planning of local tourism activities. In addition, Inayah (2012) suggested that the cultural dimension of social capital provides an important contribution to development processes. Social capital appears to play a crucial role in the functioning and strengthening of modern society.

The Quebec Declaration (UNEP and WTO 2002) stated that the community, as part of social component, comprises the human capital which plays the role and have responsibility in determining ecotourism development. Therefore, ecotourism development in a certain location should not rely solely on that location's natural and cultural assets and attractions. Hakim and Nakagoshi (2008) suggested that, unlike natural and cultural capital, which has often been discussed in ecotourism literature, a community's participation and involvement in ecotourism activities has not received much attention. The importance of identifying social capital in ecotourism development arises from the fact that the success of such development is determined by the balance of environmental, economic, and social culture aspects Krstić et al. (2008). Therefore, identifying the community's social capital at GPS was expected to provide input, for those managing GPS, to development of ecotourism at the site, and improvement of the community's ability to plan and implement ecotourism activities; thus helping to ensure the success of ecotourism in GPS.

## Method

This research was designed as a survey exploring existing social problems and social capital phenomena in relation to ecotourism development in GPS. Data on social capital dimensions, general conditions of the location, socioeconomic conditions, and ecotourism conditions, were all collected through in-depth interviews, field observation, literature and document review, and miscellaneous other interviews. A modified questionnaire, based on the Social Capital Assessment Tool (Khrisnaq and Shrader 1999), was used to obtain data on social capital from the community. Data was descriptively analyzed according to key elements of social capital, i.e., trust, social networks, social norms, relevant proactive action, and concern toward each other and the environment.

## Results and Discussion

### *Stakeholders Involved in the Management of Gunung Padang Site's Ecotourism*

There were three main stakeholders involved in the management of GPS, i.e., Forum Peduli Situs Gunung Padang (literally meaning 'a forum which cares for GPS: the Forum'), Juru Pelihara/Jupel for GPS (the official representative of the local government's Tourism Office), and Komunitas Penggerak Pariwisata/Kompepar (which literally means 'community tourism driver', a group established by the village government). There were 270 households in the community around GPS which were involved as members of the GPS Forum. The programs implemented by this forum were those arrived at by mutual agreement of community members. The community had a sense of distrust towards the Kompepar. The former felt that since the Kompepar and the Forum had the same area of concern, the Kompepar should not have been established. Only one community member defended the existence of Kompepar, while the others had resigned from Kompepar. Other stakeholders involved in GPS included village officials (head of village, heads of hamlet, and heads of neighborhood), the ustadz (Islamic religious leaders), the teachers, and the traditional leaders.

Several nongovernmental organizations (NGOs) had also been involved, i.e., Lokal Mala, Komunitas Budaya Sunda (literally meaning the Sundanese Traditional Community), and Komunitas Badui dan Kampung Naga (the Badui and Kampung Naga Community). The Lokal Mala was involved in cooperation to establish the mini arboretum. The Komunitas Budaya Sunda, i.e., Sasanggrahan Sunda, Kalakai, and Sanggar DKC Cianjur, were involved in joint training of art. The Komunitas Badui dan Kampung Naga was involved in sharing, and comparative study, of cultural tourism management.

## *The Social Capital of the Community Around Gunung Padang Site*

### Trust

#### (a) Trust toward GPS management

The majority of people of Karyamukti Village had trust in the Forum Peduli Situs Gunung Padang, since Forum members were from the community around GPS. The level of trust could be considered high, as there had never been any conflict between the community and the Forum, and the Forum had always kept to their agreements with the community. The Community's trust in the Jupel could also be considered high. The community would help the Jupel during periods of high visitor numbers. They also considered the Jupel, which was appointed by the local government, to be composed of people who were knowledgeable about the history of GPS. The Jupel was looked upon as the caretaker of the site. There had never been any conflict between the Forum and the Jupel, as there was a clear understanding of task, role, and profit between the two organizations. However, the community's level of trust of the Kompepar was low. The Forum, which consisted of community members, had previously been in conflict with the Kompepar. The conflict apparently arose from the similarity of vision, objectives, and task between the two organizations. The Forum assumed that the Kompepar would work only when there were incentives for them to do so.

(b) **Trust toward community leaders**

The community's levels of trust of their leaders, both formal (village officials) and informal (religious leaders and teachers), were high, with the exception of feelings towards their village head, distrust of whom was mainly due to the latter's personal problems; the intrusion of such personal matters being considered inappropriate by the community. In comparison with people's trust of village head, the people put more trust in village officials, since the officials were the ones that helped in solving conflicts which arose in the community. The community tended to trust religious leaders and teachers since they considered the leaders as knowledgeable persons.

(c) **Trust engendered by traditional practices (customary law) and rituals**

The majority of Karyamukti villagers still engaged in a small number of traditional practices, i.e. traditional ceremonies/rituals related to rice harvesting, the birth of babies and pregnancy. However, the role of traditional leaders had been decreasing. Only a small number of community members were still conducting traditional rituals at the GPS. This could nonetheless provide a cultural attraction for visitors.

(d) **Trust toward each other**

The people of Karyamukti Village still had low levels of trust toward each other. There was a sense of jealousy among the people in relation to job opportunities in GPS. This condition would likely produce conflict among the people and hamper the development of ecotourism in the area. The management should consider providing transparently equal opportunities for community involvement in the management of GPS, thus reducing jealousy, and promoting the development of trust among the people. As Baksh et al. (2013, p. 5) suggested, "Trust should be built through cooperation, or networking among, and participation of the local community themselves". As Jones (2010) proposed, people who tended to trust other members of a community were more willing to participate in solving environmental problems. It was expected that when people around GPS trusted each other, they would be more willing to participate in the maintenance of the site's environment as an asset in the development of ecotourism.

(e) **Trust toward outside parties (visitors and NGOs)**

The community tended to be trusting of outsiders. Hence, they readily accepted visitors in an open and friendly manner. This would have been beneficial in achieving visitor satisfaction toward tourism services in GPS. However, a particular incident, arising from miscommunication between a researcher and the people, marred the latter's trust toward visitors. A researcher used an explosive to identify stone composition and depth. The people who heard the explosion were panicked and some were so enraged that they took action against the researcher, sufficient to result in charges against them going to trial. Since then, all researchers have had to report to village officials before conducting any research at GPS.

The people still had high degrees trust toward NGOs. They had cooperated with several NGOs in the ecotourism development of GPS; such cooperation taking the form of training, and experience-sharing activities in relation to tourism development.

(f) **Trust toward local (district) government**

The degree of trust toward local government was evidenced by the community's willingness to allow their land to be used by the local government for tourism development. They trusted that the local government would provide appropriate compensation for their relocation, and that this relocation was intended for their mutual benefit. The community believed that tourism development in the site would provide job opportunities for the people. Therefore, they were willing to support the development, by allowing the use of their lands for the development, taking care of the cleanliness of their environment, and joining experience-sharing activities, and others such activities. This was in line with findings that "higher levels of institutional trust indicate that citizens tend to rely on the effectiveness of institutions" (Jones 2010, p. 132). Jones (2010) further asserted that institutional trust was connected to individuals' tendency to act collectively and protect the environment, and considered that provision of information towards such actors (local government) will facilitate the resolution of environmental problems.

## Social Networks

Cooperation between the community and local government involved a grant from the National Program of Community Empowerment (PNPM). The funds involved were used by five groups within the community: the lodging group, art group, culinary group, handcraft group, and tourist guide group. The government had not conducted any extension program for the community in relation to tourism development. 'Support' from the village government was limited to a request that the people should be friendly to visitors. The Head of hamlet had made this request to the community.

The people had a good cooperative bond with the Forum, as one element of GPS management. This was indicated by people's willingness to join the five GPS tourism-supporting groups established by the Forum. The membership of these groups had reached 110 persons, with the highest number in the art group (60 persons). Individual interactions among group members were high. The existence of a chairman in each group who coordinate the members and facilitate the interaction of members in their groups had contributed to this high interaction among group members. There was an equal division of roles among members. Each group had its own function, role, and objective for achieving mutual benefit, and there had not therefore, been any conflicts between groups.

There was a social network that was directed towards tourism development in Karyamukti Village. The people had started to develop social links with various groups, such as NGOs, the Sundanese Culture Forum, and the Local Government Tourism Office. They were also in the process of developing a 'tourism village', involving all tourism objects and attractions found in the village.

Initiatives from three members of the community had succeeded in establishing the Forum Peduli Situs Gunung Padang. The initiators had wanted to improve community welfare through tourism activities with GPS as the main object. To date, there had been 270 households involved in the Forum. People involved were active in managing and maintaining the forum network, while other people outside the forum tended to rely on the Forum's members.

Karyamukti villagers also tended to be open towards outsiders. They were willing to develop relationships with outsiders for the promotion of development. Besides Karyamukti Village, people from Cibeber village also benefited from tourism activities at GPS. Some of Cibeber village people had become motorcycle taxi drivers, and had formed their own group. The bonds formed among people with the same interest had created various levels of interaction, i.e., between individuals of the same group, between individuals from different groups, between individuals and groups, and between groups.

Cooperation, and the bonds created by such cooperation, is important for increasing people's participation in tourism development. Jones (2010) suggested that social network and civic participation significantly influenced environmental awareness and activity (environmental activation). Networking was also found to be the most dominant variable affecting tourism development in Tambaksari Village (Baksh et al. 2013).

## Social Norms

The people of Karyamukti Village held to certain social norms, i.e., traditional norms and more common norms related to dress code, avoiding infidelity, respecting elders, and respecting other people's opinions and beliefs. Norms, as Baksh et al. (2013) suggested, do not have a direct effect on ecotourism development, but can help in strengthening people's networking and participation. This would in turn affect ecotourism development. Bakhs et al. (2013) further asserted that formal rules, as part of the norms, are important in protecting natural resources and minimizing negative impact on the environment. Therefore, the norms held by Karyamukti Village people would indirectly affect ecotourism development in the area. For example, by respecting other people's opinions, conflicts that commonly arise from differing opinions, could be avoided. In this way, people would tend to be more comfortable with each other, creating a supportive social atmosphere.

## Proactive Action

The people of Karyamukti Village were accustomed to 'gathering activities', known as *ngariung*. They tended to gather in the afternoon, telling stories and sharing information and experiences with each other. They also practiced various arts together, i.e., silat (a martial art), kecapi suling (traditional musical instruments), and marawis (a combination of Middle Eastern and Batavian music; strongly influenced by religion).

There were also social organizations such as religious study groups for women and men, in the community, and the people actively participated in such groups. The groups had become their means to interact and share information with each other. Such groups could be considered to provide effective means of fostering a sense of kinship among the people.

The people of Karyamukti Village also showed high levels of participation in other social organizations, such as the Pemberdayaan Kesejahteraan Keluarga-PKK (a social organization which empowers women to participate in development), August festivals (held to celebrate Indonesian independence day), the Taman Pendidikan Qur'an (an informal educational organization, seeking to teach children about the Qur'an).

In summary, proactive action in sharing experiences would influence networking (people's participation in network).

## Concern for Each Other, and the Environment

### (a) Concern for each other

Concern for each other was still high among the community of Karyamukti Village. The majority of community members in some way related to each other as family (they had familial relationships with each other/kinship). They helped each other in dealing with unfortunate events, visiting and nursing sick people, and organizing weddings, and funerals. They had high tolerance of group differences, and respect toward community subgroups that performed traditional rituals or pilgrimage activities at GPS. Such concern would provide opportunities for collective action in achieving mutual objectives, and should be used to the full for ecotourism development of the site.

### (b) Concern for the environment (environmental concern)

Environmental concern was evidenced by people's proactive actions in maintaining the environment. The people of Gunung Padang Hamlet (the hamlet adjacent to GPS) understood the importance of the environment for their livelihood. They participated in activities to clean up the road and make gutters when there were landslides and heavy rains. In contrast, people in Gunung Mas Hamlet tended to show low environmental concern. They disposed of waste resulting from gold mine processing, into the river, polluting this river with mercury. As a result, clean water was hard to obtain, and the people had to buy clean water from elsewhere (known by the people as refill water). The situation indicated that environmental knowledge affected people's attitudes towards the environment, and hence affected ecotourism development in the area. In line with this, Zhang and Lei (2012) found that environmental knowledge affects people's attitude toward ecotourism, and consequently determines their decision to participate (or not) in ecotourism.

Apparently, the social capital of the community in Karyamukti Village should be increased, particularly in relation to their trust toward each other, and their environmental concern. Empowerment is a determinant of social capital (Narayan and Cassidy 2001). In order to improve social capital, the local government should create and implement, for the people of Karyamukti, empowerment programs aimed at increasing people's environmental knowledge. The people have to recognize the value of their environment as the main asset for tourism. Involvement of more people in tourism activities would lessen jealousy and distrust among competing people. However, the management has to ensure a clear role, and benefit sharing, for those involved in tourism activities and development. There are several ways to stimulate people's involvement, as stated by Zhang and Lei (2012), i.e., through suitable management strategies aimed at increasing people's environmental knowledge, encouraging positive attitudes toward tourism, and promoting people's affinity for and with local attractions.

## Conclusions

The findings indicate that the people of Karyamukti Village has positive social capital that could be exploited for ecotourism development related to GPS. They trust GPS management, with an exception of the Kompepar. They also trust their informal leader. There are a small number of people who still practice customary/traditional law, which could be beneficial as tourist attraction. The community tends to be open towards visitors and nongovernmental organizations. In term of networking, the people participated in various social networks, and have started to develop links with other stakeholders. They have taken proactive action in relation to experience sharing, participation in social organization, and the environment. This would also be beneficial to ecotourism development in the GPS. However, there was also a negative side to their social capital: in that they still retain some jealousy of each other in relation to job opportunities. This could pose a threat to the success of any tourism development in that there likely is unfair competition by individuals or subgroups, which in turn could affect tourism operations in the location. Therefore, an empowerment program should be implemented to enhance the people's social capital, and further ensure the success of ecotourism development in the area.

## References

[UNEP] United Nations Environment Programme, and [WTO] World Tourism Organization (2002) Quebec declaration on ecotourism. World Ecotourism Summit, Quebec City

Baksh R, Soemarno Hakim L, Nugroho I (2013) Social capital in the development of ecotourism: a case study in Tambaksari Village Pasuruan Regency, East Java Province, Indonesia. J Basic Appl Sci Res 3(3):1–7

Hakim L, Nakagoshi N (2008) Planning for nature-based tourism in east java: recent status of biodiversity, conservation, and its implication for sustainable tourism. ASEAN J Tour Hosp 7 (2):155–167

Hasbullah J (2006) Sosial capital (Menuju Keunggulan Budaya Manusia Indonesia). MR United Press, Jakarta

Inayah (2012) Peranan Modal Sosial dalam Pembangunan. Ragam Jurnal Pengembangan Humaniora 12(1):43–49

Jones S (2005) Community-based ecotourism: the significance of social capital. Ann Tour Res 32 (2):302–324

Jones N (2010) Environmental activation of citizens in the context of policy agenda formation and the influence of social capital. Soc Sci J 47:121–136. Available online at www.sciencedirect.com

Krishna A, Shrader E (1999) Social capital assessment tool. [Paper]. In: Conference on social capital and poverty reduction. The World Bank, Washington DC

Krstić B, Jovanović S, Milić VJ (2008) Sustainability performance management system of tourism enterprises. Facta Universitatis Series: Economics and Organization. 5(2):123–131

Narayan D, Cassidy MF (2001) A dimensional approach to measuring social capital: development and validation of a social capital inventory. Curr Sociol 49(2):59–102

Pongponrat K, Chantradoan NJ (2012) Mechanism of social capital in community tourism participatory planning in Samui Island, Thailand. Tourismos Int Multi J Tour 7(1):339–349

Pretty J (2003) Social capital and the collective management of resources. Science 302:1912–1914

Pretty J, Ward H (2001) Social capital and the environment. World Dev 29(2):209–227

Pretty J, Smith D (2003) Social capital in biodiversity conservation and management. Conserv Biol 18(3):631–638

Spellerberg A (2001) Framework for the measurement of social capital in New Zealand. Statistics New Zealand Te Tari Tatau, Wellington

Zhang H, Lei SL (2012) A structural model of residents' intention to participate in ecotourism: the case of a wetland community. Tour Manage 33:916–925

# Chapter 18
# The Contributions of Enclave Tourism to Community Wellbeing in Mauritius: Host Community Perspective

**Perunjodi Naidoo, Philip Pearce and Richard Sharpley**

**Abstract** Despite tourism being recognised as a promising growth sector in many Small Island Developing States (SIDS), its contribution to community wellbeing has been neglected. The current study is set within the context of tourism development in a SIDS and examines the positive and negative outcomes of enclave tourism to community wellbeing in Mauritius. International tourism to Mauritius has developed significantly over the past 30 years with the hotel industry being the most dominant form of tourism enclave. Although, large resorts provide the environment for attracting tourists to destination communities, it remains unclear how this form of tourism contributes to community wellbeing in Mauritius. Moreover, although past studies have been conducted on enclave tourism in SIDS, there is no study which has specifically been undertaken on the contributions of enclave tourism to community wellbeing. Therefore, the aim of this research is to assess the host community's perceptions regarding the contributions of enclave tourism in Mauritius. Given the lack of information on the topic, a theoretical sample with in-depth interviews was used amongst key informants of the destination. The findings reveal mostly similarities amongst the interviewees and identify that enclave tourism contributes to community wellbeing in seven domains namely: Economic, Social, Environmental, Cultural, Leisure, Educational and Political. The findings reflect that although the economic benefits of enclave tourism has a greater potential to contribute to community wellbeing, better policy and planning are needed to unleash the opportunities for entrepreneurship which are currently limited. Moreover, the findings show that

P. Naidoo (✉)
School of Sustainable Development and Tourism,
University of Technology, Mauritius, Port Louis, Mauritius
e-mail: pnaidoo@umail.utm.ac.mu

P. Pearce
College of Business, Law and Governance,
James Cook University, Townsville, Australia

R. Sharpley
School of Sport, Tourism and the Outdoors,
University of Central Lancashire, Preston, UK



A. Saufi et al. (eds.), *Balancing Development and Sustainability in Tourism Destinations*, DOI 10.1007/978-981-10-1718-6_18

tourism development in Mauritius occurs on an ad hoc basis with short-term gains, which limit the potentials of and negatively affect community wellbeing.

**Keywords** Community wellbeing · Enclave tourism · Small island developing state · Mauritius

## Introduction

Government decisions have a wide ranging impact on the lives of individuals in a community. Governments of Small Island Developing States (SIDS) perceive tourism as a panacea for development (Sharpley 2002) and pursue enclave tourism as a viable option for current and future development. However, politicians are accustomed to their success or failure being evaluated according to measures of economic growth, although an increase in GDP growth does not necessarily achieve development goals such as reducing economic inequalities, empowering individuals, providing meaningful employment or better health to the community (New Economics Foundation 2015). Increasing, evidence show that communities around the world believe that well-being is an important government goal to pursue. For example, 81 % of the population in the UK perceive that the government should bring "greatest happiness" instead of focusing on "greatest wealth" (BBC 2006). The survey also showed that Britain was less happy than in the 1950s despite being three times richer. In France, it was found that 86 % of the population believed that community wellbeing should be a goal for the future (Fabrique Spinoza 2015). Moreover, the first World Happiness Report reveals that communities are increasingly requesting for happiness to be part of government policy (Helliwell et al. 2012). There is increasing agreement that success of communities should be supplemented with holistic assessments of wellbeing which considers human progress (OECD 2013). However, there is no consensus on the definition of community wellbeing although it is frequently defined as the quality of people's experience of life (OECD 2013), life satisfaction and happiness (Diener 2009; Veenhoven 2012).

Despite enclave tourism being the most dominant form of tourism development in SIDS, research has not been undertaken into the specific contribution of enclave tourism to community wellbeing. Enclave tourism refers to specific types of tourism products, such as holiday resorts, casinos, safari lodges, cruise ships and theme parks (Edensor 2001; Liu et al. 2007), of which resorts are the most important category in warm water SIDS destinations. Resorts are self-contained structures, where tourists are encouraged to remain within as their departure to other locations would mean potential loss of revenue (Wall 1996). This type of tourism is characterised by a physically, socially and economically self-contained structure, segregated from the local community. In this study, enclave tourism refers to resorts with self-sufficient infrastructure and amenities which provide tourism services in a closed environment, such as large hotels and Integrated Resort Schemes (IRS).

They provide tourists with facilities during their holidays, such as accommodation, food and beverage, entertainment, shopping and leisure. Although some studies have been carried out on enclave tourism in SIDS (Hernandez et al. 1996; McElroy and De Albuquerque 1998; Prayag et al. 2010), no studies were found on the contributions of enclave tourism to community wellbeing. Due to the scant information available on this topic, in-depth qualitative interviews were conducted with ten Key Informants from the host community referred to as respondents H1 to H10 in this study.

## Results and Discussion

### *Community Wellbeing*

It was important to understand the perceptions of community wellbeing from the host community in Mauritius since it is believed that perceptions of stakeholders cannot be separated "from the social situations in which they are routinely embedded" (Weed 2005: 359). Perceptions of these individuals are influenced by the contextual situation and the ontological assumption of the research is that different people perceive events in different ways (Guba 1990; Hardy 2005). Thus, it becomes important to comprehend how individuals perceive the world (Weed 2005) with regards to tourism development in Mauritius.

The respondents were first asked to explain their understanding of community wellbeing. They stated that community wellbeing is about having a sense of satisfaction with life where the community is happy. For example, respondent H1 defined wellbeing as:

> to be content with conditions, what we have and what we earn on our jobs and also a sense of satisfaction. Not only making more and more money but you need to have money of course for the basic necessities but wellbeing is to be happy with what we are doing and what we have in addition to the job that we do. It's not related to the amount of money that we collect from our job every month. We feel complete in terms of salary, job, health, spirituality, we feel happy as a human being.

Another respondent explained that wellbeing for the community consists of several aspects such as "good health, an appropriate physical environment, access to infrastructure and transportation, and access to services such as health care at reduced costs and banking facilities (H6)". Moreover, respondents stated that community wellbeing is reached when the "needs of the family are met including day to day issues and in terms of education and having access to schools which develop the child in a variety of ways such as arts, sports and not only provide academic knowledge". For entrepreneurs interviewed, there is a sense of wellbeing when the business is successful and there is no debts,

> when tourists come and buy, it contributes to our wellbeing. I am at ease, I do not have debts. If every day I sell in the morning and eat in the evening, this is not wellbeing. When

you have money in your hand, when you do not have stress, then there is wellbeing. We live tranquil. When we see our stock going, it is a joy and it gives us the inspiration and motivation to continue, we feel good. When we do not sell, we do not have courage (H2).

Wellbeing happens at an individual level and then becomes community wellbeing. The community is a group of people of different backgrounds, origins and traditions living in an area, and practicing different activities for their own individual well-being. The findings support that community wellbeing is achieved when tourism improves the lives and environment of all individuals within the community. It brings life satisfaction and happiness based on a balance of several factors such as money, personal development and access to resources. Community wellbeing constitutes the common thread that provides people in a community that lifestyle that they are looking for, the resources and peace of mind that are required for individuals and their family to grow in the best possible conditions. Community wellbeing is relative in time as it is dependent on the needs and lifestyles of individuals in a community at a particular point in time.

The next section discusses the specific contributions of enclave tourism to community wellbeing based on seven domains.

## *Economic Dimension*

The findings reveal seven important domains to community wellbeing. The most important one was the economic contribution of enclave tourism. The host community stated that enclave tourism enhances community wellbeing through creation of direct and indirect jobs:

there will be employment to taxi drivers who will get jobs, people in the villages will get jobs, there is a paid salary to these people, there can be development of handicraft. Those who are involved at food processing at village levels, they can sell their products to tourists. Some construction of shops will also contribute economically, and then of course the employment to fishermen sons and daughters will also be economically important (H1).

Direct employment is also available closer to the homes of residents and a respondent stated that:

our relatives work in the hotel industry, they are a bit tired but it is good. It is more remunerated I think and it is close to their home. Big hotels bring more work, when there are jobs, there is money. Hotels also have more capacity to employ even though the working hours are long (H3).

Past studies have also revealed the contributions of enclave tourism to employment generation in developing countries (Sharpley 2003; Mbaiwa 2005; Prayag et al. 2010), however, the host community were also critical of the conditions of employment in enclave tourism, revealing that job creation alone did not contribute to community wellbeing unless it was through the provision of meaningful job opportunities with attractive salaries, reasonable working hours and conditions. Previous studies also confirmed that enclave tourism provided mostly low-skilled jobs with low salaries (Mbaiwa 2005; Sharpley and Ussi 2012).

Interviewees stated that economic opportunities were restricted due to the nature of this form of tourism and all-inclusive holiday packages, which encouraged tourists to remain inside resorts instead of interacting with local entrepreneurs. Respondents felt that resorts restricted her wellbeing since tourists only stayed in inside and access to the hotels for entrepreneurs were fee paying:

> We would also like to sell our products in hotels, but the price is outrageous. The price is around Rs 3000 [USD $85] per day and we must make a profit out of this! Big hotels do not give us a chance to get contact with tourists. Those who are inside will never know us, we have to rely only on those that go out on their own. Tourists that go with tour operators will also not come here as they go to places where they get commissions (H2).

Moreover, small entrepreneurs did not have the means to pay commissions to tour operators and thus they remained excluded from the tourism circuit.

> … we have not felt the benefits of the tourism industry at all… We are on 'stand by' mode here. Other artisans have already come and gone. It has been 1 year since we are here. It was a government based project where shops were available to artisans if we were interested. We thought that there would be lots of tourist tours like they have said. The ministry is not doing its job properly… We are not making profit at all…. Tour operators' buses are not coming… (H2).

The study shows that there is also a breakdown of trust between host community and the government. To overcome the reliance on tour operators and improve wellbeing of the locals, respondent H2 stated that there is a need to work with enclave resorts as this is where most of the customers were located. The respondents provided an example of how hotels could work with small entrepreneurs.

> The hotels could organise a handicraft market. For example…during each week-end, they could take, let's say 5 different artisans from the village to participate in the market inside the hotel. This would bring added value for us and for the tourists. Even if they put us under the coconut tree it's ok [laughs] but the hotel directors aim too high and they do not think of the small at all. This results in our exclusion (H3).

In terms of power, the host community felt that the enclave tourism industry is controlled by the private sector, leaving little opportunities for the locals. As noted by Mbaiwa (2005), enclave tourism often failed to establish linkages with locals to encourage the development of small and medium local enterprises. Conversely, the host community members who directly benefitted from enclave tourism were less critical of this form of tourism development suggesting that residents might evaluate tourism development based on the benefits and costs received as per the social exchange theory (Andereck et al. 2005). It was also perceived that enclave tourism had increased the cost of living with higher prices of land, property and goods. The findings support previous studies where tourism expansion has increased land prices resulting in more economic struggles for the locals (Andriotis 2003; Prayag et al. 2010; Lorde et al. 2011).

## *Social Dimension*

The social dimension of community wellbeing derived from enclave tourism is also important to the host community. They argued that enclave tourism contributes to community pride since the locals are proud to have well-reputed and international award-winning resorts in the community. Enclave tourism enhances public infrastructure, public utilities and services in the community, concurring with the results of past studies (Mbaiwa 2005). Wellbeing was also enhanced through other social aspects such as communication and interaction between hosts and guests. One of the respondents gave an example by referring to her encounter with a tourist couple at a bus stop in Port Louis, the capital city,

> …we continued the conversation in the bus and these South Africans stayed with us and today we are friends. I have made a lot of friends like them. They wanted to stay with us to go to the airport easily. The tourists like our Mauritian culture. We are happy that when we play a Sega [local song and dance] it creates a great ambiance. During Maha Shivaratree [local hindu religious pilgrimage] they took the bicycle and went there and stayed overnight…the fact that they appreciate our culture increases our wellbeing (H2).

The interviewees also perceived that the social contributions of enclave tourism to community wellbeing have largely been due to Corporate Responsibility (CSR) initiatives. Enclave resorts contribute 2 % of profits to CSR in different areas, such as education, leisure and environment (see also Prayag et al. 2010; Sharpley and Naidoo 2010). However, the physical boundary of enclaves reflects the barrier with the community, lack of interaction and subsequently acts as an obstacle in spreading community wellbeing.

## *Cultural Dimension*

The cultural domain was identified as an important component of community wellbeing. Enclave tourism was perceived to distort local culture to please tourists. A respondent stated the following:

> The other day I saw tourists coming from a tour bus at one of the religious pilgrimage sites. I was wondering what they had on their foreheads as I had never seen that before. Hindus in Mauritius usually wear a mark or spot between the eyebrows, but the tourists had a strange shape on their foreheads which is not at all part of our culture! It seems that this tour did not provide them with a genuine experience of the local culture. Culture has become something to sell to tourists and at times this is reflected in the exaggerated display of culture which does not reflect our reality. How far are enclaves such as tour operators giving the right explanation and educating tourists? (H4)

The culture of the community presented by this tour operator was not genuine and it affected wellbeing as there was a lack of value and respect of the local culture. The respondent also provided another example:

> The other day I was watching a cultural show in a hotel and was shocked to find that it displayed more sexual connotations rather than local culture. The way the culture is interpreted in enclaves does not add value to our local traditions and thus in the long-term it will not contribute to our wellbeing and we will lose our identity (H4).

This supports the view that tourism commodities culture and destroys the authenticity of traditional cultures (Cohen 1988; Shaw and Shaw 1999). Nevertheless, the lack of contact with locals in enclave resorts was also positively viewed by one respondent as locals feel that they were not being taken over by tourists. This can help manage conflicts between residents and tourists as well as safeguard local cultures (H5).

## *Environment Dimension*

A green environment also enhances community wellbeing. Enclave tourism has contributed to an improvement in the environment in the vicinity of hotels, but mostly on and around the routes adopted by enclave tourists and not for locals. The host community expressed concerns for the green environment which is being lost due hotel developments that have destroyed the coastal landscape and degraded the natural environment. However, respondent H3 stated that the green environment is being lost to the development of enclave resort, "we do not have nature anymore. Trees are being replaced by hotels…ecologically, we are in trouble…Owing to the demand for large stretches of land for the construction of tourism infrastructure (Andriotis 2008), enclave resorts have "taken" the best coastal land, now they are moving more inland with the IRS and RES (H8). "We will become tenants in our own island (H4)" as even the middle class will not be able to afford purchasing a house or a plot of land as has been the practice in the past. "With this fast development, our natural landscape is being reduced and will have serious consequences on our wellbeing (H4)".

## *Education Dimension*

The study revealed that education is an important component of community wellbeing. Studies have also shown that those who are better educated are more liable to have a higher degree of life satisfaction (BBC 2012). Education enhances wellbeing due to its economic and social benefits (Hill and King 1995). All respondents agreed that due to enclave tourism, formal tourism education is provided through several institutions such as universities, vocational institutions and in-house training in enclave resorts. In addition, there is evidence that enclave tourism has contributed to a high level of communication and foreign languages skills for locals without any formal education. Thus, enclave tourism has contributed to personal development through exposure to continuous informal learning opportunities. Respondent H9 stated "… I know hawkers who sell necklaces to tourists on the beach. I personally know one of them who speaks very good Italian but he has never been to school or to Italian classes!". Moreover, stakeholders stated that enclave tourism could enhance the personal development of the individual through travel opportunities.

## *Leisure Dimension*

Leisure has often been closely related to the health component of quality of life (Ferrans 1996; Liang et al. 2013). The local community expressed that there is poor leisure opportunities for locals apart from those available through CSR which can also take the form of leisure activities such a football ground. However, these facilities are only provided in the vicinity of hotels and the community at large will not benefit. Enclave tourism has reduced leisure spaces through restricted access to beaches and coastal roads. There is a loss of wellbeing due to restricted areas for relaxation, sport and socialisation. "…there is the danger as we have seen in two or three cases where the roads are closed. There is no access for the Mauritians but you have to go more in the interior as coastal roads and coasts are handed over to hotels. H1". The promoters take the best spots to develop products and thus tourists are offered the best of the island as stated by respondent H7. Some residents perceive that facilities and services provided to tourists far exceed what is made available to them.

## *Political Dimension*

The political domain is important in community wellbeing. Past studies have also shown that political factors affect community wellbeing (Barber et al. 2012). The respondents expressed that although, the construction of infrastructure for tourists and associated services might result in economic benefits, it is uncertain as to whether the benefits of enclave development will be equitably distributed to the local community. Respondents were also rather skeptical regarding the economic benefits spreading to the grassroots in the long-term due to poor tourism planning. Respondent H2 stated that "every government in power puts forward a new policy. For example, a government says that we need to target 2 million tourists. If life continues to become expensive, there will not be tourists". Moreover, respondent H5 stated that natural resources have been exploited for resorts to be economically viable. "We need to know where to stop though and identify what we want as tourists and tourist numbers. We should not have overexploitation of resources with enclaves". There is also a lack of consistency in tourism development as there is no long-term planning due to the new agenda of newly elected governments and the lack of continuity resulting in poor strategy for tourism development.

## Conclusions

Enclave tourism in Mauritius contributes positively to community wellbeing mainly through the economic, social and educational domain. Although enclave tourism contributes to employment, there is a lack of economic opportunities for small entrepreneurs who expressed their daily hardships and lack of control over tourism environment. The host community also expressed that enclave tourism had resulted in costs and sacrifices. They were highly concerned with the all-inclusive packages, which seriously threaten the economic benefits to spread to the grassroots. Thus, this study confirms the findings of Sharpley (2002) that tourism development may reinforce imbalances in power since enclave tourism perpetuates power of elites instead of reducing socio-economic gaps. Moreover, access to leisure was restricted for locals. Although the Mauritian government promotes tourism and sustainable development, the residents criticised the government for its lack of strategic planning and sound policy. It is suggested that the government has to play a more active role by providing a suitable political environment for enhancing community wellbeing. A stakeholder approach with inputs of the host community amongst others is highly desirable to further understand the role that enclave tourism can play to harness community wellbeing.

## References

Andereck KL, Valentine KM, Vogt CA, Knopf R (2005) Residents perceptions of community tourism impact. Ann Tour Res 32(4):1056–1076

Andriotis K (2003) Tourism in crete: a form of modernisation. Curr Issues Tour 6(1):23–53

Andriotis K (2008) Integrated resort development: the case of Cavo Sidero, Crete. J Sustain Tour 16(4):428–444

Barber BK, McNeely C, Spellings C (2012) Role of political factors in wellbeing and quality of life during long-term constraints and conflict: an initial study, vol 380, Supplement 1, Available at http://www.sciencedirect.com/science/article/pii/S0140673613601993#. Accessed on 30 June 2015

BBC (2006) Britain's Happiness in Decline, Available at http://news.bbc.co.uk/2/hi/programmes/happiness_formula/4771908.stm. Accessed on 30 June 2015

BBC (2012) BBC News—Well-being and education "go together". Available at http://www.bbc.co.uk/news/education-18724993. Access July 2013

Cohen E (1988) Authenticity and commodification in tourism. Ann Tour Res 5:371–386

Diener E (eds) (2009) Assessing wellbeing: the collected works of Ed Diener. Springer, Illinois

Edensor T (2001) Performing tourism, staging tourism: (re)producing tourist space and practice. Tourist Stud 1(1):59–81

Fabrique Spinoza (2015) Actualités. Available at http://fabriquespinoza.fr/wp-content/uploads/2012/03/CSA-pour-La-Fabrique-Spinoza-1200-205-Les-Fran%C3%A7ais-et-le-bonheur.pdf. Accessed on 30 June 2015

Ferrans CE (1996) Development of a conceptual model of quality of life. Res Theor Nurs Pract 10 (3):293–304

Guba E (1990) The paradigm dialog. Sage, Beverly Hills, CA

Hardy A (2005) Using grounded theory to explore stakeholder perceptions of tourism. J Tour Cult Change 3(2):108–133
Helliwell J, Layard R, Sachs J (eds) (2012) World happiness report. Available at http://www.earth.columbia.edu/sitefiles/file/Sachs%20Writing/2012/World%20Happiness%20Report.pdf. Accessed on 30 June 2015
Hernandez S, Cohen J, Garcia HL (1996) Residents' attitudes towards an instant resort enclave. Ann Tour Res 23(4):775–779
Hill A, King E (1995) Women's education in developing countries: barriers, benefits, and policies. Johns Hopkins University Press, Baltimore
Liang J, Yamashita T, Brown JS (2013) Leisure satisfaction and quality of life in China, Japan, and South Korea: a comparative study using. J Happiness Stud 14:753–769
Liu A, Liu J, Liu M (2007) Progress in enclave tourism study of overseas: a literature review, Chinese. J Popul Res Environ 5(3):76–81
Lorde T, Greenidge D, Devonish D (2011) Local residents' perceptions of the impacts of the ICC Cricket World Cup 2007 on Barbados: comparisons of pre- and post-games. Tour Manage 32 (2):349–356
Mbaiwa JE (2005) Enclave tourism and its socio-economic impacts in the Okavango Delta, Botswana. Tour Manage 26:157–172
McElroy JL, De Albuquerque K (1998) Tourism penetration index in small Caribbean Islands. Ann Tour Res 25(1):145–168
New Economics Foundation (2015) the importance of measuring wellbeing. Available at http://www.neweconomics.org/blog/entry/the-importance-of-measuring-wellbeing. Accessed on 30 June 2015
OECD (2013) Measuring wellbeing for development, global forum on development. Available at file:///C:/Users/Jodi/Documents/articlejcu/wellbeing/Session%203.1 %20-%20GFD%20Background%20Paper.pdf, Accessed on 30 June 2015
Prayag G, Dookhony-Ramphul K, Maryeven M (2010) Hotel development and tourism impacts in Mauritius: hoteliers' perspectives on sustainable tourism. Dev South Afr 27(5):697–712
Sharpley R (2002) Rural tourism and the challenge of tourism diversification: the case of cyprus. Tour Manage 23(3):233–244
Sharpley R (2003) Tourism. Modernisation and development on the island of cyprus: challenges and policy responses, journal of sustainable tourism 11(2&3):246–266
Sharpley R, Naidoo P (2010) Tourism and poverty reduction: the case of mauritius. Tour Hospitality Plan Dev 7(2):145–162
Sharpley R, Ussi M (2012) Tourism and governance in small island developing states (SIDS): the case of Zanzibar. Int J Tour Res. doi:10.1002/jtr.1904
Shaw BJ, Shaw G (1999) Sun, sand and sales: enclave tourism and local entrepreneurship in Indonesia. Curr Issues Tour 2(1):68–81
Veenhoven R (2012) Cross-national differences in happiness: cultural measurement bias or effect of culture? Int J Wellbeing 2:333–353
Wall G (1996) Rethinking impacts of tourism. Prog Tour Hospitality Res (3/4):207–215
Weed M (2005) Research quality considerations for grounded theory research in sport and exercise psychology. Psychol Sport Exerc 10:502–510

# Chapter 19
# An Exploratory Study on the Internal Impacts of Hotel CSR Activities in Bali

**Trianasari and Gede Adi Yuniarta**

**Abstract** Corporate Social Responsibility (CSR) has been a central topic within both the academic area and business practice. Academically, this topic has gained much attention through continuous research. Similarly, an increasing practice of CSR by organizations has been evident. However, while it is important that companies demonstrate their concern on CSR activities, it is essential to understand if there are impacts for the companies internally. Limited study has viewed CSR from this point, especially within the hotel context. This paper reports on the results of an initial study that was part of a larger study aiming to examine the relationship between CSR activities and brand image. At its preliminary stage, the research elicited the respondents' views on CSR activities. Their views were then compared to identify if there are differences. Furthermore, an exploration was undertaken on the internal impacts of CSR program that the hotels face. The study was conducted in Bali, involving twenty hotel managers of 3–5 star hotels. Data were collected using interview method and were analyzed using qualitative approach. The results of the study showed that all hotel managers emphasized the importance of CSR, although their degree of understanding and knowledge varied especially between different hotel star ratings. The majority of the respondents emphasized that CSR activities were programmed in order to meet the government's requirement. Furthermore, there were eight themes representing the internal impacts of CSR obtained from the data. All of the themes are considered to be favorable impacts and most of them give new insight to the CSR literature as have not been published in the previous studies. They are trust, image, protection, improved closeness to, understanding and knowledge of nature, environment, and society, pride, mouth-to-mouth advertisement, recognition, and profit. The most frequent theme was mainly in terms of increasing brand image. The majority of the themes have not

Trianasari (✉) · G.A. Yuniarta
Faculty of Economics and Business,
Ganesha University of Education, Bali, Indonesia
e-mail: nanatrianasari01@gmail.com

G.A. Yuniarta
e-mail: gdadi_ak@yahoo.co.id

 
A. Saufi et al. (eds.), *Balancing Development and Sustainability in Tourism Destinations*, DOI 10.1007/978-981-10-1718-6_19

been found in the CSR literature, suggesting a new understanding. Having these results, the follow-up study which was designed as the main study is to be undertaken. That is to statistically examine the link between CSR and brand image. Future research may also investigate the relationships between CSR, brand image, profitability, and sustainability.

**Keywords** CSR · Environment · Hotel · Brand image · Marketing · Society

## Introduction

Corporate Social Responsibility (CSR) seems to be a familiar term for both researchers particularly within the area of social science and business practitioners. Indeed, academia and managers have given increasing attention on this topic particularly related to the implementation of CSR (McWilliams et al. 2006). Within the academic realm, CSR has been considered an important area of research (Lindgreen and Swaen 2010). A great amount of research has been conducted within various business sectors. However, despite the growing interest in this topic, limited studies have focused on CSR within the context of hotel.

Similar to the academic point of view, from the business practice aspect, organizations have more concerned and commitment to implementing CSR, although many face challenges in their efforts (Lindgreen and Swaen 2010). One of the possible reasons of why many organizations struggle in implementing CSR may relate to their finance ability. Indeed, it may be arguable that higher profitability is likely to be in line with higher commitment on CSR activities. That is, a more profitable organization may spend more money on CSR activities compared to those with less profitability. Interestingly, there have been a few studies that have examined the correlations between CSR and finance performance in a different direction. As an example, McWilliams and Siegel (2000) have investigated how CSR activities affect financial performance. However, their study showed mixed results which are negative, positive, and neutral correlations. Such results suggest that the influence of CSR on profit or financial aspect is inconsistent and thus, remains unclear.

Looking into CSR from a slightly different angle, this study aimed to explore the impacts of CSR activities internally, within the hotel sector in Bali. It first answered the question of how hotel managers perceive CSR. Exploring the degree of importance of CSR in the eyes of hotel manager combined with leadership style leads to further understanding of the strategic planning an organization of CSR activities (Waldman et al. 2004). It may not ignore the possibility that some managers have negative personal view of CSR such as a burden, especially when it is implemented by the government. Second, the study explored whether the increasing number of organizations paying attention to CSR activities are solely the result of the Government's regulation. In other words, this study helps to understand whether hotel CSR actions are merely a fulfillment of the government obligation or there are other purposes or agenda. If so, it may not entirely serve the real purpose of CSR. That is, the hotel CSR activities may implicitly focus on marketing campaign or improving a brand image. If that is the case, the genuinity of helping community and preserving culture and the environment may be questionable. Thus, up to this point, the pure motivations of CSR implementation can be concluded. Next, this study elicited understanding of the impacts of CSR internally within the organization. The results of this study are used as the foundation of a larger study focusing on the relationship between CSR and a brand image. The rationale of this study is that CSR research within the context of a hotel has been lacking. Among the limited studies, Trianasari and Yuniarta (2014) explored the implementation of hotel CSR in Bali which has not answered the impacts of CSR on a brand image. Thus, to a certain degree, this study enhances the current understanding of CSR within hotel contexts.

The remainder of this paper presents briefly a review on relevant literatures, followed by research methodology, findings and discussion, and limitation and conclusion.

## Literature Review

This section provides a brief overview on CSR literature. It starts with the definition of CSR followed by its impacts.

**What is CSR?** CSR has become a central topic within the literature of social sciences, although it has been introduced for several decades (see for example, Friedman 1970). As largely known, the term stands for CSR. In the literature, there are various definitions of CSR which in fact have similar meanings. One of the simple ways to define CSR is by referring to the meaning of each word in it. That is, the responsibility of corporate for social aspects. In other words, it is a support an organization offers to society to show its responsibility. In a more profound manner, CSR is described as “situations where the firm goes beyond compliance and engages in ‘actions that appear to further some social good, beyond the interests of the firm and that which is required by the law.” (McWilliams et al. 2006). CSR is one of the ethical applications within the business operations that should address the society, environment, or nature.

As occurring globally in Indonesia, CSR has gained increasing popularity following the inclusion of CSR programs in the Indonesian government rules and regulations. That is, organizations are obliged to perform CSR activities. However, it has not yet been uniformly implemented by all organizations registered in Bali. Such an obligation originates from the fact that organizations can involve in the positive development of a society where the organizations are located. At a certain point, the existence of an organization gives impacts to the society although the degree of their influences may vary from one organization to another.

## *The Implementation of CSR*

Previous research has shown that the implementations of CSR vary from one organization to another. Among the limited studies, Trianasari and Yuniarta (2014) have classified seven areas in which hotels in Bali implement their CSR program. They are environment, energy, health, and safety in the workplace, labor, product, community involvement, and miscellaneous. Little attentions were given on the perseverance of culture, and religious and historical tourist objects. Furthermore, it was found that the implementation of CSR program was mostly held during the anniversary of hotels.

However, while it is essential for an organization to show social responsibilities, the core issue may not be about the obligation to perform CSR program. Most notably, the question of how far an organization should implement CSR program should be addressed (Untung 2014). In other words, the key point in CSR is about how much an organization should spend or how deep an organization should care and involve itself in supporting society and environment.

## *The Impacts of CSR*

As presented earlier in this paper, organizations seem to experience difficulties in their implementation of CSR (Lindgreen and Swaen 2010). Perhaps, such experiences relate to some consequences of CSR implementation. The first and foremost impact may be the reduction of shareholders' wealth (Friedman 1970 as cited by McWilliams et al. 2006). This impact is likely to already be anticipated following the organizations' commitment. In other words, when organizations commit themselves, it can be assumed that they are ready to allocate some CSR funds. However, some authors have attempted to study the relationship between CSR and financial performances in a different direction. For example, McWilliams and Siegel (2002) have found mixed results of neutral, positive, and negative relationship between CSR and financial performance. As such, these findings are considered inconsistent and need further empirical studies. In their previous work, these authors stressed that ideally, the level of CSR should refer to the cost–benefit analysis (McWilliams and Siegel 2001). Furthermore, in the long term, CSR has been found to have relationship with profitability (Robledo and Batle 2002).

Next, the impacts of CSR can be seen from other managerial aspects. For example, when supported by political strategies, firms can use CSR strategies as sustainable competitive advantage (McWilliams and Siegel 2002). Furthermore, for certain organizations, their CSR toward environment may constitute more resources or capability that leads to sustainability (Hart 1995). A good environment management helps to protect and sustain resources and thus, the business itself. Moreover, in terms of marketing, there is a possibility to create or improve a positive image (Lansing and De Vries 2007). It can be expected that having more publications of kind and caring acts to society via CSR program, a company may in turn receive benefit such as a good or better image. The improvement of an image is likely to increase the public or potential customers' awareness and thus, business. However, another question may need to be addressed, that is, whether CSR is a pure or just a marketing campaign (Lansing and De Vries 2007). Furthermore, organizations can attract a socially responsible customer (Baron 2001). However, it may be hard to determine what types of CSR attributed to be put in a product that such customers are willing to pay (McWIlliams et al. 2006).

## Research Design

The study employed a qualitative approach in which a semi-structured interview was used as the method of data collections. There were 20 respondents participated in the study. They were Hotel Managers and Human Resources Managers at several hotels in Bali. The ratings of the hotels range from 3 to 5 star hotels. Consents were obtained prior to their participation. The data were analyzed using qualitative method. First, the raw data were coded. The main purpose of the coding was to identify themes related to the research questions. The themes derived were then grouped based on their relevance or similarity or closeness in terms of their meaning. The data analyses were started by searching for patterns and ideas represented by the patterns. During the analysis process, themes developed were modified as more ideas found. Results of the data analysis were interpreted, defined, and described.

The next section presents the findings of the study and discussion.

## Findings and Discussion

There were two research questions that this study addressed, they are: "How hotel managers perceived CSR?" and "What were the impacts of CSR that the hotels receive internally?" Findings of the study are discussed below.

### *Managers' Personal View of CSR*

The managers' personal view on CSR was sought in order to understand if there were differences especially between respondents of different hotel star rating. Then, their views were contrasted with the implementation of CSR activities. As the role of manager is important in the planning, organization, and implementation of CSR program, it is arguable that, to certain degree, the implementation of CSR depends on the understanding of the managers of CSR (Waldman et al. 2004). The analysis revealed that all respondents stated the importance of CSR activities. The majority of the respondents demonstrated sufficient knowledge and understanding of CSR. To be more specific, the respondents who represented 5-star hotels showed deeper knowledge than those of lower star hotels. Using similar expressions, some respondents explained that CSR constituted actions performed by organization to show their responsibilities for society and environment that has been implemented by the Indonesian Government. The field notes taken during the interview suggest an emphasis on the government's rule and regulation. This part of findings may imply the relatively strong consideration of government. In addition, it was found that the 5-star international chained hotels have a more clear guidance within their organization and more variation of CSR activities. Furthermore, there seems to be a link between CSR and financial abilities. A female respondent working at a 3-star hotel emphasized that the CSR program depended on the hotel's budget. She further explained that when the business is low, the hotel may decide not to spend money on CSR programs. Similar comments were obtained from other respondents. This finding is in line with the work of McWilliams and Siegel (2001). Also, it partly supports the work of Friedman (1970) in the way that CSR program requires financial resources. The inability to steadily allocate CSR funds can result in inconsistent implementation of CSR. Such situation reflects the statement of Lindgreen and Swaen (2010) with regard to the difficulties to implement CSR.

The implication of this finding is that CSR activities are important. Especially, with the government regulation, CSR cannot be neglected, although there was still an unclear guidance about consequences if hotels fail to do so and about the ideal amount of resources (financially) the hotels should invest. Next, with the understanding and knowledge of CSR the manager acquired, the implementation of hotel CSR seemed to have served the purpose. However, as also found in the work of Trianasari and Yuniarta (2014), the selection of areas to be included in hotel CSR programs needs to be guided by the government policy. As such, overlapping support for certain areas on one hand and lacking support for other needy areas on the other hand, can be minimized or avoided. In other words, the distribution of CSR supports can be more equalized.

## *The Internal Impacts of CSR*

In this study, the operational definition of internal impacts is the consequence of CSR activities that the hotels under study bear internally. There are eight themes gathered from the analysis of the data. All of the themes are considered to be pleasant impacts and most of them give new insights to the CSR literature as they have not been published in the previous studies. They are trust, image, protection, improved closeness to, understanding and knowledge of nature, environment, and society, pride, mouth-to-mouth advertisement, recognition, and profit. Each of the themes is described as follows.

### Image

Image is the most frequent theme elicited in the interview data. All respondents especially stated that this theme as being an important impact of CSR performance. This finding is in line with the work of Lansing and De Vries (2007). In fact, positive image can be directly stored in the mind of society who receives CSR support, whereas for potential customers, public, or other stakeholder, the image can be built indirectly via advertorials or publications in the printed or electronic media. Thus, it can be expected that the more CSR activities performed and the more advertisement created will result in a more positive image.

### Trust

Some respondents put forwarded trust as being one of the impacts the hotels face following CSR activities. This theme can be obtained from the local people, the government, employees, as well as potential customers, other stakeholders, and the public. It was explained that the more CSR program undertaken, the more trust will be gained by the hotels. Furthermore, one respondent commented that hotels that perform regular CSR activities may be seen as being considerate, well organized, and trust worthy.

### Protection

A few respondents emphasized that hotels often gain protection from the local society through their CSR program. They think that at a certain point, CSR activities give room for increasing level of the local people's sense of belonging. That is, the local people with whom the hotels engaged in CSR program would become more familiar with the existence of the hotel and may consider that the hotels are parts of their belonging. An example that respondents further described was that the local people help hotels when there are unusual situations occurring such as accidents, fire, or thieves entering the hotel area. Thus, improvement of security can be gained with the help of the local community.

### Improved Closeness/Relationship, Understanding and Knowledge of Nature, Environment, and Society

Through the CSR activities, respondents stated that the hotel employees increase their closeness or relationship to the nature, environment, and local society, especially when the CSR actions become a traditional program for the same objects or recipients. The respondent further explained that the feeling of closeness to a certain degree improves their understanding and knowledge. This improvement may lead to higher awareness and protection which can be demonstrated in the employees' daily and personal and professional lives. For example, they may save power, energy, and water when not used or they adopt the reducing, reusing, and recycling concept when it comes to garbage, and many more. This finding is consistent with the work of Hart (1995).

### Pride

Pride is another theme drawn from the data. Respondents explained that society acting as the object of CSR program often show their pride as being considered by the hotel. For example, children in a village who receive shoes, school uniforms, hats, and other packages, commonly express that they are proud of their new stuff. Another example is the village leaders often show off their new facilities such as a public toilet, library, and many more that they received from hotels. Such feeling can be taken as a compliment by the hotels. In light of pride and its relationship to CSR, no study has specifically shown it in the literature. Thus, this finding enhances the understanding of CSR, although it may require further empirical work.

### Words of Mouth Advertisement

The respondents described that this type of advertisement can be voluntarily aired by an institution, the local government, and society, who receive support from CSR activities. However, the impact of such promotional method may not provide immediate results. In the literature, there are no studies specifically investigating the relationship of CSR to mouth-to-mouth advertisements. Thus, this finding gives new insights within the CSR literature.

### Recognition

Although not commented by all respondents, recognition is another theme that emerged from the data. By performing CSR activities, the hotels may gain recognition especially from those who receive the CSR support such as the society, hotel schools, and the government. For example, the hotels may be recognized as obeying the rules and regulations specified by the government. Moreover, CSR activities may increase the interest of hotel schools not only in Bali, but also outside of Bali and Indonesia to set out an agreement with regard to students apprentice program or job opportunity.

### Profit

The majority of respondents stressed profits as being one of the CSR impacts the hotels were likely to enjoy. They rationalize that the advertisement of CSR activities may attract more customers especially those who have interests in ethically operated companies. Although it may affect in the long term, this impact cannot be ignored. This part of the findings partly supports the study undertaken by McWilliam and Siegel (2002) and is consistent with Robledo and Batle (2002), although, empirical studies using a quantitative approach is required to test its significance.

To summarize, the implication of these results is that CSR programs and activities provide favorable impacts on the hotels. Thereby, while it is important to show social and environmental responsibilities in the existence of the hotels, it is also a benefit to bear the pleasant impacts of CSR performance. Indeed, CSR actions are reflections of the idea to operating business in an ethical manner (Trianasari et al. 2013). However, most notably, a commitment to engaging in CSR activities involves the shares of financial aspects which hotels should plan in advance.

## Limitation

There are limitations of this study to be acknowledged. First, the results cannot be generalized and applied into wider population as the nature of a qualitative approach. The next limitation is in terms of sample. That is, the hotels employed as sample were not in the same categories of star ratings. The major difficulty admitted here was in collecting data especially from 4 or 5 star hotels. Nevertheless, in qualitative study, focus is often given more to the authenticity of the data rather than to its reliability (Silverman 2001). This study served as an initial type of study in order to mainly elicit understanding if there are impacts of CSR internally. Thus, qualitative study is deemed suitable for such a research purpose.

## Conclusion

The results of the study reported in this paper have improved the understanding of how CSR activities give impacts to hotel internally. There were eight themes serving as the impacts of CSR derived from the study. They were trust, image, protection, improved closeness to, understanding and knowledge of nature, environment, and society, pride, mouth-to-mouth advertisement, recognition, and profit. More specifically, respondents perceive favorable impacts from CRS activities particularly on their hotels' brand images. While such result gives a new insight, an examination using quantitative approach is yet essential to test whether the impact is significant or not. Thereby, a follow-up work is required. Furthermore, future study may focus on investigating real motives of organizations' CSR activities. Also, a study that integrates the relationship between CSR, brand image, profitability, and sustainability may be advantageous to deepen the understanding of CSR especially within the hotel setting.

## References

Baron D (2001) Private politics, corporate social responsibility and integrated strategy. J Econ Manage Strategy 10:7–45

Budi U (2014) CSR dalam Dunia Bisnis. Penerbit Andi, Yogyakarta

Friedman M (1970). The social responsibility of business is to increase its profits. New York Times Magazine, 13 Sept

Hart S (1995) A natural resource-based view of the firm. Acad Manage Rev 20:986–1014

Lansing P, De Vries P (2007) Sustainable tourism: ethical alternative or marketing ploy? J Bus Ethics 72:77–85

Lindgreen A, Swaen V (2010) Corporate social responsibility. Int J Manage Rev 12(2):1–7

McWilliams A, Siegel D (2000) Corporate social responsibility and financial performance: correlation or misspecification? Strateg Manage J 21(5):603–609

McWilliams A, Siegel D (2001) Corporate social responsibility: a theory of the firm perspective. Acad Manage Rev 26:117–127

McWilliams A, Siegel D (2002) Corporate social responsibility and financial performance: correlation or misspecification? J Manage Study 21:603–609

McWilliams A, Siegel DS, Wright PM (2006) Corporate social responsibility: strategic implications. J Manage Study 43(1):1–18. doi:10.1111/j.1467-6486.2006.00580.x

Robledo MA, Batle J (2002) Re-planning for tourism in a mature destination: a note on Mallorca. In: Voase RN (ed) Tourism in Western Europe: a collection of case histories. CABI Publishing, Wallingford

Silverman D (2001) Interpreting qualitative data: methods for analysing talk, text and interaction, 2nd edn. Sage/Thousand Oaks, London/New Delhi

Trianasari, Yuniarta (2014) The implementation of hotel CSR funds: an exploratory study in Bali. In: Paper presented at the international conference on business, marketing, and management, 8–12 June 2014, HongKong

Trianasari, Yudiaatmaja F, Dharma A (2013) Konsep Tri Hita Karana dan kaitannya dengan Etika Bisnis: sebuah kajian konseptual. In: Presented at the 6th annual doctorate colloquium and conference. Gajah Mada University, Yogyakarta

Untung B (2014) CSR dalam dunia Bisnis. Yogyakarta, Andi

Waldman D, Siegel D, Javidan M (2004) CEO transformational leadership and corporate social responsibility. In: Working paper, Rensselaer Polytechnic Institute

# Chapter 20
# An Exploratory Study of Tourists Photo Taking Behavior at Nglanggeran Ancient Volcano, Gunungkidul, Indonesia

**Handayani Rahayuningsih, Kwartarini Wahyu Yuniarti and Tri Kuntoro Priyambodo**

**Abstract** "No pict is hoax" is a statement usually used by young people, which means that pictures need to be taken to provide a sense of meaningfulness to their stories. Generally, tourists use pictures to share their experiences from the tourist destination with their relatives, friends, and communities. In this digital era, there is no real need to share the photographs in printed version. Through the Internet, people can share their moments with all of their acquaintances in social media. This paper reports on research undertaken into tourists photo taking behavior at Nglanggeran Ancient Volcano. It finds that respondents take some pictures as part of self-actualization. Tourists mainly use camera in their mobile phones in preference to other photographic devices, which they will typically share via social media. The result shows that most respondents take pictures and share them in a single device that is mobile phone.

**Keywords** Tourists behavior · Photography · Mobile phone camera · Social media · Pictures · Travel evidence · Picture sharing

---

H. Rahayuningsih (✉)
Universitas Gadjah Mada, Yogyakarta, Daerah Istimewa Yogyakarta, Indonesia
e-mail: handayani.rahayuningsih@ugm.ac.id

K.W. Yuniarti
The Faculty of Psychology, Universitas Gadjah Mada, Yogyakarta, Daerah Istimewa Yogyakarta, Indonesia
e-mail: kwartarini_yuniarti@yahoo.com

T.K. Priyambodo
Graduate School, Universitas Gadjah Mada, Yogyakarta, Daerah Istimewa Yogyakarta, Indonesia
e-mail: mastri@ugm.ac.id

 
A. Saufi et al. (eds.), *Balancing Development and Sustainability in Tourism Destinations*, DOI 10.1007/978-981-10-1718-6_20

## Introduction

Destination is the most important part of tourism. Whether the place is iconic or recommended by friends, the activity of traveling requires visitors to essentially make a trip to the destination, otherwise it may not be called "to travel" (Urry and Larsen 2011).

Photography and tourism are intrinsically connected (Sheungting et al. 2011). Pictures can be a document and a form of experience during the journey (Haldrup and Larsen 2003; Larsen 2008). It may also provide the opportunity for tourists to share experiences with others (Groves and Timothy 2001). In addition, tourists take pictures to produce a memory that can be seen and enjoyed after the trip. Through pictures, tourists try to make the experience of traveling last longer (Urry and Larsen 2011).

Since its discovery in the 1840s (Urry and Larsen 2011), photography has always been associated with traveling. Consequently, photography can no longer be separated from tourism (Larsen 2006). For tourists, photography can connect the pictures of the tourist destination with the memory of visiting the destination. Therefore, for many people, traveling is not complete without carrying a camera. This is especially true nowadays with the idea of "no pict = hoax" trending among young people. The idea argues that the absence of pictures to support one's story implies that the story lacks of validity. The picture is the evidence that tourists use to describe their experience of visiting a destination (Ntarangwi 2000).

Photography plays different roles in tourism. The role that we can see the most is to prove that a person has directly been to a destination (Hillman 2007); and to provide evidence that a person has experienced various forms of life that are authentic, exotic, and different from the daily routine (Chalfen 1979).

Urry (2002) reveals a theory about people gaze. People gaze is a picture of a destination that is formed before seeing it in person. According to him, when traveling, people tend to refer to the brochure or postcard with the aim of taking the same picture that attracts them to visit the destination. This indicates that most tourists are concentrated on a very limited area.

In contrast to the theory put forward by Urry (2002), Haldrup and Larsen (2003) examine family gaze, offering tourists an example that photography can produce not only personal pleasure, but also social relationships. The family used a camera to show success, togetherness and love, capturing memories, personal narrative, and identity. These authors argued that the most important part of taking pictures is to build, to strengthen, and to show family relationships.

With the development of technology, there are currently many different types of cameras that can be used to capture the unforgettable moments, such as digital camera, video cameras, and Polaroid cameras. Tourists no longer had to wait for the pictures to be printed to see and share their moments, as modern technology has allowed us to directly see the camera shots, delete unwanted pictures, and share them through their mobile devices. Social media such as Facebook, Twitter, Path,

Instagram, blogs, etc., have enabled users to instantly select which pictures they would like to share with their peers and family members.

Indonesia too has begun to embrace the power of photos to market new emerging tourist destination through the Internet. An example of destination with special interest is available in Gunungkidul, Yogyakarta, namely the Nglanggeran Ancient Volcano. Many people have visited the area since the site manager posted attractive pictures of the ancient volcano in the social media. Nowadays, the social media is followed by thousands of tourists and most of the pictures of the Nglanggeran Ancient Volcano available on the Web are uploaded by them.

Unfortunately, there is a limited number of studies that reviews and proves the existence of a close relationship between tourist destination development and photography, especially in terms of tourists photo taking behavior. Therefore, it was incredibly crucial to conduct this study as such information is invaluable in providing assurance and consideration for the stakeholders in formulating development decisions.

## Theoretical Framework

There have been many studies revealing that photography is the main attraction in the field of tourism (Gillespie 2006; Gillet et al. 2013; Hunter 2007; Larsen 2005). One study that reviewed the behavior of tourists in photo taking is conducted by Prideaux and Coghlan (2010) located at the Great Barrier Reef, Australia. Prideaux and Coghlan explained the differences in the photo taking behavior of socio-demographic tourists, as well as how the digital pictures taken by tourists can become a marketing opportunity for tour operators. However, these marketing opportunities are more likely done by tourists, as the author did not mention how the tour operator or manager may utilize such opportunity.

Sheungting et al. (2011) linked tourism and tourists photography by focusing on the online picture sharing activities and experiences among Hong Kong residents. This study used a quantitative approach to profiling method. The results showed that 89 % of the population takes pictures while traveling, and 41 % of them upload their pictures to the internet.

A study of tourism located in Nglanggeran Ancient Volcano done by Masruri (2014) examined the feasibility of Nglanggeran Ancient Volcano Ecotourism Region as destination (79.21 %), with 43.71 % satisfaction rating (somewhat satisfied). The formulation of development strategies were based on SWOT analysis.

Although the research located in Nglanggeran Ancient Volcano has already been done, exploratory study on tourists photo taking behavior is the first study in this location. Moreover, while the previous study used a quantitative approach to calculate the feasibility level of Nglanggeran Ancient Volcano as a tourist attraction, this study uses a qualitative approach to examine tourists' photo taking behavior.

## Method

This research is a qualitative research in the form of exploratory study. Exploratory study is employed as the researcher has limited information about tourists' photo taking behavior due to the lack of research examining this issue. This study is required in order to understand the phenomenon of tourists' photo taking behavior at Nglanggeran Ancient Volcano.

The research took place in one of the special interest attractions in Gunungkidul, namely Nglanggeran Ancient Volcano. The ancient volcano is selected as the research location because the destination's popularity is highly supported by the pictures uploaded by local residents in social media. In addition, there is a mini gallery exhibiting some interesting pictures collection of Nglanggeran Ancient Volcano, which manages to attract more visitors.

Respondents in this study are the informants who testify through group interviews, observation, and documentation. Selection of respondents is conducted by purposive method to obtain credible subject, in accordance with predetermined criteria (Strauss and Corbin 1990).

Primary data is obtained by group interviews, observation, and documentation. In this study, the participant observation method is carried out by following the activities of the respondents and observing the respondents' activities. Through group interviews, respondents have the opportunity to listen to the contributions of other respondents, as it can trigger a new understanding and cause the subjects to develop their ideas more clearly (Ghony and Almanshur 2012).

## Tourists' Motivation in Taking Pictures

Based on the research results, there are only 5.9 % of the respondents who work professionally in the field of photography. This shows that most tourists who take pictures during the field research at Nglanggeran Ancient Volcano do not have any occupational, professional, or job-related purposes.

The field research found there are 20.6 % of respondents with a hobby in photography. This proves that the reason visitors take pictures in the tourist attraction has less to do with the activity of photography itself and indicates other motivations that need to be examined further.

The results reveal a deeper understanding on the tourists' behavior as more than half of them (55.9 %) do regularly take pictures in everyday life, despite not having any interest in photography. In relation with photo taking behavior in the destination, it can be argued that visitors take pictures of unusual and interesting experiences that are different from their daily routines.

Almost all respondents agree that the pictures taken at the site are to be shared to their friends and families, in addition to keeping them for personal documentation as evidence that they have visited the site, capturing the moment, testing photographic techniques, and acquiring photo stock.

## Tourists Photo Taking Behavior

There are four types of cameras used by the respondents, namely pocket camera, DSLR/Mirrorless, Go Pro, and mobile phone cameras. The result from the field research shows that mobile phone camera is the type of camera mostly used by respondents, i.e., 26 people out of 34 respondents (76.5 %).

The main reason for the respondents to take pictures with the mobile phone camera is because of its practicality. Mobile phones have small size and therefore are highly portable and appropriate with the trip up to the Nglanggeran Ancient Volcano. In addition, the mobile phone camera is fairly easy to use and has a variety of features that help users to get the desired image.

There are two properties revealed in pre-research and field research, namely the tripod and monopod. Usually, tourists in a group bring a tripod or monopod so that they can take pictures for all group members. However, most respondents did not use any property to help them take pictures.

The results show that 32.3 % of respondents prefer to capture the landscape as the object; while 26.5 % prefer a combination of landscape and people; and 41.2 % prefer taking pictures of not only other people, but also of themselves.

In regard to the landscape of the site, despite many interesting spots that can be photographed, the peak is the most preferred tourist spot for photo taking. From the top, visitors are able to see the full view surrounding Nglanggeran Volcano, including Mount Merapi in the north, Wonosari in the east, and the Imogiri hills in the south. Moreover, the peak is the best spot to watch the sunrise.

A theory on people gaze by Urry (2002) applies seamlessly to the Nglanggeran Volcano case. Urry (2002) argues that visitors who have previously seen brochures or postcards of the tourist attraction spots tend to avoid copying exactly what they see in the promotional photos, as they prefer to create their own perspectives. On the contrary, the theory proposed by Albert and James (1988) stating that brochures and postcards have made visitors project a certain expectation on the pictures they take is not entirely relevant to the current study, as proven by how the selection of objects the tourist prefer tend to be incidental.

According to several respondents, there are particular moments or conditions that they have marked in order to produce the best pictures in Nglanggeran Ancient Volcano. They recommend to take pictures on sunrise and sunset; at about 8 o'clock in the morning when the weather is nice and sun can be seen through the clouds; when there are few people; at night with background of the city lights; and at the peak. Moreover, it is also important to take into account when to avoid certain times to ensure a good photograph in Nglanggeran Ancient Volcano, such as during rainy seasons or other unfavorable weather conditions, and on the peak season where the site is packed with crowds of visitors.

## Things to Do After Taking Pictures

There are 5.9 % of respondents or 2 people who prefer to not share the pictures taken with their relatives, while among the respondents who share the images with others, 88.2 % will upload their pictures to social media.

There are several social media that the respondents use as a means to upload the pictures, including *Whatsapp, Line, BBM, Facebook, Path, Instagram,* www.1X.com, www.viewbug.com, www.flickr.com, and www.pixoto.com. Moreover, it is imperative to point out how social media reflect the respondents' occupations. Professional photographers and photographers with deep interest in photography upload their shots to photography discussion forum such as www.1X.com photography, www.viewbug.com, www.flickr.com, and www.pixoto.com in order to obtain critics and encouragements so they can improve their knowledge and quality of the photographs. Meanwhile, other respondents prefer to upload their pictures on social media such as *Whatsapp, Line, BBM, Facebook, Path, and Instagram.*

There are 44.1 % of the respondents who prefer to edit pictures before uploading them to social media in order to make them look more appealing, while 55.9 % preferred a non edited pictures to show a more natural, untouched quality of the scenery.

## Factors Behind Tourists Photo Taking Behavior

30.8 % of the male respondents prefer the landscape as the object, while the other 30.8 % prefer people. The rest of the male respondents prefer both, whereas most female respondents (19.3 %) prefer people as the object. This indicates the selection of the object based on gender Table 20.1.

The differences also occur in the selection of objects respondents at different age levels. In the age range 19–24 years and 30–34 years, most of the respondents prefer people as the object, while in the age range 25–29 years; most of them love both landscapes and people. Those who are beyond 34 years prefer the landscape as object photography. This suggests that different age range is significant to the objects selected (Table 20.2).

**Table 20.1** Photography object choice based on socio-demography

| Socio-demography | | Photography object (%) | | | | | |
|---|---|---|---|---|---|---|---|
| | | Pre-study | | | Field research | | |
| | | Landscape | People | Landscape and people | Landscape | People | Landscape and people |
| Gender | Male | 12.5 | – | 50 | 30.8 | 30.8 | 3.8 |
| | Female | 12.5 | – | 25 | 3.8 | 19.3 | 11.5 |
| Age | 19–24 | 12.5 | – | 12.5 | 19.3 | 26.9 | 7.7 |
| | 25–29 | 12.5 | – | 62.5 | 3.8 | 3.8 | 7.7 |
| | 30–34 | – | – | – | 7.7 | 19.3 | – |
| | >34 | – | – | – | 3.8 | – | – |
| Occupation | Student | 12.5 | – | 62.5 | 19.3 | 30.8 | 15.4 |
| | Employee | 12.5 | – | – | 3.8 | 19.3 | – |
| | Entrepreneur | – | – | – | – | – | – |
| | Lecturer | – | – | 12.5 | 3.8 | – | – |
| | Photographer | – | – | – | 3.8 | – | – |
| | Doctor | – | – | – | 3.8 | – | – |

Based on the occupation, most of the students prefer to photograph people as objects. As in the preceding discussion, the object in question is itself due to the need to maintain the self-existence. Self-existence can be proved by showing or showing off the picture of themselves to others. Similarly, students, respondents who work as private employees also prefer themselves as the object. In contrast to students and employees, other respondents who work as a lecturer, photographer, and doctors prefer the landscape as the object.

Students, private sector employees, teachers, and doctors mainly use the mobile phone camera as a tool for taking picture. The device is much preferred not only because of its portability but also it does not require a high technique and easy to use. Moreover, mobile phone devices usually come with picture editing features so that users can easily edit pictures and can directly upload them to social media without transferring the pictures to another device such as laptop.

However, unlike other professions, respondents who work as photographers prefer to use the type of DSLR cameras. This type of camera is preferred to produce good quality images. Photographers who have mastered the technique of photography select the technology that supports the ability to produce the desired image.

According to most respondents, both male and female do not like edited picture and prefer the natural picture without any editing process. However, among a particular age range (25–29 years old) respondents do take time to consider if the picture needs to be edited before it is uploaded to social media. According to their professional backgrounds, only photographers prefer to edit pictures (Table 20.3).

**Table 20.2** Camera choice based on socio-demography

| Socio-demography | | Camera choice (%) | | | | | | | |
|---|---|---|---|---|---|---|---|---|---|
| | | Pre-study | | | | Field research | | | |
| | | Pocket camera | DSLR/Mirrorless | GoPro | Mobile phone camera | Pocket camera | DSLR/Mirrorless | GoPro | Mobile phone camera |
| Gender | Male | – | 12.5 | – | 50 | 7.7 | 7.7 | 3.8 | 46.2 |
| | Female | 12.5 | – | – | 25 | – | 3.8 | – | 30.8 |
| Age | 19–24 | – | – | – | 25 | 7.7 | 3.8 | 3.8 | 38.5 |
| | 25–29 | – | 12.5 | – | 62.5 | – | 7.7 | – | 11.5 |
| | 30–34 | – | – | – | – | – | – | – | 23.1 |
| | >34 | – | – | – | – | – | – | – | 3.8 |
| Occupation | Student | – | 12.5 | – | 62.5 | 7.7 | 7.7 | 3.8 | 46.2 |
| | Employee | – | – | – | – | – | – | – | 23.1 |
| | Entrepreneur | – | – | – | – | – | – | – | – |
| | Lecturer | 12.5 | – | – | – | – | – | – | 3.8 |
| | Photographer | – | – | – | – | – | 3.8 | – | – |
| | Doctor | – | – | – | – | – | – | – | 3.8 |

**Table 20.3** Picture editing necessity based on socio-demography

| Socio-demography | | Edit picture (%) | | | |
|---|---|---|---|---|---|
| | | Pre-study | | Field research | |
| | | Yes | No | Yes | No |
| Gender | Male | 50 | 12.5 | 23.1 | 42.3 |
| | Female | 12.5 | 25 | 15.4 | 19.2 |
| Age | 19–24 | 12.5 | 12.5 | 19.3 | 34.6 |
| | 25–29 | 50 | 25 | 15.4 | 3.8 |
| | 30–34 | – | – | 3.8 | 19.3 |
| | >34 | – | – | – | 3.8 |
| Occupation | Student | 62.5 | 25 | 26.9 | 38.5 |
| | Employee | – | – | 7.7 | 15.4 |
| | Entrepreneur | – | 12.5 | – | – |
| | Lecturer | – | – | – | 3.8 |
| | Photographer | – | – | 3.8 | – |
| | Doctor | – | – | – | 3.8 |

Most respondents share the pictures they take in Ancient Volcano Nglanggeran through social media. Social media as a means to distribute pictures to family, friends, and colleagues is widely used regardless the gender, age, and occupation, in spite of slight differences. Respondents who work as photographers tend to prefer social media where they can obtain advice and criticism from fellow photographers about the pictures they take, such as photography forum websites, e.g., www.1X.com, www.viewbug.com, www.flickr.com, and www.pixoto.com.

## Conclusions

Based on the results of the research, it can be concluded that,

- Tourist motivation in taking pictures is not job-related or hobby purposes, but because of the interesting things as they find in daily life. The motivations are as a showcase, for personal documentation, as evidence has ever been to Nglanggeran Ancient Volcano, to capture the moment, to test photographic techniques, and to get photo stock. The motivation is mostly related to have an evidence of the experience, to have personal documentation, and capture the unforgettable moment, which are in accordance with previous studies (Haldrup and Larsen 2003; Larsen 2008; Hillman 2007). However, this study find another motivation that has not been yet revealed previously, namely showcase, to test photographic techniques and to get photo stock.
- Tourists photo taking behavior at Nglanggeran Ancient Volcano are using mobile phone camera as a tool for taking pictures without property, most favorite object is people (themselves); the most favorite spot to take picture is

the peak, and people do not prefer to copy exactly the picture seen in brochure or postcard. While in previous research (Prideaux and Coghlan 2010) most respondents use digital cameras. This research reveals that many more tourists prefer to use mobile phone camera because it is practical and easy to use. While landscape is the most photographed object in the previous research (Prideaux and Coghlan 2010), in this study, the most photographed object is people. They take more pictures of themselves to maintain the social media presence and to follow the trend selfie.

- Things to do tourists after taking pictures are show them to relatives, upload to social media, and they do not edit pictures before uploading to social media. In the previous study in Sheungting et al. (2011), there were 41 % of respondents who upload pictures to the internet. In this study (2015), 88.2 % of respondents upload their pictures to the internet.
- Factors behind tourists photo taking behavior are the gender, age, and occupation. Similarly with previous study (Prideaux and Coghlan 2010), this study agrees that socio-demographic factors affect tourists behavior in taking pictures. For instance, while previous study said that a woman takes picture of her family, now she takes picture of herself. Moreover, while the factor age used to affect the choice of technology for sharing pictures, now it affects the act of editing pictures which is well-known in social media platform.

**Acknowledgments** Tourism Studies, Graduate School, Universitas Gadjah Mada

## References

Albert PC, James WR (1988) Travel photography: Amethodological approach. Ann Tourism Res 15:134–158

Chalfen R (1979) Photography's role in tourism: some unexplore relationships. Ann Tourism Res 6(8):435–447

Ghony M Djunaidi, Almanshur F (2012) Metodologi penelitian kualitatif. Ar-Ruzz Media, Jogjakarta

Gillespie A (2006) Tourist photography and the reverse gaze. Wiley on behalf of the Am Anthropol Assoc 34(3):343–366

Gillet S, Schmitz P, Mitas O (2013) The snap-happy tourist: the effects of photographing behavior on tourists' happiness. J Hospitality Tourism Res 20(10):1–21

Groves DL, Timothy DJ (2001) Photographic techniques and the measurement of impact and importance attributes on trip design: a case study. Loisir et Societe 24(1):311–317

Haldrup M, Larsen J (2003) The family gaze. Tourist Stud 3:23–45

Hillman W (2007) Travel authenticated?: postcards, tourist brochures, and travel photography. Tourism Anal 12(2):135–148

Hunter WC (2007) A typology of photographic representations for tourism: depictions of groomed spaces. Tour Manag 29:354–365

Larsen J (2005) Performativity of tourist photography. Space dan Cult 8(4):416–434

Larsen J (2006) Geographies of tourism photography: choreographies and performances. Dalam: Diedit oleh Jespern Falkheimer dan Andre Jansson (eds) Geographies of communication: the spatial turn in media studies. Nordicom, Hal, Goteborg, pp 243–261

Larsen J (2008) Practices and flows of digital photography: an ethnographic frame work. Mobilities 3(1):141–160

Masruri NW (2014) Studi Pengembangan Objek Daya Tarik Wisata Kawasan Ekowisata Gunung Api Purba Nglanggeran Kabupaten Gunung Kidul. Universitas Gadjah Mada, DIY, Tesis

Ntarangwi Mwenda (2000) Education, tourism, or just a visit to the wild? Afr Issues 28(1/2):54–60

Prideaux B, Coghlan A (2010) Digital cameras and photo taking behaviour on the great barrier reef —marketing opportunities for reef tour operators. J Vacation Mark 16:171–183

Sheungting Lo I, McKercher B, Cheung C, Law R (2011) Tourism and online photography. Tour Manag 32:725–731

Strauss AL, Corbin J (1990) Basic of qualitative research: grounded theory procedures and technique. Sage Publication Inc, Newbury Park

Urry J (2002) The tourist gaze. Sage Publications, London

Urry J, Larsen J (2011) The tourist gaze 3.0, Sage Publications, London

# Chapter 21
# Exploring the Public Participation Practices in Planning for Sustainable Tourism in Malaysia

**Arni Abdul Gani, Khairil Wahidin Awang and Abdullah Mohamad**

**Abstract** Natural environment in Malaysia has become one of the major attractions for tourists. Participation in recreation and tourism in natural areas can bring many benefits such as family cohesiveness, relaxation, and provide green spaces for people to appreciate the aesthetical values of the park. These activities help to spread the awareness, disperse knowledge, and foster a sense of environmental appreciation. In addition, the development of tourism in these areas provides employment and economic growth of the local community. The increasing popularity of parks and protected areas as tourism destination definitely calls for special attention to its planning and management. It is crucial to ensure planning and development of tourism infrastructures and programs will not create extensive damage to the environment. One principle of sustainable tourism is incorporating public views and ideas through public participation practice. The study explores respondents' opinions on the current pattern of public participation, with reference to participation in planning for tourism in natural settings. The results indicated some lacking, especially along the overall practice, on the methods used in the drill, the extent of information presented, the opportunity for the public to share views and the thoughtfulness of the public's opinions in decision-making. These findings encourage for thoughtful consideration by the respected authorities to improve the current practice of public participation and calls for further studies to investigate further on the issue.

---

A.A. Gani (✉)
Department of Tourism Management, Faculty of Hotel and Tourism Management, Universiti Teknologi MARA, Puncak Alam Campus, Shah Alam, Malaysia
e-mail: arni@salam.uitm.edu.my

K.W. Awang
Faculty of Economics and Management, Universiti Putra Malaysia, Seri Kembangan, Malaysia
e-mail: khairilawang@yahoo.com

A. Mohamad
Faculty of Forestry, Universiti Putra Malaysia, Seri Kembangan, Malaysia
e-mail: amohd@gmail.com

 
A. Saufi et al. (eds.), *Balancing Development and Sustainability in Tourism Destinations*, DOI 10.1007/978-981-10-1718-6_21

**Keywords** Natural environment · Public participation · Protected area · Sustainable tourism · Tourism planning

## Introduction

Participation has become the essential part of any development today. Participation by the public in the development process is significant in contributing valuable information to the project. According to Shirley et al. (1995) it is imperative that many of respectable development projects to include some provision for 'participation' of the people, where this provision is most likely will become as integral to any project as 'evaluation'. World Bank (1998) indicates that participation is a process where stakeholders influence and share control over development initiatives where decisions made would affect them. There is a wide variety of definition and interpretation associated with the concept of participation. The dictionary meaning of 'participation' is 'the act of taking part in an activity or event' (Oxford Advanced Learners. 2015). Thus, to participate means to take part or become involved in an activity. It can be anticipated that participation in general is taking part or sharing something in activities, in a group with the condition of sharing commonalities among the members of the group.

The support and participation of the public are essential to the achievement of sustainable development. Sustainable development implies economic growth together with the protection of environmental quality, each reinforcing the other. Principal to this form of development is a firm relationship between human activities and the natural world, which does not lessen the prospects for future generations to enjoy a quality of life at least as good as our own. Basically, it means that people living in this world must share with each other and maintain to care for the Earth. Sustainability concept which expresses the idea that people must live within the capacity of their environment to support them are essential, especially in tourism industry, which depends on the maintenance of the environment qualities (Piagram 2000). The issues related to adopting correct and right ways toward sustainable development in tourism destinations have become crucial. Sustainable tourism development requires collective action where inputs from all parties involved and those affected by the development projects must be considered. The form and degree of their participation determine the rate at which the objectives of sustainable development are realized (ACCA 2015).

Despite its importance, the application of public participation practices in reality is just rhetoric. This happens especially in developing countries where tourism plans are usually prepared by the central authorities and are also largely market-driven, the plans usually fail to consider local issues and to provide opportunities for proper consultation with the local communities (Tosun 2000; Tosun and Timothy 2003; Wall and Mathieson 2006). In addition, local knowledge and traditional values are often being ignored in sustainable decisions made sustainable development planning (Moisey and McCool 2008) compares to knowledge of experts which is regarded as more valid and credible information in decision making process (Shirley et al. 1995). It is important to understand that poor management of tourism especially in natural environment could contribute to disaster-prone environment. Unmanaged tourism in these areas can degrade the environment, damage the sites, create severe impacts to the wildlife and the human and sparks conflicts between users. One way all these can be mitigated is through partnerships and participation by all the relevant parties during planning process (Abdullah et al. 1999). Participation by the public can help to create sense of belonging among the affected parties, becomes the platform to share ideas and views and promotes understanding on the goals of the development. The public must participate in the planning process as they are usually the most affected by the development (Arni et al. 2012). This is particularly vital as tourism development in any type of natural environment can have affects not only the surrounding areas, but can significantly affect emotions and well-being of the public.

In the case of Malaysia, the issue relating to participation practice is not new but certainly not enough researches have been conducted to explore it further. Cheuk et al. (2010) mentioned that in Malaysia the public input is mere tokenism in nature. The public are being informed on what is going on in their communities and that is only it. Further consultation normally does not occur. Sharina et al. (2011) and Dola and Dolbani (2006) further stressed that local community participation and public involvement in conservation moves in Malaysia, especially in resource management need urgent attention from all parties involved. Currently, the approach used by many government agencies in Malaysia is not proactive in ensuring true participation occurs especially at the decision making process. Their findings were then supported by Marzuki (2010), Marzuki et al. (2012) which indicated several shortcomings to participation by the public in Malaysia. These include lack of cooperation between the stakeholders, little to no access provided to the public and awareness problems.

In view of the problems, this research intended to investigate the current issues pertaining to public participation practices in planning for sustainable tourism in Malaysia. The main objective was to explore the current practices of public participation and whether the authorities are doing it right. For the purpose of the objective, a group of experts in the field were assembled and asked about their opinion on the current practices of public participation in Malaysia.

## Methodology

The study was conducted by using an expert elicitation process where a group of respondents was identified and invited to participate. They were representatives from selected government agencies, universities, and non-governmental agencies. The government agencies identified were organizations directly related to the natural environment and national park management in Malaysia, relevant ministry officers, and their agencies. As for the universities, the experts were identified based on their background and department they were attached which must be related to sustainable development, tourism, and natural environment management. As for the non-governmental agencies, these are registered agencies in Malaysia with the objectives to promote conservation and preservation of the natural environment.

For the purpose of selection of panels, the study adapted the steps as suggested by Tersine and Riggs (1976) comprised of three main procedures which include selection of the panel of experts, development of the questionnaire and analysis of the replies. The technique, which involves a set of procedure for eliciting and refining opinions of a group of experts, is intended to get consensus without the experts' face to face meetings. Of the total respondents participated in the study, 16 of them were male and 13 of them were female. The largest number of the experts was employed in the industry sector (62.1 %) while the remaining was academician (37.9 %). The experts were categorized according to three expert heading group; government, non-government and university. 41.4 % of the expert panels were categorized under government, 20.7 % were categorized under non-government agency and 37.9 % were categorized under universities. In terms of working experience, the majority has been working between 15 and 20 years (34.5 %) and only a small percentage has been working for more than 30 years (3.4 %).

They were given a set of questionnaire containing five questions with options for answers. They were based on an in depth review on the subject by the researcher and on information from previous studies conducted in Malaysia (Dola and Dolbani 2006; Marzuki 2009; Omar and Leh 2009; Sharina et al. 2011). In addition, the respondents were encouraged to write in the provided spaces. The questions asked for the experts' opinions on the approaches; the current practice of public participation, the methods suitability, how well they think it is being managed and if the current approaches has been successful in practicing true participation by the public. They were reminded at the beginning of the process that the questions referred to public participation in tourism planning for sustainable development. The data were analyzed using frequencies, presenting both responses in the percentage and percentage of cases. Any additional remarks made by the respondents were cross check with the existing choice of answers, to look for fresh terms or words used to express their opinions. However, none of the additional remarks made requires new expressions of opinions to be added to the existing answers given. The Kruskal–Wallis one-way analysis of variance was performed for testing whether samples originate from the same distribution.

## Findings

Table 21.1 summarizes the findings for all the questions asked to the respondents. For opinion on the current practice, the majority of the expert panels (34.8 %) felt that the public participation in Malaysia is only moderate in practice. They also indicate that the practice was not effective and did not exercise true participation with 13 % response recorded respectively. Large percentage was also recorded for

**Table 21.1** Experts opinion results

| | | *N* | Responses (%) | % of cases |
|---|---|---|---|---|
| Opinion current practice[a] | Practice is not effective | 6 | 13.0 | 20.7 |
| | Practice is moderate | 16 | 34.8 | 55.2 |
| | Practice does encourage participant presence | 3 | 6.5 | 10.3 |
| | It is not a true participation | 6 | 13.0 | 20.7 |
| | Practice is only a formality | 15 | 32.6 | 51.7 |
| | Total | 46 | 100.0 | 158.6 |
| Opinion on methods[a] | The banners and presentations are not effective | 5 | 12.5 | 19.2 |
| | There are lack technical advices | 13 | 32.5 | 50.0 |
| | The method using workshops is effective | 9 | 22.5 | 34.6 |
| | Methods allows dialogue and feedback | 13 | 32.5 | 50.0 |
| | Total | 40 | 100.0 | 153.8 |
| Opinion on current information delivered to public[a] | The information on the planning is not enough | 8 | 9.8 | 27.6 |
| | Lack in environmental information | 8 | 9.8 | 27.6 |
| | Information not comprehensive | 10 | 12.2 | 34.5 |
| | Information could not be understood | 6 | 7.3 | 20.7 |
| | Information not specific | 7 | 8.5 | 24.1 |
| | Allow public to view information and give comments | 14 | 17.1 | 48.3 |
| | Information are presented in sufficient time | 4 | 4.9 | 13.8 |
| | Problem to understand information due to public education levels constraint | 15 | 18.3 | 51.7 |
| | Government elaborate information and inform public | 10 | 12.2 | 34.5 |
| | Total | 82 | 100.0 | 282.8 |
| Opinion on opportunity to share views[a] | Government allows public to share views and concerns | 10 | 25.6 | 34.5 |
| | The approach helps establish trust and credibility | 2 | 5.1 | 6.9 |
| | The public has no equal rights | 12 | 30.8 | 41.4 |
| | There are lacking in active discussions | 15 | 38.5 | 51.7 |
| | Total | 39 | 100.0 | 134.5 |
| Opinion on Government consideration[a] | Government did not take opinions seriously | 12 | 20.7 | 41.4 |
| | The Government is dominant | 15 | 25.9 | 51.7 |
| | The public does not influence decisions made by govt | 17 | 29.3 | 58.6 |
| | The public does not receive satisfactory feedback from govt | 14 | 24.1 | 48.3 |
| | Total | 58 | 100.0 | 200.0 |

[a]Dichotomy group tabulated at value 1

public participation as only a formality (32.6 %) required for planning purposes. In terms of the current methods used by government and its agencies in promoting public participation, the majority of the expert panels felt that there was a lacking in technical advice (32.5 %) to facilitate understanding of relevant parties during the meetings and discussions. However, it was agreed by the majority as well that current methods used did allow for dialogue and feedback (34.5 %). 22.5 % of the responses indicate that the workshops were the most effective methods to be used in public participation. While they felt that the methods used were effective, they were concerned with promotion wise, where banners and other promotions used such as presentations to disseminate information were not effective (12.5 %).

Information dissemination is very important to ensure that the public can understand the development plans and able to take part successfully. The expert panels felt that the government did play their part in attempting to provide information for the public (12.2 %) as well as allowing the public to view and give comments (14 %). This shows that government has made an effort to incorporate public participation in the process. However, with reference toward the management of information, the majority of the expert panels indicated that the information for the public were not comprehensive (12.2 %), were lacking in environmental information (9.8 %), were not enough to facilitate proper understanding (9.8 %), information were not specific (8.5 %) and the information could not be understood (7.3 %). Major constraint to this was the level of education held by the public (18.3 %) which could hinder through understanding of the information provided by the government. Only a small percentage of the panel experts felt that sufficient time (4.9 %) was allocated to present this information to the public. This could mean that longer times are needed for workshops and public displays and presentation of development plans to the public. In terms of whether the current practice allows the opportunity to share views of the public, the majority of the expert panels felt that while there were opportunities given (25.6 %) by the government, there were obvious lacking in active discussions between relevant parties (38.5 %). They also felt that the public has no equal rights (30.8 %) in terms of making influence in decision-making. This is why public participation in Malaysia was treated as 'tokenism' in nature (Wong 1995), which participation was just a matter of fulfilling the requirement. Nonetheless, some of the experts felt that the current practice did help to establish some form of trust and increase credibility (5.1 %) of the relevant parties.

The expert panels were also asked their opinion on the level of consideration given by the government to include public opinions and views in decision-making. The majority of them felt that the public did not influence decisions made by the government (29.3 %) and that the government were dominant in their decisions (25.9 %). In addition, they felt that the public was not given a satisfactory feedback by the government (24.1 %) as well as not serious to take proper consideration on public opinions and views when making decisions (20.7 %). In conclusion, this indicates that there were obvious lacking in how the government treated the information they received from the public pertaining to any development plans in Malaysia. While there were definitely efforts to encourage public participation, the

**Table 21.2** Experts opinion on the success of public participation in Malaysia

| | Frequency | Percent |
|---|---|---|
| Disagree | 10 | 34.5 |
| Neutral | 14 | 48.3 |
| Agree | 5 | 17.2 |
| Total | 29 | 100.0 |

management and execution of such process were still questionable. The final question asked the expert panels their opinion whether the current practice of public participation in sustainable tourism planning is a success or otherwise (Table 21.2).

The majority of the experts (48.3 %) rated neutral on their opinion when asked if public participation in Malaysia is a success. 10 experts representing 34.5 % rated disagree while only 5 experts (17.2 %) rated agree. Based on these findings, the expert panels felt that the public participation in Malaysia is not seen as a success yet as there are many issues need to be addressed by the respective authorities. A Kruskall-Wallis H was performed to examine if the opinion on success between the expert panels differs. The test showed that there was a significant difference in opinion on success between the expert headings, $X2$ (2) = 6.759, $p = 0.034$, with a mean rank 16.70 for government agencies, 7.33 for non-governmental agency and 14.00 for university academicians.

## Conclusions

Public participation in Malaysia was not seen as a requirement that need profound consideration by relevant authorities. While efforts were made to include public participation in the planning process in Malaysia, the expert panels still thinks that overall practices were yet to achieve its intended purpose. Essentially, the findings from this study suggested that public participation for such intention is unsuccessful drawing similar results from prior studies such as (Dola and Dolbani 2006, 2012; Sharina et al. 2011).

Despite being allowed to participate in the planning, the public's contribution is often very limited with most of their involvement is only at the initial stage of planning but never surpassing to the decision-making process. Attention needs to be drawn also toward the ineffectiveness of public participation that is performed only to satisfy some obligatory requirements. While such practice may encourage the public presence, the approach used is exorbitantly in need of an upturn for more genuine participation. The public needs to be empowered by including their concerns, validation, and ideas in planning for sustainable tourism. This is significantly crucial because empowerment is apparently one of the main concerns that need to be highlighted especially when the planning would affect the public.

Results from the findings may bring to a close that the limitations for true and effective public participation in planning for sustainable tourism is the upshot of weak practices by the governing agencies. The analysis gathered has able to prove that information on technical advice and development plan is either absent or insufficient for the public to review. Unfortunately, they are also given very limited opportunities in sharing their views and concerns, let alone taken into consideration on matters which would be taken in the decision-making. These findings call for an urgent means to improve public participation in the planning of sustainable tourism in Malaysia. This is timely since the country is keen in its efforts to develop, implement and promote the concept of sustainable tourism.

## References


Abdullah M, Yaman AR, Jamaluddin MA (1999) Recreational opportunities for public use in Ayer Hitam Forest: setting the stage and park management approach. Pertanika J Trop Agri Sci 22 (2):161–166

ACCA (2015) Public participation in sustainable development administrative centre for China's agenda 21. Retrieved 18 June 2015, from http://www.acca21.org.cn/chnwp20a.html

Arni AG, Khairil WA, Zaiton S, Abdullah M (2012) A discourse on sustainable tourism development and human dimensions of public participation in protected areas. In Zaiton S, Syamsul Herman MA, Norfaryanti K (eds) Sustainable forest utilisation: concept and practices. Universiti Putra Malaysia Press, Serdang, Malaysia

Cheuk S, Liew-Tsonis J, Ing GP, Razli IA (2010) An establishment of the role of private and public sector interests in the context of tourism transport planning and development: the case of Malaysia. Int Bus Econ Res J 9(2):59–67

Dola K, Dolbani M (2006) Public participation in planning for sustainable development: operational questions and issues. ALAM CIPTA, Intl J Sustain Trop Design Res Pract 1(1):1–8

Marzuki A (2009) A review on public participation in environmental impact assessment in Malaysia. Theor Empirical Res Urban Manage (12):126–136

Marzuki A (2010) Tourism development in Malaysia: a review on federal government policies. Theor Empirical Res Urban Manage (17):85

Marzuki A, Hay I, James J (2012) Public participation shortcomings in tourism planning: the case of the Langkawi Islands, Malaysia. J Sustain Tour 20(4):585–602

Moisey RN, McCool SF (2008) Sustainable tourism in the 21st century: lessons from the past, challenges to address. In: McCool SF, Moisey RN (eds) Tourism, recreation and sustainability: linking culture and the environment. UK, Cabi

Omar D, Leh OLH (2009) Malaysia development planning system: Kuala Lumpur structure plan and public participation. Asian Social Sci 5(3):30–36

Oxford Advanced Learners (2015) Oxford learners dictionaries. Retrieved 6 Jan 2015, from http://www.oxfordlearnersdictionaries.com/definition/english/participation?q=participation

Piagram JJ (2000) Tourism and sustainability: a positive trend. In W. C. Gartner & D. W. Lime (Eds.), Trends in outdoor recreation, leisure and tourism. Oxon: Cabi Publishing 373–382

Sharina AH, Hood S, Mustaffa O (2011) Engaging the local community in participatory resource management through learning: the experience from Langkawi Island, Malaysia. Kajian Malaysia 29(1):125–139

Shirley AW, Nair KS, Ascroft J (1995) Participatory communication. Sage Publications, London

Tersine RJ, Riggs WE (1976) The Delphi technique: a long-range planning tool. Bus Horiz 19:51–55

Tosun C (2000) Limits to community participation in hte tourism development process in developing countries. Tour Manag 21:613–633

Tosun C, Timothy DJ (2003) Arguments for community participation in the tourism development process. J Tour Stud 14(2):2–15

Wall G, Mathieson A (2006) Tourism: changes, impacts and opportunities. Pearson Education Ltd, England

Wong LP (1995) Defining leisure and recreation in Malaysia. (Doctoral Dissertation, Oregon State University). Retrieved from http://ir.library.oregonstate.edu/xmlui/handle/1957/10405 Proquest Database

World Bank (1998) World bank participation sourcebook. World Bank. Retrieved 3 June 2011, from http://www.worldbank.org/wbi/sourcebook/sb0100.htm

# Chapter 22
# Land and Tourism in Post-war Sri Lanka: A Critique on the Political Negligence in Tourism

**Iraj Ratnayake and Mahesh Hapugoda**

**Abstract** This critique argues that the mechanism of land utilization under post-war tourism in the Northern and Eastern provinces of Sri Lanka was instigated by marker-driven political negligence. It discovers that the introduction of *post-catastrophic tourism* (Zizek 2014) has been influenced by immediate commercial motives and no attention has been paid to the political significance of the traumatic historical memory of the community affected by a lasting conflict. The post-war tourism has generated a phenomenon called *dark tourism* in the area concerned, where the curiosity to see the 'monuments' of the fierce battle could be exploited for recreational purposes. The victorious southerners used their political hegemony to legitimize their commercially motivated invasions by converting the once battlefield to a tourism zone of 'leisure tourism practices' (Pieris 2014: 266). For this purpose, the most valuable lands located in potential of nature tourism have obviously been grabbed by the Sri Lankan military depriving of those who claim the right to those lands. In the meantime, the apparent mass tourism promotion by the government between 2009 and 2014 too has significantly disregarded the fundamental principle that careful attention must be paid to, 'the visitation to places, where tragedies or historically noteworthy death has occurred and that continue to impact our lives' (Tarlow 2005: 48) before obvious profit motives. In the context of this paper, the above aloof attitude in neglecting geo-political sensitivity of the community by the then politico-administrative agents, who were blinded by their own development paradigm, is identified as a dangerous humanitarian mistake from a wider universal sense. While acknowledging tourism as an agent of development, this paper understands that the trivialization of land rights and the disregard towards

I. Ratnayake (✉)
Department of Tourism Management, Sabaragamuwa University of Sri Lanka, Balangoda, Sri Lanka
e-mail: iraj@mgt.sab.ac.lk

M. Hapugoda
Department of Languages, Sabaragamuwa University of Sri Lanka, Balangoda, Sri Lanka
e-mail: hapugoda@gmail.com



A. Saufi et al. (eds.), *Balancing Development and Sustainability in Tourism Destinations*, DOI 10.1007/978-981-10-1718-6_22

the significance of a historical memory and the ownership of a community, which has turned 'the suffering into a leisure experience for contemporary tourists' (Smith et al. 2010: 38) cannot also be easily overlooked.

**Keywords** Post-catastrophic tourism · Post-war politics · Tourism development · Land uses · Sri Lanka tourism

## Introduction

Tourism is widely considered 'one of the most remarkable socio-economic phenomena of the twentieth century' (Neto 2003: 1). Perhaps as the largest industry in the world, it functions as a change agent of development (UNWTO 2013). Therefore, governments, especially those in the third world with inherent natural heritage, are unhesitant to include tourism as a main source of income in their economic policy manifestos. In doing so, they often tend to forget the subsequent socio-cultural and political concerns that can, in turn, have drastic impacts on their communities. It is acknowledged that some of the potential tourism attractions have to be promoted to uplift the living conditions of the communities concerned. Most importantly, in the precise case under review of this paper, the communities which were subject to a thirty-year civil war between Sri Lankan government and the Tamil secessionists, tourism could have been a major modernizing factor to be considered under the post-war development strategy. When it comes to the utilization of land for any development concerns, not only the ownership of land of those community, but the above geo-political attachment as well have to be kept in mind for a better community integration, since the very eruption of the conflict ignited from the very geo-political misunderstandings themselves. In the context of land use and management, this study focuses on how the introduction and development of recreational tourism has neglected the politico-ideological players that can be in conflict with historical and geo-political concerns of the community.

To illustrate the mentioned negligence, the study identifies the manner of acquisition and management of lands for tourism purposes in certain post-war areas as potential empirical evidence. It ascertains that land has been the most prominent grievance for the beginning of the ethnic conflict in the North-Eastern Sri Lanka, but the manner in which same aspect is exploited can be noticed as a 'focal target of obliterating the ethnic identity' of the inhabitants in the area. In this very regard, the introduction of tourism has functioned as *a signifier* of the dominant ethnic group whose authority in political decision-making has contributed to reaffirm its hegemonic status as *the Master* than an agent of sustainable development. Hence, in a restricted sense, both politically and culturally, tourism has not been able to establish *trust* in the community in a reciprocal sense 'as a rewarding relationship in the communities concerned' (Hunter 2004: 308). In addition, though the tourist arrivals impressively increased during 2010 and afterwards, as Tourism Watch observes, the Sri Lanka tourism industry has created 'new inequality and is further increasing the conflict rather than bringing peace and cross-cultural understanding' (Mangalassery 2010). As a result, in contrast to the promotion of mass tourism, Lokuhetti and Jayawardene (2013) suggest the necessity of a new approach for long lasting peace and a sustainable form of tourism in match with the higher expectation of 2.5 million tourists in 2016. Deviating from civil government agencies such as Tourism Authorities, Provincial Councils or Local Government Bodies, the military has used its discursive 'security concerns' to forcefully acquire land for tourism especially over looking general resistance for tourism and right for land use. Hence, from the very outlook, negotiations or compromises between the community and political agencies are denied, and the type of tourism that is facilitated and promoted by the military in these areas seems alienated from the community igniting further resistance.

Empirically, it is observable that in most of the controversial lands acquired and secured by the military, for instance *Peanut Farm* in Panama and *Thalsevana* in Kankesanthurai (KKS), the resorts are commercially managed by the military themselves. The militarized environment in the KKS site with heavily guarded checkpoints, security interrogation in the entrance to the above resorts, and constant surveillance in the area restricts authentic tourists' behaviour creating a grotesque sense of unnaturalness. By analysing the empirical evidence of land use in tourism and reviewing the policy implementation strategies in initiating mass tourism in nature tourism attractions, this study examines how the denial of a true political articulation based on sustainable development paradigm unfortunately relinquished the opportunity of ethnic harmony and nation building. Further, on the ground that the thirty-year war itself was a product of geo-political sentiments of ethnic nature, the study suggests that the land use policies for mass tourism should be handled with more cautiousness and transparency rather than not being subject to the demand-driven tourist motivation.

## Objectives

By deriving empirical evidence from the sites subject to forceful land acquisition and tourism operation and reviewing existing literature about the said phenomenon, this study aims to examine the nature of the tourism product generated in the post-war phenomenon. It also intends to observe how the absence of civil authorities and channels (both political and administrative) and the presence of military in their place have aggravated the situation in which the communities have been deprived of land rights and fundamental rights. The paper looks into this artificial phenomenon of lucrative tourism operation and reveals newly emerged controversies and de-humanization. In the context of lasting historical geo-political grievances that sparked a thirty-year war, the study critics that the short-sighted politico-administrative decisions taken by the then governments in implementing tourism as a 'change agent' in the post-war development has further intensified misunderstanding between government agencies and affected communities. The study substantiates the possible rise of high-end tourism in ethnically sensitive geo-political entities in Sri Lanka results in community resentments and a possible rise of disaster capitalism (Klien 2007).

## Methodology

The information used for this research was based a review of the literature and documents relating to the land use decisions of state sector, institutions and as well as private operators of different scales varying from small to medium scale operators in concerned coastal communities. State sector institutions include those who are with direct involvement in land administration and semi-structured interviews were conducted with officers of those institutions. Informal interviews with local community groups and representatives from non-governmental organizations were also conducted. In addition, participant observation was used to collect primary data and also allowed the researchers to determine the accuracy of the information gained in interviews. The researchers spent a total of 18 days during June and July, 2015 on-site at Arugam Bay, Panama, Pothuvil in the Eastern Coast and Jaffna Peninsula in the North.

## Empirical Evidence

This study is based on the empirical evidence derived from two military controlled sites in the Northern and Eastern Sri Lanka. *Shastrawela* and *Ragamwela* from the Eastern Province and *Kankesanthurai* from the Northern Province are the two locations that are chosen out of many such examples scattered around the coastal belt in Sri Lanka. Shastrawela in Panama carries a history of land grabbing controversies starting from the Tourism Master Plan in 2005, while Kuchchaveli and Kankesanthurai has been focused on by many recent reports in relation to land grabbing.

### *Shastrawela and Ragamwela Case (South of Arugam Bay)*

The East coast of Sri Lanka offers infinite opportunities for all kinds of active and passive water based recreation. Arugam Bay is a fishing village located in two and half kilometres south of the small town of Pothuvil, which is situated in the dry zone of Sri Lanka's remotest southern end of the Eastern coast. Arugam Bay is 320 km away from the capital Colombo and it is one of the most beautiful bays in the island—a popular tourist destination with perfect surfing conditions. It is probably the best surf destination in Sri Lanka and one of the world's top ten surf locations. Hence, the area is of a great attraction especially for the surfers from all over the world.

In late 2004, the Tsunami devastation had severely affected the people in the area and, as a result, most of super-infrastructure facilities were destroyed. Due to its popularity among tourists, the area has managed a slow recovery and the community has developed tourism with their own hard work and inspiration. The situation has drastically changed after the conflict. Considerable increase in tourism demand with sustained peace the government has adopted a new approach to occupy those customary lands of locals. BBC reports , '… more than 350 families in Ragamwela and Shastrawela villages have been forcibly evicted from their lands by the military' (as cited in UNHRC, 2012: 3). The government forcibly appropriated these lands in 2010 and handed over to the Navy and the Air Force. Peanut Farm is the only remaining plot of privately owned land located in-between the lands occupied by air force and the navy. The present owner who is from Panama Village shared his experiences with us, "I have been operating here since 1997. Tsunami affected my property in 2004. However, I managed to start operations back. The situation was changed after the war as the government exerted pressure on us to leave these lands. I was threatened over the phone and asked to give up, but I was the only one who could survive after all" (Respondent 3: Owner/Peanut Farm Surfing Point, personal communication, June 27, 2015). The presidential secretariat was directly involved in the land grabbing. This incident alone created a colossal unrest among villagers.

After this appropriation, the land use decisions were abruptly taken by the military. These decisions were ad hoc, where the land has been used for the construction of a hotel and an intentional conference centre. These projects have been carried out by Navy and the Air Force, respectively, disregarding tourism development protocols. For instance, relevant governmental institutions have not been consulted from the outset. "Air force did not allow Coast Conservation Department (CCD) to enter the site for inspection and navy too did not obtain proper permission from us to initiate their construction within the area under our custody" (Respondent 4: Planning Assistant/CCD, personal communication, June 29, 2015). Responded further asserted that, CCD had to obtain a court order to oversee the construction work by air force, which had to temporarily suspend and navy was started negotiating and coordinating with the CCD. "This case is typical evidence for ad hoc nature of decisions taken by the developers, "is a violation of existing legal framework of the country" (Respondent 4: Planning Assistant/CCD, personal communication, June 29, 2015). Due to continuous pressure exerted by the villagers the present government has decided to release the land except 25 acres that occupied for the construction of an international conference centre by the air force (Divaina 2015). The whole process is unforeseen. "Nothing was transparent. Initially we were told that this land is for a military base and later to construct a housing scheme for war heroes, but after all they did something different" (Respondent 2: Community Representative, personal communication, June 21, 2015).

## *Kankesanthurai and Kuchchaveli Case*

Kankesanthurai (KKS) is one of the prominent examples of forceful land occupation by military after the war, where the area has been declared as a high security zone (HSZ) on the northern tip of the Jaffna peninsula. This much controversial land grabbing that occupies some 5000 acres of heredity lands of locals has not been exclusively occupied for military purpose, but also for recreation facility development called *Thalsevana* Holiday Resort formally known as Harbour View Hotel. It illustrates the case that new phenomenon in the post-war period, where the recreational business has been added onto military portfolio.

Several important observations were made during the first visit to Thalsevana during early 2015. First, the visitors had to register themselves at a security point just before the HSZ. And the security inquired about 'early reservation' which sounds suspicious and welcoming at all. The researchers drove further about few kilometres passing several military and police points, which were on both sides of the road up to the resort. It is observed that a prime plot of land right in the heart of the HZS of KKS—with a strip of beach and immediate waterfront has been used for the recreation tourism. It had typically followed the same construction and operational process commonly available elsewhere in resort management. The guests are being served by all in-service military personnel and as a result locals do not find any employment opportunities here in these resorts. At time of the site visit, several Navy crafts were patrolling out in the sea was no addition to the ambience of the resort, but might registering a concern in the minds of guests whose aspirations were otherwise.

## Discussion

The thirty-year civil war was ended in 2009 and the urgency for development was the common hope of everyone. In the atmosphere of the necessity to 'next massive leap forward' the much stagnated economy due to war and to transform it as an influential economy in Asia (Ministry of Finance and Planning: *Mahinda Chinthana* 2010), Sri Lanka was expected to experience a rapid development in the tourism industry by increasing the number of tourists visits. The vision of the rapid economic development of the then government has been, according to the above *Mahinda Chinthana* policy manifesto, 'is to transform Sri Lankan tourism sector, by 2020, to be the largest foreign exchange earner in the economy' (164). This vision was coupled with its mission to make Sri Lanka a strategically important economic centre in the world and making it also a commercial hub of Asia. Since tourism was identified as a main source of national income in the anticipated economic growth that was believed to 'rebuild' Sri Lanka, it was given a prioritized area in sustainable economic development encompassing everything else.

The Minister of Economic Development, the most powerful Minister during the 2009–2014 eras, was assigned to handle the tourism industry with great anticipations. Together with a wide promotional mechanism, the Ministry started parallel activities which targeted to attract mass tourists. To accelerate the functionality of tourism and to build the necessary infrastructure the Ministry of Defence too contributed in various face-lifting activities in the central capital and peripheral zones, where tourism was deemed to operate as a major commercial contributor to the national economy. The military interventions in post-war civil activities were controversial and subject to criticism. But with the effectiveness of the military victory and the rising totalitarian characteristics, there had been very little protest for their intervention in public (non-military) activities. The government tactically used the symbolic achievement of the military forces to 'silence' the public in sensitive issues such as commercially benefitting land-related matter, commercially productive projects, where community resistance can spring up. For examples, Urban Development Authority used the military charisma for removing residents from congested city areas forcefully restricting community resistance. It is during this season that the Sri Lankan military forces got directly involved in tourism-related activities in some of the most controversial areas. However, the function of the military in the field of post-war tourism in Sri Lanka can be identified under two major sections.

(a) Its trained manpower was effectively and cheaply utilized to enhance the infrastructure and face-lifting activities both in North and East as well as in the central capital.
(b) The victorious character of the military forces was used to acquire commercially valuable lands by 'showing' strategic military significance by successfully thwarting community resistance for tourism.

In addition to these attributes, it was observable that a degree of 'dark tourism' was emerging in Northern and Eastern Sri Lanka, between 2009 and 2012 in which people from South started travelling to see the 'monuments' of war. This trend of post-catastrophic tourism (Zizek 2014) focused on areas where the final stage of the civil war took place. They were the first group of visitors who visited the areas which were occupied by Tamil Tigers for nearly three decades. The Sri Lankan military started small cafeterias and rest houses alongside the A9 road to cater the unmanageably pouring crowd to Northern Peninsula to witness the remnants of devastating war. Later, the military forces went into the lucrative tourism business by putting up luxury hotels and resorts near the beaches or any other nature tourism attractions. From 'dark tourism' now it has gained momentum to mass tourism in many areas where it has been declared as High Security Zones (HSC), there are resorts and hotels that are maintained by the Sri Lankan military.

The Tourism Master Plan that came about somewhere in the early 1990s aimed to develop Arugam Bay as 'a model for an up-market, boutique tourism destination though tourism in the village was historically low budget' (Robinson and Jarvie 2008: 640). The Task Force TAFREN further suggested that 'the local population was to be relocated from the 200-m protected zone. Although this land is privately owned and falls outside the government-regulated 200-m limit, the area was to be redeveloped for hotels and tourism infrastructure' (641). The then Chairman of the Tourist Board had made it clear in one of the post-tsunami Arugam Bay meetings, 'we are looking about the high level tourists, not the 5 dollar tourist' (cited in Braunmuhl 2006). In advanced rhetoric, as quoted by Naomi Klien, Sri Lanka Tourism Board saw the tsunami disaster as a blessing in disguise, 'In a cruel twist of fate, nature has presented Sri Lanka as a unique opportunity, and out of the great tragedy will come a world class tourist destination' (Klein 2007: 11). 'Promoting high-end tourism seems to be one of the driving forces of TAFREN, which seems geared towards encouraging high-end tourism, export-oriented agriculture and manufacturing, and large scale fisheries. It clearly does not include small-scale fishing, subsistence farming or community based tourism' (Fernandoas cited in Klien 2005).

It is clear that the post-war tourism had learnt straightaway from the initial objectives of the Tourism Master Plan in 2005 which aimed tore build the tsunami affected community in Arugam Bay. The 100-m Tsunami Buffer Zone which was meant for safety measure was not applied for tourism. 'On the surface it made sense, but there was a glaring problem with that rationale—it was not being applied to the tourism industry. On the contrary, hotels were being encouraged to expand into the valuable seafront, where fishing people had lived and worked' (Klien 2007: 388). She further states, 'the federal government had commissioned a team of international consultants to develop a construction blueprint for Arugam Bay, and this plan was the result it called for Arugam Bay to be levelled and rebuilt, transform from a hippie-charming seaside town into a high-end boutique tourism destination' (388). According to Klien, this kind of plan was initiated with the motivation of the United State government backed by theories adopted by Chicago School of Economics (392). As she examines, 'The US government was so enthusiastic about Sri Lanka's potential as a high-end tourism destination, with all its possibilities for resort chains and tour operators, that USAID launched a program to organize the Sri Lankan tourism industry into a powerful Washington style lobby group' (392). The land that was targeted under the 2005 Master Plan, the stretch of the coastal belt consisting of 150 acres from Panama to Pothuvil, was indeed what was at present taken into a naval military base wherein a tourism resort is managed by the military themselves.

As evidenced by Klien (2005) and Robinson and Jarvie (2008), *Shastrawela* in Panama has been aimed for the utilization of hotel and resorts targeting mass tourists. In this case, the land that was recommended under the 2005 Tourism Master Plan for its potential for tourism has now been occupied by the Sri Lanka Navy. The negative impacts of having such a state monopoly seem to far outweigh the benefits mentioned above. With increasing population, rural people have been compelled to encroach on state lands through sheer necessity. Adding to this challenge are the economic, social and environmental impacts of continued degradation of Sri Lanka's land resources from misuse and over-exploitation. Therefore, the clearing of lands for other uses such as tourism is a big problem.

Moreover, competition for land among various users including tourism will, inevitably, become a more pressing problem. Under this context, military was an effective mean to take control of the commercially conducive lands for tourism purpose and to silence the community resistance after the war. The arbitrary use of military leads to various other issues due to lack of guidance and supervision by tourism experts who can function as 'intermediaries' between the government agencies and communities. The military has no ideas how they disregard the universally accepted tourism and hospitality protocols, civil and human right conventions, land use policies, feasibility standards for resort construction and management incorporating other government bodies such as tourism authorities, enforcement agencies and external expertise.

For the Southerners who visited to see the 'remnants of victories' of war, the Sri Lankan military built tourism resorts in their former battle grounds. The indigenous Tamils may feel that it is this 'conquered' battle field that is converted for entertainment and recreation. In a multi-ethnic and multilingual country, where there had been historical battle for political identity, the government has to be careful in implementing commercially feasible projects, which can hurt the historical sentiments of another community. It is said that the original owners of these lands cannot even see those while the outsiders can go and enjoy in these resorts built in those lands. Hence it is viewed that the true 'invasion' has taken place not in terms 'dark tourism' (Pieris 2004) made of timely inspired Sinhalese visitors right after the war, but in terms of commercially driven mega scale tourism projects that made way to these areas and are here to stay forever.

Military involvement in tourism has largely ignored the necessary consultation of other governmental institutions that have direct stake in land use decisions. It has been clearly evident that the military has by-passed all civil protocols that need to be followed, which has resulted in disturbing social order by and large. From site development to operations, lack of transparency in the development agenda has created frustration among the deprived people. Public access has been restricted particularly to the beach at locations occupied by the military. Function of a tourist resort has been 'redefined' as tourism operations in these locations do not stick to conventional resort management techniques. For instance, one must declare his or her identity by producing travel documents at the security checkpoints. We have personally experienced that walk-in guests are not welcome at those resorts, as it was questioned often for the confirmed reservation (personal observation). The premises are full of military personnel, who regularly move to and from for various purposes. Customer service operations are also carried out by military personnel themselves.

The ultimate tourism product that has been triggered out of all these controversies does not, on the other hand, represent the accepted standards of quality maintenance and supervision. Except the comments given in the *Trip Advisor* and other web sites, the quality of service given to customers is not made according to the standards laid out by tourism industry or any other relevant authorities, who are responsible for consumer care. The resort planning and designing, quality of food, service quality and delivery, safety measures, etc., are all decided and delivered by the respective military authorities, mostly Sri Lankan Army and Navy. The exclusive nature of the product may sometimes inspire those, who love exotics military tourism but it will surely decay the acceptable tourism and hospitality standards.

## Conclusion

The study finds that the areas considered in the study have been subject to serious land acquisition inequities and controversies. Despite the rhetoric of development, lack of transparency in the development agenda has also created frustration among the deprived people. The presence of military in the place of civil administrative bodies in land acquisition and later utilization for tourism operation has caused misunderstanding in the communities concerned. The relationship that has emerged in the war-tone coastal belt between land (mis)-used and the lucrative tourism operation has led post-war Sri Lanka to significant controversies and de-humanization. In the context of lasting historical geo-political grievances that sparked a thirty-year war, the study finds that that the short-sighted nature of politico-administrative decisions taken by the contemporary governments over the implementation of tourism as a 'change agent' in the post-war development in Sri Lanka has further intensified misunderstanding between government agencies and affected communities. Instead, the study substantiates the possible rise of high-end tourism in ethnically sensitive geo-political entities in Sri Lanka results in community resentments and disaster capitalism (Klien 2013).

## References

Braunmuhl CV (2006) Tsunami: a study on disaster response in Sri Lanka. Media International, Colombo

Divaina (2015) *Panam Patthtu weK and ulu Matha Hotal* (in Sinhala). Divaina E-news paper, p IX. Retrieved from http://epaper.divaina.com/index.php?option=com_flippingbook&view=book&id=2299&page=1 on 19 July 2015

Hunter WC (2004) Trust between culture: the Tourist. In: Williams S (ed) Critical concepts in the social sciences. Routledge, London, pp 279–296

Klein N (2007) The shock doctrine: the rise of disaster capitalism. Metropolitan Books, New York

Lokuhetti A, Jayawardene C (2013) Developing a shared vision for tourism in post-war Sri Lanka. Emerald 5(5):486–494

Mangalassery S (2010) After the civil war: Sri Lanka relies on tourism. Retrieved from http://www.tourism-watch.de/en/content/after-civil-war-sri-lanka-relies-tourism on 2 Jun 2015

Ministry of Finance and Planning (2010) Mahinda Chinthana: vision for the future Sri Lanka: the emerging wonders of Asia. Ministry of Finance and Planning, Colombo

Neto F (2003) A new approach to sustainable tourism development: moving beyond environment protection. United Nations, New York

Pieris A (2014) Southern invasions: post-war tourism in Sri Lanka. In: Postcolonial studies, vol 17, no 3. Taylor and Francis, London

Robinson L, Jarvie JK (2008) Post-disaster community tourism recovery: the tsunami and Arugam Bay, Sri Lanka. In: Disasters, vol 4, no 32. Overseas Development Institute, Oxford, Mercy Corps, Journal Compilation, Blackwell, pp 631–645

Smith M, Macleod N, Robertson MH (2010) Key concepts in tourist studies. Sage, London

Tarlow P (2005) Dark tourism: the appealing dark side of tourism and more. In: Niche tourism, contemporary issues trends and cases. Elsevier, Oxford, pp 47–58

UNHRC (2012) Written statement submitted by the Asian Forum for Human Rights and Development, a non-governmental organization in special consultative status: Human Rights Council, Nineteenth session, Agenda item 4, Human rights situations that require the Council's attention. Retrieved from, http://www.sangam.org/2012/03/AFHRD_Statement.pdf on July 16, 2015

UNWTO (2013) Sustainable tourism for development guidebook: enhancing capacities for sustainable tourism for development in developing countries. World Tourism Organization, Madrid

Zizek S (2014) Living in the end times. Verso, London

# Chapter 23
# Implementation of a Cluster-Based Model for Waterfront Urban Tourism Development in Manado, Indonesia

**Bet El Silisna Lagarense and Agustinus Walansendow**

**Abstract** Tourism Development is one of regional development priorities for North Sulawesi, Indonesia. Manado Waterfront Development (MWD) is one of city development programs to enhance the image of Manado as a waterfront city. Manado waterfront is being developed to strengthen opportunities for both tourism and other businesses. There are many buildings that are used for trade and business purposes. The spatial distributions of tourism, commercial, and residential land uses overlap. The previous study had developed a clustered based model for Manado waterfront area. This study is an evaluation on implementation of the model and strategy proposed in the previous study for waterfront urban tourism in Manado Indonesia. This study is aimed to examine to what extent the model is successfully applied and how stakeholders perception and public participation can contribute to development program for the city tourism. The study applies descriptive approach using survey, in-depth interview, and Focus Group Discussion (FGD) to get inputs and recommendation on the implementation of tourism development model for Manado as a waterfront city based on previous research. Field research consisted of a survey questionnaire, interview and observation on study sites. The results show that stakeholders such as government, tourism industries, local residents at and around the waterfront, city planners, and academics all agree to include the model into the city planning and implement it into practice. The study proposed a formulation of strategies and policies for waterfront tourism development and proposed to local authorities for implementation.

**Keywords** Cluster-based · Model · Waterfront · Urban tourism

---

B.E.S. Lagarense (✉)
Tourism Department, Manado State Polytechnic, Manado 95252, Indonesia
e-mail: bet_lagarense@yahoo.com

A. Walansendow
Business Administration Department, Manado State Polytechnic Indonesia, Manado, Indonesia
e-mail: walansendowagus@gmail.com



A. Saufi et al. (eds.), *Balancing Development and Sustainability in Tourism Destinations*, DOI 10.1007/978-981-10-1718-6_23

## Introduction

Manado as Waterfront City with the abundant ocean tourism potential available, the government of North Sulawesi is giving special attention in developing tourism sector. The major responsibility to realize the program of the provincial government lies in the hands of the government of Manado City. As the provincial capital, Manado City is the host which is responsible for promoting tourism of the area particularly its marine tourism. The efforts for building branding destination, the master plan of the spatial city planning has focused on the city with waterfront image and develop the coastal area with a well planned as well as aesthetically managed and controlled both river and coastal waterfront (Manado spatial city planning 2007–2027). The City of Manado consists of a large area, comprising land, sea areas, and islands in the northern part. The combination of land and sea results in a special natural tourism potential, such as coastal areas with beautiful panoramas, a wealth in sea ecosystems, a wealth in coastal society cultures, a city with a mountainous panoramic views, lakes and other elements that are the basis of tourism of the region.

Coastal tourism plan for Manado City is focused on analysis of marine tourism potentials and tourists arrivals because Manado is in the strategic position as an entry gate and perceived as a display room for it serves the North Sulawesi visitors at the first place before visiting other places within the province. Physically, Manado has tourism product and market segment potential for marine tourist events; core events, major events and supporting events. One of the most popular annual events in Manado is coastal festival in July each year. All the attractions are conducted along Manado coastal areas; Boulevard, Karang Ria, Malalayang and Bunaken island. The event is conducted and delivered to celebrate Manado City anniversary which provides activities such as industry expo, cultural festival, traditional culinary festival, and traditional marine sport tourism festivals.

## History of Waterfront Development

McGovern (2008) points out that the waterfront in Philadelphia is situated at the edge of the central city and adjacent to a refurbished historic district and serves as a center of commerce, tourism, and recreation. Waterfronts have unique characteristics and, therefore, provide special opportunities for urban development (Malone 1996). Waterfront development has occurred when capital has been available and, in recent years, in response to changes in the global economy (Amin and Thrift 1992). The history of waterfront development is well documented in the academic literature. For the last 200 years, waterside locations have been used for port facilities, manufacturing industry, boat building, repair and maintenance, drainage and sewage treatment plants (Craig-Smith 1995). Redevelopment of waterfront areas in the United States began in the late 1950s under the Urban Renewal Program. In the 1970s, urban regeneration of waterfront areas emerged as an area of study by architects, planners, and urban geographers in North America (Craig-Smith 1995). Ten years later, political scientists, geographers, and economists fostered study of waterfronts in Europe (Hoyle et al. 2002 in Craig-Smith 1995).

Contributions to the waterfront development literature have also emerged from Australia. Bradbourne (1989) traced the international nature of the waterfront revitalization processes. In United Kingdom, waterfront development has been undertaken by Urban Development Corporations (UDGs). The development and redevelopment of Manado City are currently concentrated along the coastline of Manado Bay. This can be seen in the development activities in the reclaimed areas where new land has been created. The land reclamation along Manado Bay, where MWD is taking place, is a huge activity which has been taking place for more than a decade. Thus, the area of reclamation has changed over time. However, the waterfront development has raised growing criticism and high concern from various parties, such as environmentalists, NGOs, and academics regarding shoreline revitalization, particularly the creation of new land for waterfront development. Given this strategic issue, it is important to critically review what has been done in terms of protection and improvement of the environment leading to sustainable development.

On the one hand, waterfront development and the considerable tourism attraction potential attached to it have created good opportunities to gain economic benefits through regional and community development. On the other hand, environmental degradation has gradually increased within and surrounding the area. In such a case, there is a need to reconsider the balance between these two important aspects of development to make sure that as many stakeholders as possible share in the benefits. Therefore, in the process of planning and development, it is essential to have active participation from different types of groups and institutions to seek their insights and to incorporate them into the development program.

## Urban Tourism

Urban tourism has been used to create economic growth by providing new employment opportunities and by increasing business capacity. This has led to an increase in planning for tourism in cities based on existing resources and the creation of new products, providing a challenge to develop urban tourism in a sustainable manner. According to Dieke (2005), tourism planning refers to the methods policymakers adopt to achieve tourism development objectives. Such planning can occur at national, regional and local levels, incorporating the following components: (1) analysis of demand; (2) analysis of the availability and quality of tourism assets; (3) forecasting of visitor demand; (4) costing and financing of the tourism plan; (5) human resource development issues; and (6) marketing. Dieke stated that implementation of urban tourism plans require a plan of action.

When the plan is accepted (usually by government), it should also have incorporated three additional components: (1) an implementation strategy (action plan); (2) a monitoring procedure (Is the plan meeting the objectives and/or have unforeseen problems or other difficulties arisen?); and (3) an evaluation function which relates to an assessment of whether objectives have been achieved, need to be modified or discarded. Such a process should ensure that the plan and its implementation are constantly monitored so that they can be altered as necessary to meet changing market conditions or priorities. To plan effectively for urban tourism development while lessening its negative effects, planners need to understand the multiple sectors that exist in cities and their relationship to tourism and how these have been changing over time and space. Tourism is often considered as a challenge to the sustainability of urban environments, just as it is within wilderness and rural environments (Hinch 1996).

Urban tourism in Manado has the potential to attract. This has led to an increase in its tourism profile through product development based on tourists' demands. Tourism is seen by local authorities as a significant sector to generate economic growth. This has led to an increase in planning for tourism in the city based on existing resources and the creation of new products. Tourism in Manado is in an expansion phase and tourism is being used as a development catalyst. At the same time, there is a need to invest in other tourism resources, such as heritage and/or historical attractions and infrastructure, in order to enhance the tourism image leading to competitive advantages for the city. However, urban tourism development provides many challenges for Manado City if urban tourism is to be planned and developed in a sustainable manner. Urban tourism in Manado has emerged as a result of intensive development of tourism infrastructure and product development that has required a process of tourism planning. However, the forms of urban-based tourism require further studies in order to understand the phenomenon and the complexity of urban functions which will influence tourism development in the area. Waterfront development, as a part of urban tourism, has been adopted to support the growth of the city. Heavy investment by the local authority in providing infrastructure for tourism facilities, including through the development and redevelopment of the waterfront, requires integrated planning for the overall urban tourism development in the area and integration with broader urban development concerns.

## Cluster-Based Urban Tourism Development

Urban tourism models for less developed countries have received limited attention. Therefore, it is necessary to consider: (1) how the urban tourism concept should be applied (Singh 1992), and (2) the evolution of urban tourism space (Weaver 1993) with specific reference to case studies in less developed countries. Blank (1994) identified five major factors that characterize cities as tourism destinations: (1) high populations, which attract high numbers of tourists who are visiting friends and relatives; (2) major travel nodes that serve as gateways or transfer points to other destinations; (3) focal points for commerce, industry, and finance; (4) concentrations of services such as education, government/administration center, health, and others; (5) places that offer a wide variety of cultural, artistic, and recreational experiences. Wall (2006) argued that complexity, which is inextricably melded into the nature and structure and of urban tourism, gives rise, at the same time, to many challenges and opportunities.

Fainstein and Judd (1999) asserted that the complexity of urban tourism can be addressed through three elements that involve (1) the tourist, (2) the tourism industry, and (3) the city. According to, Singh (1992), in the case of cities such as Lucknow in India, tourism should exist in the city as part of a concern to secure its heritage assets. In addition, the unique characteristics of cities in LDCs countries provide a different perspective on the adoption of tourism from that in developed countries. Urban tourism in LDCs has involved intensive development of tourism infrastructure and the adoption of a process of tourism planning. The spatial distributions of tourism, commercial, and residential land uses overlap as can be seen in the complexity of urban functions.

## Methods

Manado waterfront was selected as the site for this study because it is a location that is undergoing massive land reclamation in an area that has important tourism resources in Manado. For development purposes, Manado waterfront has been divided into three clusters that are to be developed consecutively: cluster A, cluster B, and cluster C (Fig. 23.1).

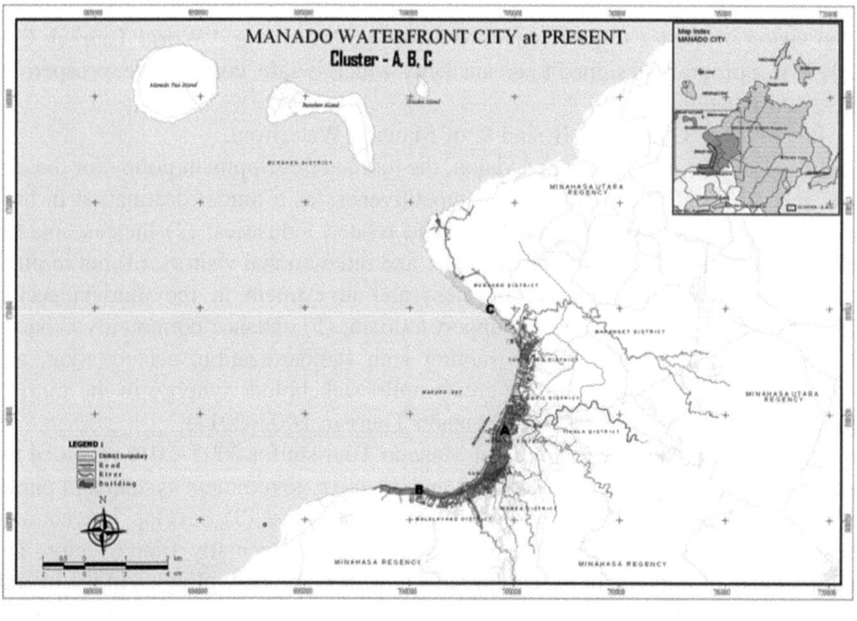


**Fig. 23.1** Map of Manado Waterfront City (Author, 2014)

However, for the study purposes, cluster A was selected as the study area. This site was selected due to the high level of use by the local community. Changes in waterfront uses to date have primarily taken place in cluster A. Consequently, massive environmental impacts have arisen as economic and social benefits have been sought from MWD. Division of Manado waterfront into cluster A, cluster B, and cluster C is shown on the map in Fig. 23.1. This research focuses on the case of waterfront development in Manado, Indonesia, which has been planned since 1991. Case studies are a qualitative strategy through which the researcher explores in-depth a program, event, activity, process, or one or more individuals.

The case(s) are bounded by time and activity and the researcher collects detailed information using a variety of data collection procedures over a sustained period of time (Cresswell 2009). The advantage of a case study approach is that it permits a particular case to be explored in-depth. However, there is no guarantee that a particular case is typical and this limits the extent to which findings can be generalized to other cases. This study examines the MWD program, process, and the people involved in the project. City government of Manado has introduced waterfront development to the public over a period of almost two decades. This case study involved fieldwork comprising questionnaire surveys, interviews, and on-site observations to provide evidences of waterfront development in Manado.

## Research Results

### *Result from On-Site Observation and Documentation*

The vision of the local government to designate Manado as "World Tourism City 2010" is a program designed to create a city which is safe, competitive, prosperous, just, and prestigious.

Figure 23.2: Cluster A, B, and C of Manado Waterfront.

Based on the aforementioned vision, the tourism development policy for the city has several goals to: (1) maintain competitiveness as a tourist destination in high quality of service, products, facilities, and related industries; (2) increase market share through marketing to both domestic and international visitors; (3) put in place an incentive system to attract business and investment in the tourism sector; (4) develop the infrastructure to support tourism; (5) enhance community capacity building through education and training with standardization, accreditation, and competence certification; and (6) create multi stakeholder synergy in the tourism program designs (Master Plan of Manado Tourism 2007–2017).

Furthermore, the City Planning of Manado Tourism for 2005–2016 recorded the strategies to (1) implement an efficient and effective government system and public services; (2) design tourism-based spatial distributions; (3) develop international scale infrastructure; (4) create a pleasant and environmentally friendly city. The efforts made to implement these programs have become very obvious in the overall city development, especially focused on the waterfront. Cluster A, B, and C of

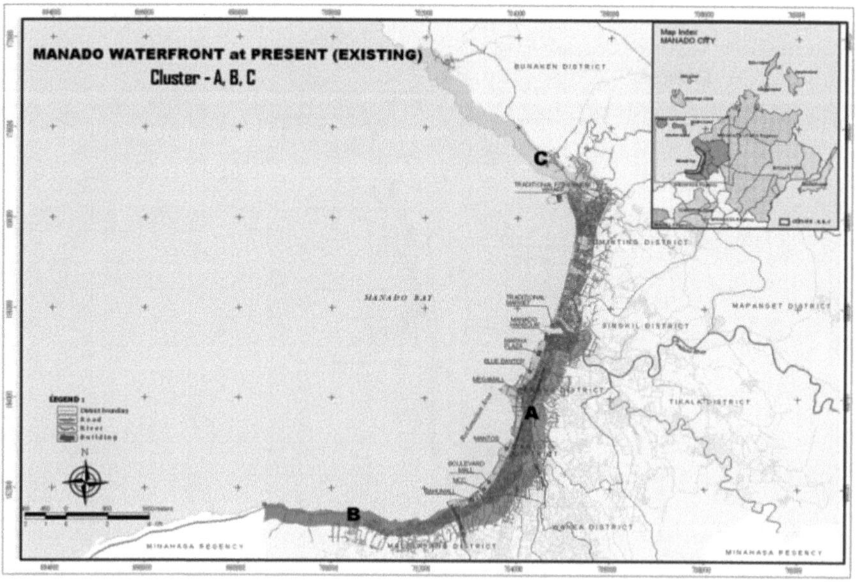


**Fig. 23.2** Cluster *A*, *B*, and *C* of Manado Waterfront (Author, 2014)

Manado Waterfront development (MWD) (Fig. 23.2) occurs along the coastline of the city to serve the communities with their various needs and interests. Manado waterfront with its various kinds of tourist attractions is being developed to enhance and strengthen opportunity both for business purposes and as a tourist destination. Waterfront development has been extensively and urgently expanded for several reasons. The waterfront area especially cluster A is designated as a center for trade and business, leisure, lifestyle, and various tourism needs and activities. The plan is to equip the city with various business and tourism facilities as an integrated part of the city development plan which has currently and extensively being implemented.

## *Cluster A of Manado Waterfront in the Past*

MWD especially cluster A has been established since 1991 with the length of about 3 km where land reclamation existed. While MWD has been unofficially divided into cluster A, B, and C, cluster A has been enormously expanded for business and trade center for the whole city. This can be observed from the field that all extended services and products provided within the areas are mostly for trade and private enterprise expansion. However, this research revealed that cluster A has currently developed rapidly with multi purposes at the gradual stages of empirical developmental stage. In the past, the majority of Manado coast area has been utilized, accessed, and dwelled by traditional fishermen (Fig. 23.3). The map illustrates the physical environment of Manado waterfront where traditional fishing occurred in

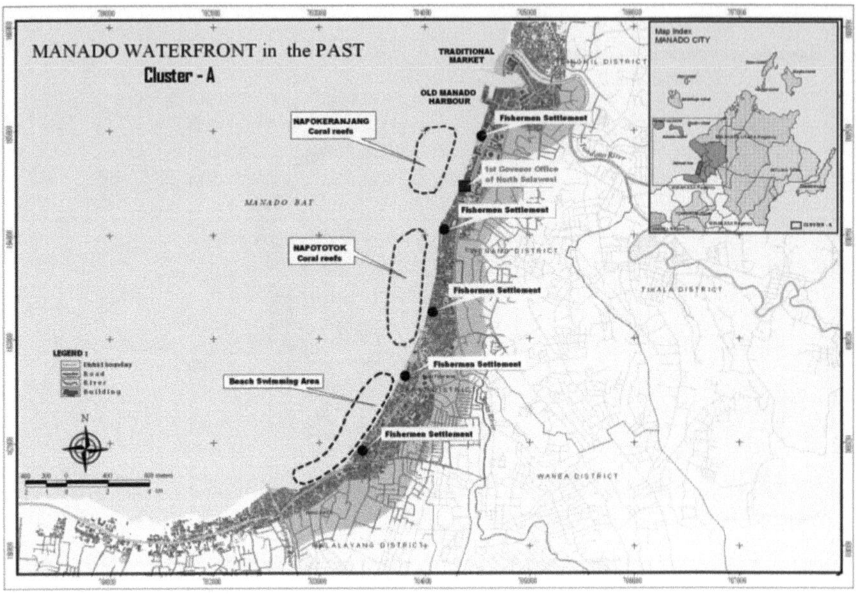


**Fig. 23.3** Manado waterfront in the past (Author, 2014)

the past before land reclamation. The map shows that Manado City is mostly located along Manado coast and resided mostly by local fishermen.

The first governor office was also located right at the coast. Manado harbor is at the end of the study site with dive spots surrounding where abundant of coral reefs emerged. This was the place where local community depend their lives on. As the reclamation is taking place, rapid changing and massive coral destruction on sites are unavoidable to serve the need for creating trade, business, and enterprises area for economic benefit gains purposes. Reclamation within the area has recently created bizarre disaster to the environment. Waterfront development within cluster A has failed to consider environmental protection for future generation. Historically, this area has empirically served various needs such as.

## Fishing Ground for Traditional Fishermen

Manado waterfront in the past was mostly used as traditional fishing ground for local fishermen. Traditional fishers in the past have extensively utilized the coast and the sea as their place to support their living. They had free access to the sea for their entire fishing activities from the dawning to the end of the day. However, this has changed due to the land reclamation along the Manado Bay. Since then, fishermen became apathetic and marginalized to the two end sides of the reclamation area of cluster A. Consequently, there have been extensive criticisms by the fisher society that the developers and government have failed to protect them from the marginalization as a result of reclamation along Manado Bay. This has currently created very contradictive issues on MWD where the elites are allowed to operate their business in the area while ignoring to protect traditional fishermen's sociocultural environment who have lived there for generations. On one hand, the public admitted that the fishermen should be protected and facilitated to be removed to an appropriate place. On the other hand, the city government has no courage and goodwill to discuss this issue in a public discussion to solve problems. Ironically, the land reclamation along Manado Bay within cluster A has currently kept moving to further stages for expansion. This has created unending debates among stakeholders for years. The agreement has never been reached as many interests and concerns have to be taken into account.

## Fishermen Wharf

As the fishermen had their activities as local fishermen from whom the local residents buy fish, they traditionally used the coastal area especially around Manado port to anchor their boats. Conventionally, Manado coastal was once used as a temporary fishermen wharf and boat terminal where fishermen sold their catch to local people (Fig. 23.4).

**Fig. 23.4** Traditional fishermen's boats along Manado Coast (Author's file, 1990)

Apparently, this area was historically very popular for the local residents as a place to buy fresh fish with reasonable prices to meet their daily needs. Fish market has temporarily and locally existed in this place in the morning and late afternoon. For this purpose, the fishermen had access to Manado coast closely and freely. This was the place where they earned money for their living to maintain their family' lives.

## Beach Swimming Area for Local Residents

For recreational activities, the Manado coast within Manado bay was also used as a swimming area with warm water and clear vicinity. It was a place for local people to have recreational beach swimming in the morning or in the afternoon while enjoying the sunset and the mountain '*Gunung Manado Tua*' panoramic views. Swimming activities had traditionally occurred within the area for the local residents. Easy access to the opened beach by the local people to have free swimming sport might raise an awareness to protect the natural environment for future generation.

## Meeting Points for Local Communities

Historically, people had meeting along the coast for several purposes; social gathering, marine boat festivals, and government meeting along the Manado Coast especially where the old harbor had existed at this place. Accessibility to the waterfront was very popular for many activities and interests. Sense of belonging to maintain the place as a central focus for residents' leisure interests emerged.

## Old Manado Harbor as Local Sea Transport Center

In the past, Manado waterfront particularly at the old harbor, the sea, and the coast has had enormous influences on the local transportation for goods and passengers (Fig. 23.5). Vast majority of transporting materials by the Manado port has improved access to marine settings. Traditional boats and other manual marine-based vehicles were used and terminated at the harbor. The old harbor had historically been a transportation center for the city residents and visitors to travel to and from Manado to other nearby island destinations. With respect to the new inventions of the means of transportation, the use of Manado port has changed gradually. Today, a wide variety of machines and transportation equipment is used to extent safety and convenience for easy access to the sea. Unfortunately, the result is an increasingly diverse range of chaotic transport activities at and surrounding Manado harbor and Manado waterfront settings.

Fig. 23.5 Chaotic mixed uses of Manado port (Author, 2009)

## High Public Access

Manado waterfront is a core place for local fishers to fish and relax while watching sunset for free. Prior to land reclamation, Manado Coast was exploited by local residents to enjoy sunset during a relaxing time. Numbers of city residents came to approach this site just for doing occasional recreational activities such as fishing, rowing, or just sitting along the coast. Since the reclamation is taking place, the free access has reduced dramatically and changed the number of people visiting and approaching the coast. A massive increase of land reclamation in the area was inevitable and occurred over the last few years. Today, the increased use of the area for reclamation has limited the public access to Manado waterfront. The boulevard is improved and the malls, shophouses, and business centers are constructed within the area. These facilities served to increase the popularity of the Manado waterfront as a business center while concerns on public access to waterfront are denied.

**Boulevard for Pedestrians**

Due to the easy and free access to Manado waterfront, the Boulevard was used for most pedestrians (Fig. 23.6). The interaction between people and waterfront was high and the local populations have easy access to the favored areas particularly coast or beaches. The pedestrians were able to walk close to the shore line. Nowadays, this area has been blocked by buildings between the boulevard (road) and the shore lines. As a result of the rapid development and change of uses of the Manado waterfront, people's access to the waterfront is hindered.

## *Cluster A of Manado Waterfront (Existing)*

The creation of new land along Manado waterfront (Fig. 23.7) has currently operated by six developers such as Megasurya Nusalestari Ltd (36 ha), Multicipta Perkasa Nusantara Ltd (6 ha), Bahu Cipta Persada Ltd (7.5 ha), Papetra Perkasa Utama Ltd (5,325 ha), Sulenco Boulevard Indah Ltd (9 ha), and Gerbang Nusa Perkasa Ltd (10 ha) (City Spatial Document, 2008). As the growth of land reclamation area has become widespread, the large numbers of developers show that significant environmental, economic, and sociocultural damage results from the MWD. MWD on Manando Bay has currently continued to exist.

As the stage of waterfront development continues, resulting in an increasing mode of reclamation, increasing numbers of losing habitats, increasing development of the sites infrastructure to cope with business and trade needs while decreasing environmental quality. Since the land reclamation in 1991, significant

**Fig. 23.6** Boulevard for pedestrians (Anton, 1992)

**Fig. 23.7** Land reclamation for Manado waterfront development (Author's file, 1996)

extent of traditional land uses within the area has rapidly changed. Consequences emerged particularly in the heavily used Manado coastal areas by six developers where this research focused on. There are significant numbers of issues which currently illustrate the rapid development of the land reclamation along the existing Manado waterfront. Based on field observation, such development has been intended for long planned business and trade locations and has relevance to economic benefits gain purposes

## Chaotic Manado Port

Manado waterfront especially on the harbor area has a mixed use for transporting goods and passengers including tourists. Historically, this area is actually a heritage from the Dutch Indies government. Therefore, the Manado Harbor area is a historical remnant that must be maintained because it held an important role in the historical development of Manado City. The role of the area as the only harbor in Manado City is strongly related to the development of the surrounding trading area as well as Manado City and other areas within North Sulawesi Province. For the city tourism development especially on the waterfront, this historical aspect is considered one that has a selling point as tourism object. However, a description of the existing Manado harbor seems to be like a chaotic uses comprising of unsafe guarantee of sailing and the shipping services include domestic shipping and public shipping.

## Unorganized Traditional Local Market

Right beside the Manado port area, there is a traditional market called "*Pasar Bersehati,*" the oldest and the biggest traditional market in the city. This traditional local market provides fruits, vegetables, fish, and meat which are mostly all locally produced. Daily commodities at reasonable prices can be found in this market. The city harbor of Manado City is just 100 m down the street. However, the chaotic market condition and environment appear to be the problem. Agricultural products, vegetables, foods and beverages, various fruits, various cattle and pottery, meat, fish, sugar and cake, fabrics, and handicraft are sold here without building blocks for selling grouped different goods.

## Boulevard on Business (B on B)

Boulevard has been designated as the center of business or Central Business District (CBD) where the city residents are doing various business activities. This place offers opportunity to establish business and trade for economic benefits purposes. Commercial enterprises such as shopping malls, banking, cinemas, entertainments, hotels, convention centers, trades, and office areas occurred within the area to serve the basic needs of the city residents. The area is very popular with the term B on B (Boulevard on Business).

## Extensive Construction of Malls and Shophouses

An important concept that illustrates the dangerous impacts to natural environment posted by massive land reclamation along the Manado waterfront is the 'economic successions.' Because of the higher demands to live surrounding the waterfront and the majority of the city residents do shopping on boulevard, the development of the malls and shophouses show a corresponding increase. Figure 23.8: Megasurya Nusalestari Area

As the biggest developer, Megasurya Nusalestari has so far established 386 units of shophouses (Fig. 23.8) consisting of Mega Style (181 units), Mega Bright (25 units), Mega Profit (24 units), Mega Smart (156 units) (Company Profile of Megasurya Nusalestari, 2008). It has been observed that this area becomes the central development of Manado waterfront for cluster A as the development sites are discovered and used for businesses and trading companies. Due to the rapid development of malls and shophouses businesses, the Reef Restoration Group has cynically mentioned that the vision for Manado as a waterfront city might be parallel and associated to Manado as a mall city. This group argued that is based on recent trend and the population growth, the projected future of the land reclamation within Manado Bay is for chasing economical development and failed to take care of the coastal environment.

## Chaotic City Transportation, Traffic Congestion, and Pollution (Water, Air and Noise)

The other environmental impacts of MWD have been a significant increase in pollution (water, air, and noise) and traffic congestion. There are a couple of transportation options to get within Manado City. Private taxis and public transport vehicles called '*mikrolet*' are the most common means of transportation for city residents and visitors. Since Manado City's center is on the waterfront along the boulevard, it is easily accessed by the city residents as well as boats from islands destinations. The reclamation and building construction activities have absolutely created city chaotic transportation along the boulevard area, at the heart of Manado City. The most common disturbance for physical use of roads along the boulevard is an increased queuing for local transport. The traffic demand surrounding the B on B area is great resulting vehicles slows the speed of the traffic stream and congestion is incurred. This condition creates the extreme traffic congestion within the area as the vehicles are fully stopped for periods of time then the traffic jam occurred. Ultimately, land reclamation within Manado Bay has significantly contributed to harmful air, water, and noise pollution. Besides, pollution from cars and trucks to and from the reclamation areas has affected coastal residents living near busy boulevard and impacted to local air quality.

**Fig. 23.8** Megasurya nusalestari area (Author's file, 2008)

## Center for MICE Tourism

There are still a few of exhibitions and fair centers that have been built in Manado City. However, MICE tourism has recently grown in Manado City to serve the need for international events held within North Sulawesi. Manado Convention Center (MCC) on boulevard area becomes a famous venue for the events (Fig. 23.9).

Manado has a long coastline with tourism potentials, hotel, and resorts. There are services for conference and convention in accordance with eating, relaxing, and city tour program services. Exhibitions and business and commercial fairs are held in MCC on the Manado waterfront. This provides opportunities for MICE tourism to be developed in the area. A number of regional and international cultural events and exhibition fairs were held successfully in MCC such as World Ocean Conference (WOC) and Coral Triangle Initiative (CTI) in May 2009 as well as Sail Bunaken in August 2009. The events created many cooperative chances in Ocean science and technology including tourism. The WOC held on May 11–15, 2009 in Manado was a formal and very important meeting of heads of the states that have coastal and marine territories. Scientists, NGO's, journalists, private sector, and other stakeholders attended the meeting to discuss current issues in the marine related to climate change and degradation of marine resources. This was aimed to achieve international commitments in sustainable development of marine resources and the

**Fig. 23.9** Manado convention center (Author, 2009)

prosperity of mankind. The event was attended by about 1000 participants from 111 countries. Other side events in WOC were presented such as International Ocean Science, Technology and Policy Symposium, as well as the Technology and Industry Exhibition. The 33 provinces of Indonesia and all regencies within North Sulawesi have taken part in the exhibition. Other local events such as Manado expo, marine festivals, music and cultural festivals, and international choir competitions have also taken place in MCC (Manado Convention Center).

#### Limited Access for Fishermen

Traditional fishermen who were marginalized as a result of reclamation activities move on. They found to resettle at two ends of Cluster A, nearby Malalayang River and Tondano River. At the end by Malalayang river, right at the border line between cluster A and cluster B of the Manado waterfront resettled traditional fishermen as a result of a rapid and huge reclamation within cluster A. Furthermore, public space to enjoying free sunset as well as city park and green space are no more available as consequences follow the massive reclamation activities on the waterfront. To conclude, the overall result of economic succession within cluster A of Manado waterfront is a gradual development of facilities and infrastructure and gradual loss of pristine environment and environmental quality. Assessing the impacts of the intensive reclamation and development on Manado waterfront has been a challenge for sustainable development concept and its implementation.

### *Results from Questionnaire-Based Survey*

#### Roles of the Stakeholders in Decision Making Process for MWD

Table 23.1 and Fig. 23.10 present kinds of meeting and rates of contribution to the MWD with the average score where informal meeting (2.45), multi stakeholders meeting (2.40), public consultation (2.34), consultant meeting (2.31), and workshop (2.22).

Interestingly to point out here that informal meeting played the most significant roles in the participation for MWD (Fig. 23.10). This indicates that the city residents as the users of Manado waterfront are initiative, responsive, and highly concerned on decision-making process for MWD.

Furthermore, the highest rates of respondents who valued their involvement in the meeting are very important (44 %). They have a high level of self-confidence that their engagement in the decision-making process through meetings will contribute and determine MWD in the future. It is important that the city residents be aware on the importance of their participation that will bring their insights to be taken into account during the process of MWD.

**Table 23.1** Kinds of meeting and rates of contribution

| Meeting and rates of contribution | Not significant | Somewhat significant | Very significant | Mean score* |
|---|---|---|---|---|
| | % | % | % | |
| Workshop | 9 | 7 | 16 | 2.22 |
| Multistakeholders meeting | 9 | 6 | 22 | 2.40 |
| Public consultation | 7 | 5 | 17 | 2.34 |
| Consultant meeting | 8 | 4 | 17 | 2.31 |
| Informal meeting | 7 | 3 | 21 | 2.45 |
| Others | 0 | 2 | 0 | 2.00 |

*Source* Survey, 2014
(*3 = Very significant; 2 = Somewhat significant; 1 = Not significant)

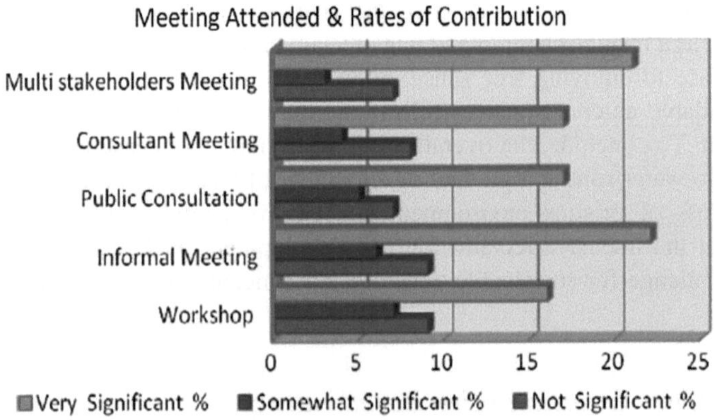


**Fig. 23.10** Meeting attended and rates of contribution (*Source* Survey, 2014)

## Manado Waterfront as Part of Urban Tourism Planning

Manado has been designated as a waterfront city for it is mostly located along the coastal line. Tourism was expected to be the leading sector within the region and it is considered as having important roles in the city development. Figure 23.11 shows the respondents ideas on Manado waterfront as an important part of tourism planning where 53 % strongly agree and 41 % agree.

Therefore MWD as a part of urban tourism planning is expected to become the center of economic activities for city development. For this concern, Manado has been declared to be the tourism city by government and this vision has been well recognized by communities. However, it is evident that the MWD has been so far prioritized for business and commercial purposes only. As a matter of fact, the priority for city government to include tourism in the regional development program has been blurred without detailed guidelines in the implementation.

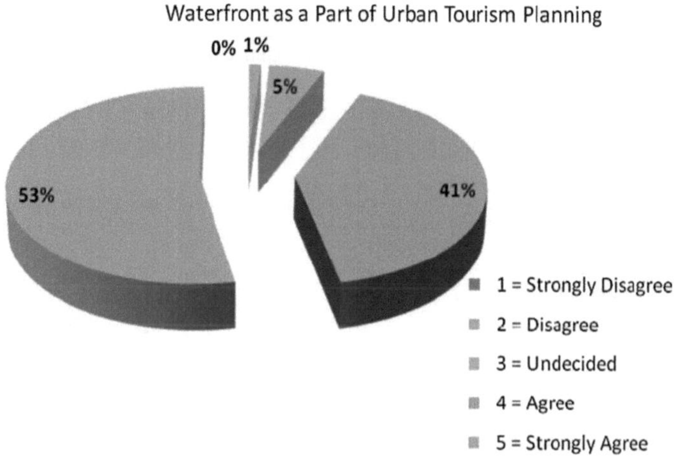


**Fig. 23.11** Waterfront and urban tourism (Survey, 2014)

## Impacts of MWD

The general public sample expressed their thoughts regarding the impacts of MWD (Table 23.2). More than half made strongly favorable responses regarding positive aspects of MWD; (1) It was seen as an important part of city tourism (59 %); (2) It will improve the city's potential as a tourist destination (58 %); (3) It is good for local economic development (57 %); (4) It will bring more tourists to the city (56 %); and (5) It will have positive impacts on local business (55 %). Thus, it was widely and strongly considered that it will be positive for tourism and economic developments in Manado. The acquisition of tourism benefits were widely recognized as a development priority. However, it was also acknowledged that infrastructure, human resources in tourism, and other supporting facilities needed to be strengthened. The results also indicate that majority of the city residents and business operators at and around Manado waterfront have recognized positive impacts of MWD both for the community and region development.

Impacts of MWD is illustrated on Table 23.2 have been, respectively, arranged based on the mean scores. It is evident that the highest value is that MWD has been perceived as an important part of city tourism (4.62) and MWD improves city potential as tourist destination (4.20). This implies that MWD is highly considered as it had strongly influenced the tourism development in Manado. The general public is likely to have a strong idea for Manado as a waterfront city and a tourist destination as well. For this reason, benefits gained from tourism have been widely recognized as a priority of the city development.

**Table 23.2** Impacts of MWD

| Issues on impacts of MWD | Strongly disagree | Disagree | Undecided | Agree | Strongly agree | Mean score* |
|---|---|---|---|---|---|---|
| | % | % | % | % | % | |
| MWD as important part of city tourism | 1 | 0 | 5 | 59 | 34 | 4.62 |
| MWD improves city potent as tourist destination | 0 | 2 | 6 | 57 | 29 | 4.20 |
| MWD improves local economic development | 0 | 2 | 9 | 56 | 32 | 4.15 |
| MWD will bring more tourists in the city | 0 | 2 | 13 | 55 | 29 | 1.08 |
| MWD has positive impacts to local business | 0 | 2 | 16 | 53 | 27 | 3.99 |
| MWD helps protection land and coastal areas | 0 | 5 | 20 | 51 | 22 | 3.84 |
| MWD improves awareness on envi. protection | 3 | 3 | 23 | 43 | 23 | 3.62 |
| MWD Improves appreciation to environment | 3 | 2 | 24 | 25 | 20 | 3.59 |
| MWD increases traffic congestion | 5 | 23 | 25 | 28 | 16 | 3.21 |
| MWD creates more crowding with the area | 4 | 22 | 23 | 32 | 14 | 3.18 |
| MWD creates noise, air, and water pollution | 9 | 27 | 27 | 21 | 13 | 2.93 |
| MWD reduces people access to waterfront | 13 | 40 | 21 | 15 | 5 | 2.44 |

*Source* Survey, 2014
(*5 = Strongly agree; 4 = Agree; 3 = Undecided; 2 = Disagree; 1 = Strongly disagree)

## Conclusions

The case of MWD is typical of many land reclamations in the less developed countries. This has been popular to create new lands that have economic benefits purposes. For the case MWD, the massive and rapid reclamation along the Manado bay have been conducted to expand business and trade areas. However, it appears that environmental degradation is inevitable when MWD becomes established within the area. Waterfront development occurs along the coastline of the city to serve the communities with their various needs and interests. On the one hand, waterfront development has created good opportunities to gain economic benefits through regional and community development. On the other hand, environmental degradation has gradually increased within and surrounding the area. In such a case, there is a need to reconsider the balance between these two important aspects of development to make sure that as many stakeholders as possible share in the benefits. Therefore, in the process of planning and development, it is essential to have active participation from different types of groups and institutions to seek their insights and to incorporate them into the development program. Waterfronts have multiple uses and, therefore, are likely to be of interest to and involve a variety of stakeholders. If the interests of various groups are to be incorporated into development plans leading to greater support, then stakeholder involvement should occur and, ideally, partnerships among stakeholders should be established.

A multistakeholder approach for waterfront development could make a substantial contribution not only to the concepts and theory of user-centered designs but also to its practice, including appropriate strategies and methods. Waterfront development in Manado, Indonesia has been extensively and urgently expanded. The waterfront area is designated as a centere for business, leisure, lifestyle, and various tourism needs and activities. However, urban tourism development provides many challenges for Manado City if urban tourism is to be planned and developed in a sustainable manner. The waterfront development has raised growing criticism and high concern from various parties, such as environmentalists, NGOs, and academics regarding shoreline revitalization, particularly the creation of new land for waterfront development. Therefore, it is important to critically review what has been done in terms of protection and improvement of the environment leading to sustainable development. The forms of urban-based tourism in Manado require further studies in order to understand the phenomenon and the complexity of urban functions which will influence tourism development in the area.

Waterfront development, as a part of urban tourism, has been adopted to support the growth of the city. Heavy investment by the local authority in providing infrastructure for tourism facilities, including through the development and redevelopment of the waterfront, requires integrated planning for the overall urban tourism development in the area and integration with broader urban development concerns. As waterfront development achieves greater importance in urban setting, tourism will most likely expand its roles in the enhancement of waterfront development functions

## References

Amin A, Thrift N (1992) Neo-marshallian nodes in global networks. Int J Urban Reg Res 16:571–587

Blank U (1994) The community tourism industry imperative: the necessity, the opportunities, its potential. Venture publishing, State College

Bradbourne A (1989) A new future for Auckland's harbour edge. Plann Q 96:15–18

Craig-Smith SJ (1995) The roles of tourism in inner-harbour redevelopment: a multinational perspectives. In: Craig-Smith SJ, Fagence M (eds) Recreation and tourism as a catalyst for urban waterfront redevelopment. Praeger, London

Creswell JW (2009) Research Design; Qualitative, Quantitative and Mixed Methods Approach, Los Angeles. SAGE Publication

Dieke PUC (2005) Critical success factors for tourism development in less developed countries (LDCs). Paper presented at the 'debate on tourism in the toolbox of development projects'. University of Leuven, Belgium, 11 Oct 2005

Fainstein SS, Judd DR (1999) Global forces, local strategies and urbantourism in Judd DR and Fainstein SS (eds.), The tourist city, New Haven: Yale University Press

Hinch TD (1996) Urban tourism: perspectives on sustainability. J Sustain Tourism l 4(2):95–110

Hoyle B, Pinder D, Husain MS (2002) Urban renewal in East African port cities: Mombasa's old town waterfront. Geo J 53:183–197

Malone P (1996) City, capital and water. Routledge, London and New York

McGovern SJ (2008) Evolving visions of waterfront development in postindustrial Philadelphia: the formative role of elite ideologies. J Plann Hist 7(4):295–326

Singh S (1992) Urban development and tourism; case of Lucknow, India. Tourism Recreation Res 17(2):71–78

Wall G (2006) Tourism, environment and community: examples from Indonesia. In: Cukier J (ed) Tourism research: policy, planning and prospects. Department of Geography, publication series occasional paper, #20, University of Waterloo

Weaver DB (1993) Model of urban tourism for small Caribbean Islands. Geogr Rev 83(2): 134–140

# Chapter 24
# Sharia Tourism Development Strategy for West Nusa Tenggara, Indonesia

**Busaini**

**Abstract** Sharia tourism consists of recreational activities that are based on Islamic values. Indonesia is a predominantly Muslim country, and this is especially true of West Nusa Tenggara Province. However, the word tourism market does not associate the province with the concept of sharia travel. The Indonesian tourism ministry has designated three provinces for sharia tourism development: Nanggro Aceh Darussalam, West Sumatera, and West Nusa Tenggara. These three provinces should be in a good position to implement this policy. West Nusa Tenggara (also known as Nusa Tenggara Barat, or NTB) is well positioned to prepare for the arrival of tourist from Muslim countries and regions because Muslims comprise the majority of the population there. NTB is known as having "a thousand mosques", as well as many Islamic boarding schools and clerics. It has many places of Islamic heritage, there are many hotels and restaurants serving halal food, and its art and culture reflect the nuances of Islam. This paper assesses potential sharia tourist destination in West Nusa Tenggara using qualitative methods to recommend a tourism development strategy.

**Keywords** Tourism development strategy · Sharia · Economic development · West Nusa Tenggara · Islam · Indonesia

## Background

The most fundamental value in Islam is that the individual's noble character is put into practice. Islam teaches commendable behavior in all aspects, such as establishing a sense of human brotherhood and getting to know each other, which tourism can help with. As described by Allah in the Qur'an (chapter 49, Al Hujurat, The Private Apartments, 13): "*O mankind! We have created you from a male and a*

Busaini (✉)
Department of Economic Development, Faculty of Economic and Business, University of Mataram, Mataram, Indonesia
e-mail: busainidr@gmail.com



A. Saufi et al. (eds.), *Balancing Development and Sustainability in Tourism Destinations*, DOI 10.1007/978-981-10-1718-6_24

*female, and made you into nations and tribes, that you may know one another*" (Yusuf 2005). Therefore, for Islamic cultures, it is important to show visitors how their religious values are sympathetic with hospitality and mutual understanding.

West Nusa Tenggara (also known as Nusa Tenggara Barat or NTB) is well positioned to prepare for the arrival of tourist from Muslim countries and regions because Islamic culture is well represented there. NTB is known as "the islands of a thousand mosques" (CTO 2015). A mosque is a place of prostration and place of worship for Muslims, as well as an activity center and various Muslim social activities. The question is whether the slogan of "a thousand mosques" has anything to do with the actual behavior and culture of the Muslim community of West Nusa Tenggara. Theoretically, it has a very close relationship with the spirit of worship in praying time in the mosque, however in fact, the mosque is crowded with worshipers only on Fridays.

Sharia tourism is a recreational activity that reflects Islamic values. Sharia travel is growing worldwide and offers increasing business opportunities for entrepreneurs, including those in West Nusa Tenggara. Awareness of the benefits of the Islamic halal concept is increasing in terms of food, travel, financial services, and in other areas. West Nusa Tenggara is a province with Muslim-majority population, but it is not associated in the world tourism market with the concept of sharia travel, as compared to other places in the world, such as Turkey and Malaysia. Muslim travelers seek vacation experiences that fit their values and West Nusa Tenggara could serve as one of the main destinations for them (Bahaman 2012).

If tourism in West Nusa Tenggara is to reach the province's goal of two million tourists, the potential of sharia tourism, and the participation of religious communities, needs to be developed. Sharia development principles are highly relevant to the goals of sustainable tourism development, growth and employment promotion, poverty alleviation, environmental preservation, good governance, integrated cross-sector participation, cross-regional participation, and encouraging public and private sector partnerships. Islamic culture is deeply integrated into the lives of Muslims in West Nusa Tenggara. As such, the development of sharia tourism should receive extensive support from both from the Government and the businesses.

Questions that need to be addressed in this current condition are: first, what are the strategies to develop the potential of tourism in the West Nusa Tenggara? and second, how does the sharia tourism contribute to support the economy of the West Nusa Tenggara? This paper aims to explore and describe the sharia tourism potential in West Nusa Tenggara, and to propose development strategy that explores these potentials as well as to estimate the contribution that can be given to economic development to support a sustainable tourism.

## Methodology

The paper relies on qualitative methods to define issues and present a tourism development strategy. A description of a potential approach to sharia tourism is laid out, as well as requirements for the infrastructure necessary to support its development. Strategies are proposed based on normative descriptions from the Qur'an.

## Discussion

The potential for sharia tourism development in West Nusa Tenggara (NTB) is based on the heritage of the region. NTB has a majority Muslim population with an abundance of mosques and numerous Islamic Boarding Schools. It also demonstrates the culture and art of the Islamic religion along with halal products and Islamic festivals. The successful strategic development of these sharia tourism products required that the government of NTB guides the tourism destination in a manner that will fulfill the prerequisites of sharia tourism. Without a focused collective action in this way, creating sharia tourism destination in NTB cannot be achieved.

Strategic tourism development needs to initially focus on the availability of supporting infrastructure and accessibility improvements within tourism destinations. This includes identifying core destination areas for development and providing adequate transportation systems. For sharia tourism, human resources must also be developed so that those responsible for providing services to sharia tourism customers will have a deep understanding of the essential aspects of sharia. This can be accomplished in collaboration with local universities and other educational institutions that educate tourism operators and produce sharia travel guides. Equally important is the standardization of Islamic tourism products, such as hotels, restaurants, spas, and travel agencies. Essentially, Lombok as a tourist destination should create a friendly, safe, and convenient atmosphere for Muslims travelers and non-Muslim travelers to achieve success in developing sharia tourism.

The major potential sharia tourism destinations in West Nusa Tenggara Province include the following:

(a) Mataram (NTB's capitol city) and its surrounding areas
(b) The resort area of Senggigi, including the three Gilis (small islands), on Lombok island
(c) The emerging resort areas Kuta, Mandalika and the surrounding area on Lombok island
(d) The SAMOTA (Sumbawa) region, including the Teluk Saleh lagoon area, the Moyo Island protected area, and Tambora Volcano and their surrounding area
(e) The Alas Utan coastal district in West Sumbawa and its surrounding area
(f) The Pototano, Maluk areas in West Sumbawa

(g) The mining town of BatuHijau and the Dodo Rinti protected forest (future mining) area in Sumbawa
(h) The Hu'usurfing district in Sumbawa and its surrounding area
(i) The Wawo-Rada-Sape harbor area of West Sumbawa

Developing a diversity of sharia tourism areas around NTB, and transportation linkages between them, will allow for a greater number of visitors and avoid over capacities during peak seasons, which will improve the level of the tourist experience.

## Marketing Strategy

Because Islam is a cultural attraction for tourists, it is necessary to provide government support for cultural development as an essential element in the marketing of NTB Province. Art gallery facilities need to be improved and national and international Islamic events should be organized. In addition, guides need to be properly educated to inform tourists about the local Islamic arts and culture. In addition, more standard marketing strategies need to be adopted to develop a marketing plan for sharia tourism, including the strengthening of stakeholders and the organizing of partnerships among them, the building of cooperative promotional networks with other sharia tourism destinations at the national and international level, and cooperation with airlines that serve Islamic tourist source countries.

## Issues in Sharia Tourism Development

The development of sharia tourism needs to take into account Islamic values to be successful. In Arabic, travel is often termed *as-siyahah*, taken from the phrase *saha-maasiyahah* (running water, to thaw, or to melt). The expression is used to refer to water that flows and runs on the ground. *As-siyahahis*, when used in a human context, means traveling on the surface of the earth to worship, increase piety, or sometimes without any purpose. Travel is basically permitted in Islam, unless there is an indication that something does not allow it. The Islamic philosopher IbnuHazm (994–1064) suggested that that traveler is allowed as long as the sun has not slipped on Thursday (before the start of the Friday day of prayer). Within Islam, most travel is allowed, within the following rules:

1. Some travel is required for all Muslims to do as a religious obligation, such as the obligatory pilgrimage (the Hajj to Mecca in Saudi Arabia).
2. Some travel is *sunnah*, which is not required by Islam, but for which a Muslim is rewarded if doing it. An example of an *umrah* ("activity")that is sunnah is to visit Mecca in the non-Hajj season.

3. Travel can be *makruh*, which is an activity that is not a sin, but is better for a Muslim to not do because the reward is not worth it (like smoking cigarettes).
4. Travel can also be *haram*, or unlawful for Muslims, if is to do an evil deed.

Popular opinion among Islamic scholars is that travel for entertainment and pleasure is allowed. It is permissible to visit non-Muslim countries for purposes of safety and security and for business purposes. In addition to the quote about brotherhood from the Qur'an above, the following are statements specifically on travel from the Qur'an (Yusuf 2005):

- Al-Ankabut:20—*Travel in the land and see how (Allah) originated the creation, and then Allah will bring forth the creation of the Hereafter, verily, Allah is able to do all things.*
- Al-Fathir:44—*Have they not traveled in the land, and see what was the end of those before them- though they were superior to them in power? Allah is not such that anything in the heavens or in the earth escapes Him. Verily, He is all-knowing, all-omnipotent.*
- Saba':18—*And We placed between them and the towns which We had blessed, towns easy to be seen, and We made the stage (of journey) between them easy (saying): Travel in them safety both by night and day.*

## Sharia Tourism Development

The development of sharia tourism begins with improving capacities for creating tourism businesses and creating policies on halal activities, both food and amusements. This requires good cooperation and education between the government, religious leaders, youth leaders and the business sector, as well as local residents.

With regard to halal, basically food and drinks of any kind are permitted, with some exception. For example, Allah says in the holy Qur'an (29): "He it is Who created for you all that is on earth." Nevertheless, anything that can harm the body, religion and morals, including as pork, liquor, and alcohol, is forbidden. Pork is forbidden in Islam based on the statement in the Qur'an (Al Baqarah 173): "He has forbidden you only the Maitah (dead animals), and blood, and the flesh of swine, and that which is slaughtered as a sacrifice for others than Allah." Whereas *khamr* (Arabic) is any beverages, foods, and substances that alters, interferes with, or controls the mind and senses. Liquor is prohibited based on Allah's statement in the Qur'an (al Maidah 90): "O you who believe! Intoxicants, gambling, Al Ansab and Al Azlam (arrows for seeking luck or decision) are an abomination and is of Shaitan's (Satan's) handiwork. So avoid (strictly) that (abomination) in order that you may be successful."

To maintain a destination as sharia tourism, it must adhere to strict policies and control of the implementation of sharia travel. This includes the provision of halal foods, as well as, maintaining a clean environment that is consistent with practicing religious values, and preserving Islamic cultural values. Sharia tourism can bring contribution to the economy of the West Nusa Tenggara Province. With its majority Muslim population, tourism can provide employment for Muslims, especially for Islamic events, at Islamic venues and on Islamic tours. Increasing employment translates into increasing revenue for the Muslim population, and therefore contributing to poverty alleviation. To support sharia tourism, the destination needs to increase its development with Islamic nuances. At the same time, this approach could foster the emergence of new innovations, based on Islamic values. In the end, creating a sharia destination that complies with the teachings of Allah will bring blessings from Allah.

## Conclusions

The development of a sharia tourism destination is feasible and will bring much opportunity for West Nusa Tenggara Province, due to its majority Muslim population and culture. A long term strategic development approach should be adopted that integrates all relevant stakeholders. The teachings of Islam do not prohibit travel for good purposes. However, halal products and activities cannot be compromised because these are the provision of God. Economic benefits that a sharia tourism destination could bring include, among others, an increase in labor absorption and an increase in income for many Muslims in West Nusa Tenggara. Better relationships and increased friendship among Muslims across the world would also be an expected outcome.

## References

Bahaman FS (2012) Panduan Wisatawan Muslim (Muslim travelers guide). Pustaka Al-Kausar, Jakarta Timur (in Bahasa Indonesian)

Cultural and Tourism Office of West Nusa Tenggara Province (CTO) (2015) Calendar event of West Nusa Tenggara. Mataram Press, Mataram

Yusuf A (2005) The Holy Qur'an, text and translation. Islamic Book Trust, Kuala Lumpur

# Part III
# Balancing Sustainability in Tourism Economic Development

**Akhmad Saufi**

University of Mataram, Mataram, Indonesia

Since the end of World War II, travel and tourism has grown to be the largest service industry in the world. Many countries develop tourism for the economic benefits it can bring. Sharpley (2002) points out that tourism is an agent of development. Promoting tourism means advocating the development of economic, socio-cultural and environmental sectors. Yet, the development of tourism can bring either positive or negative impacts to a destination. As a result, studies on tourism still focus on how to balance development and sustainability, particularly with regard to how economic activities impact the local environment and culture. Liu (1998) ascertained three economic costs of tourism development. First, the tourism industry is susceptible to social and political conditions, which can result in economic instability if a place becomes over reliant on tourism. Second, tourism can create an inflation affected due to differences between tourist and host buying power. Third, tourism development may shift opportunity cost, encouraging local workers to move away from other economic sectors, resulting in their downturn.

Tourist interest in the products of local culture can stimulate local efforts to preserve their culture and heritage. This, in turn provides opportunities to gain economic benefits from tourism. For instance, cultural shows in Bali, Indonesia, are organized and promoted by small local enterprises. The operation of the cultural shows helps distribute income among local people and increase their living standards. The economic benefits that locals receive from products of their culture can increase their sense of community belonging, identity, togetherness, dignity and self-esteem. Understanding the positive and negative impacts of tourism can help to optimize its economic benefits and avoid its economic costs.

A study by Wulandari and Andilolo (Chap. 25) suggests that the implementation of cashless payment system can enhance tourist experiences in a destination. In theory, cashless payment systems can increase the quality of tourist services, can influence tourist behavior for the better management of impacts, and can increase the quality of tourist experience. Innovations of this type, however, require an entrepreneurial spirit, which can be enhanced through internal marketing. Internal marketing refers to the socialization of tourism development among tourism stakeholders, including practitioners and destination communities. The success of the internal marketing is indicated by the emergence of local tourism entrepreneurs who undertake external marketing of their destination to tourism markets. According to Walansendow and Lagarense (Chap. 26), the economic benefits of tourism can stimulate local entrepreneurs to constantly create high quality of local products, which in turn helps create a good image of a tourist destination.

Information and Communication Technology (ICT) plays an increasingly important role in tourism marketing. ICT is a platform for word of mouth (WOM) sharing and promotion in tourism, which Panchal et al. (Chap. 29) examine by looking at how young Indonesians living in Singapore perceive Derawan Island as a domestic destination in Indonesia. Their study finds WOM and social media (or eWOM) are common means used by young people to acquire travel information, and communicate and spread their travel experiences to others. It is therefore essential for increasing the awareness of tourism destinations. Afif et al. (Chap. 30), on the other hand, emphasize the important role that profesional photographers are coming to play in influencing the decision of tourists to visit destinations. ICT also helps distribute and increase income from tourism industry, as elucidated in Nugroho and Sihite's study (Chap. 31). For example, ICT can be used to increase the service quality of airlines and open business opportunity for local people in a tourist destination.

Data on tourist distributions can help in tourist destination marketing. Aktas et al. (Chap. 33) studied the spatial distribution of tourists in Turkey to help develop appropriate strategies for identifying market segments. They found that factors including motivations for travel, country of origin, and climate and environmental conditions can impact Turkey's income from tourism. In another study of Turkey, Yuncu et al. (Chap. 32) emphasize the importance of targeting domestic tourists and building facilities for them to better distribute tourist visitations across the country. Such efforts can increase the number of tourists in all destination and increase economic benefits to local communities.

Marketing can be used to increase the importance of beaches as a sustainable tourism destinations, as elaborated by Thiruman and Raghav (Chap. 36). They propose five phases of beach growth from a marketing perspective, including holiday destination, water fun destination, local beach festivals, mega event beach destination, and beach homes and urbanization. Their findings provide reference for tourism decision makers and Destination Marketing Organisations (DMO) in controlling the number of visitors and building appropriate tourist facilities and customer services (Thirumaran and Raghav).

Azmi et al.'s (Chap. 27) study elaborates how government organizations, traders, and tourism distribution channels are the key elements in the tourism supply chain in bordered areas of Malaysia. Failure in collaborations among these key players will result in poor planning and development of a tourist destination. For example, the low level of local involvement in the tourism supply chain stimulates outside players to replace locals and eventually come to control local economic transaction within the tourism sector. This creates excessive economic leakages. Hamid and Isa's (Chap. 28) study focuses on the tour operator practices as important player in the tourism supply chain. They suggest that tour operators can contribute to sustainable tourism development through corporate social responsibility (CSR) practices that intervene in the supply chain by supporting sustainable tourism and ecotourism attractions. Tour operators can, through their selection of the products they sell, help distribute income from tourism throughout the supply chain, provide employment, and open entrepreneurial opportunity for local people.

In the context of tourism destinations, competitiveness is related to the ability of one destination to compete with another. Such an ability is based on the uniqueness of the destination, including its natural landscape and local culture, that differentiate it from others places. Competitiveness also includes managerial knowledge and the skills of tourism stakeholders to manage and promote their destination, and to make it more attractive than others. According to Hanafiah et al. (Chap. 34), culture and nature are the core resources of tourism and are the key elements of destination competitiveness and performance. Strong and healthy cultural and natural resources, in turn, help create sustainable economic growth. Differences in cultural and environmental policy-making within a destination and within a country can, therefor, directly impact tourism performance. Pawitan et al. (Chap. 35), further highlight the importance of government support in the development of tourism competitiveness. Effective tourism competitiveness requires socio-political support through government intervention in the business environment, in social regulations and in human resources development. Their findings indicate a reciprocal relationship between economic impacts and the government's socio-cultural and political influences on the tourism industry.

## References

Liu Z (1998) Tourism and economic development a comparative analysis of tourism in develop and developing countries. In: Tisdell A, Roy KC (eds) Tourism and development: economic, social, political and economic issues, vol 21–37. Nova Science, Commack

Sharpley R (2002) Tourism: a vehicle for development. In: Sharpley R, Telver DJ (eds) Tourism and development: concept and issues. Channel View, UK, pp 11–34

# Chapter 25
# Preliminary Study on the Cashless Payment as an Experience-Based Marketing Vehicle in Tourism Destination

**Nuri Wulandari and Imanuella Romaputri Andilolo**

**Abstract** Tourism has undergone fundamental changes due to the vast developments in Information and Communication Technology (ICT) over the past decade. One of the most important aspects of technological development has been in the area of payment system. As we are heading toward a more digital society, governments are encouraging the adoption of cashless system to be implemented in every sector, including tourism. Cashless payment systems are promoted for their efficiency and inclusiveness. The systems are especially important for island destination where access to financial services is often limited. The development of ICT in the form of a cashless payment system can strengthen a destination's connectivity and creating a more seamless tourist experiences. A memorable tourism experience has been argued to lead to destination loyalty. Thus a cashless payment system has been considered to be the future method of payment for tourism destinations. Tourism has been a pioneer in cashless payment systems predominantly in the gaming industry. However, the study of cashless payments in other areas of tourism has been rarely explored. This study serves as a preliminary study to explore consumer behavior regarding cashless payment systems in island destinations. It argues that the cashless payment system can enhance the vacation experience, which can serve as a differential marketing value for a destination. Employing mini focus group discussions, it investigates tourist behavior in cash and noncash payments. The findings provide interesting insights on developing criteria for cashless payment systems and communication tools for cashless systems in tourism, as well as building a foundation for future research on the possible alternatives to cashless payment devices.

N. Wulandari (✉)
Research Department, Indonesia Banking School, Jakarta, Indonesia
e-mail: nuri.w.h@ibs.ac.id

I.R. Andilolo
Management Department, Faculty of Economics and Business, University of Mataram, Mataram, West Nusa Tenggara, Indonesia
e-mail: imanuelita@gmail.com



A. Saufi et al. (eds.), *Balancing Development and Sustainability in Tourism Destinations*, DOI 10.1007/978-981-10-1718-6_25

**Keywords** Cashless payment · Consumer behavior · Experience · Marketing · Tourism destination · Island tourism

## Introduction

Information Communication Technologies (ICTs) have been transforming tourism industry in a global scale since 1980s (Buhalis and Law 2008). The technological revolution experienced due to the arrival of Internet in the late 1980s has changed market condition dramatically for tourism organizations. Internet is probably the strongest driving force for the changes within the tourism industry (Werthner and Klein 1999). ICT is increasingly playing a critical role for competitiveness of tourism organizations and destinations as well as for the entire industry (UNWTO 2011).

In tourism industry development, ICTs become the central tools to connect and enable tourist experiences (Tussadiyah and Fesenmaier 2009). It is also suggested to promote increased social engagement and involve consumers to co-create experiences (McCabe et al. 2012; Sfandla and Björk 2013; Sigala 2012). Customers will be able to use technology to customize the products and personalize their experience thus creating a memorable experience of the destinations.

Although several studies have portrayed ICTs as central tool to connect and enable tourist experiences, there is lack of studies that discuss on how a payment system can contribute to the creation of experience. In addition, current literature focuses on technology availability with less perspective on consumers. Thus, there is a crucial need to explore how the system of payment in destination may affect tourist experiences in destinations.

### *Exploring Cashless Payment System in Tourism*

To investigate the area of experience and system of payment, this paper is interested to explore how cashless payment system can be implemented to enhance tourism experiences and thus can become a competitive advantage for a destination. Two main drives for the investigation on cashless payment system are first the encouragement from local government for a cashless society and second the importance of research experience in academic world.

Tourism has been the pioneer in the development of cashless payment system with its implementation in the gaming industry. Since then, the cashless payment system was also adopted for other form of leisure destination such as exclusive resort and amusement park. Recently, cashless payment system is also applied to all-inclusive island destination in the Fiji islands to cultural villages such as Maasai tribe in Kenya. It seemed that in the future the implementation of cashless system would be adopted in more and more tourism destinations as governments are encouraging a cashless society. In Indonesia the movement of using less cash is announced by the Central Bank in 2014. The implementation is not limited to cities but also to the villages and remote areas. Nevertheless, the cashless payment implementation studies are still limited and not focused to specific areas such as tourism.

Governments are driven to develop cashless society based on the benefits provided by cashless payment system compared to traditional cash payment system. The cashless payment system has been suggested to increase efficiency and inclusiveness. Since the cashless payment system is not printing money, it promotes efficiency for being a faster, cheaper, and safer method of payment. Inclusiveness means that cashless payment system could be adopted even for the least bankable individual and remote areas where financial services are limited. Furthermore, cashless payment system will enable to strengthen tourism network and inter-operability thus delivering seamless experience for tourists.

The value of experience is central to tourism activity as it is imperative in creating destination loyalty and becoming a destination competitive advantage. Customer experience has gained wide support from the academic world with the topic being one of research priorities area of the Marketing Science Institute in period 2014–2016. In line with the discussion, the future ICT development is predicted to focus on consumer-centric technologies that will support organizations to interact with their customers dynamically (Buhalis and Law 2008). Thus, development and implementation of ICT needs to be thoroughly investigated in customer experience-based perspective to ensure that the technology is relevant to its user.

## Objectives

This study is a preliminary study to outline relation of cashless payment system in creating tourism experience. To investigate the implementation possibilities of cashless payment system technology, destinations need to explore the role of payment system as a part of tourism experience. Thus, this study will investigate how consumers' responses to the cash and cashless payment system in creating their tourism experience to identify the challenges and opportunities of implementation.

The preliminary study is exploratory thus more loosely defined with a limited focus on the use of cashless payment system in leisure and tourism. The study results are expected to give insights to the nature of using cash and noncash payment in tourism setting, to identify criteria of cashless payment system and the preferred form for cashless payment device.

## Theoretical Background

The theoretical background takes root from experience studies in marketing and tourism areas. The literature review will also discuss briefly the past studies regarding ICT in general and cashless implementation in the tourism context.

Experience has always been a central component of leisure and tourism studies. Services such as restaurants and tourists have a hedonic, aesthetic, or emotional content, which professional or financial services generally lack in providing a context for examination of interesting and novel consumer behavior (Johns 1999). However, it is only recently that tourism and leisure academics have begun to focus specifically on more experiential aspects of travel.

The definition of experience can be referred to Schmitt (1999) who describes experiences as the result of encountering, undergoing, or living through situations. They stimulate the senses, the heart, and the mind. Experiences also connect company and its brand to the customer's lifestyle, and place individual customer actions and purchasing occasions in a broader social context. Experiences provide sensory, emotional, cognitive, behavioral, and relational values that replace functional values. Johns (1999) suggests that it would be of value to categorize service experiences on the basis of their emotional/hedonic content and its significance to the customer.

The typologies of experience can be divided into several parts. Pine and Gilmore (1999) divide leisure experiences into four "realms"; namely education, entertainment, escapist, and aesthetics. In line with the suggestion, the empirical research of Dube et al. (2003) finds that consumers classify pleasurable experiences into four possible types. First is sensory pleasure that is pleasant sensations induced during the experience. Second, social pleasure derived from one's interactions with others. Third is emotional pleasure borne of feelings, ideas, or mental images and lastly, intellectual pleasure from appreciating the complexities and subtleties of things around the consumer.

As the studies in tourism experiences continue to grow, there is interest in studies on technology as service experience. For businesses it has become a central endeavor to exploit the potential of technology and use it for the creation of meaningful tourist experiences (McCabe et al. 2012). In response to the practice area, a large body of work has drawn attention to the impact, role, and value of ICTs in the tourist experience (Kim and Tussyadiah 2013; Neuhofer et al. 2013). Recent work has underlined the value of smart phone applications to gather information, enrich and construct experiences (Wang et al. 2012) and the use of social networks to support and share on-trip experiences (Kim and Tussyadiah 2013).

Neuhofer et al. (2013) propose a typology to present the various parameters to enhance the experience (needs, tools, processes and results) and the degree with which ICTs can enhance the experience (depending on the intensity of the co-creation and the use of technology). While the study by Dan et al. (2014) confirms that the impact of technology on travel evolves as the traveler gains experience using new technologies and the affordances of this technology.

## Methodology

To identify the behavior of tourists regarding the payment system during their vacation experience, an exploratory qualitative inquiry was employed by means of focus group discussion. The discussion was guided by a set of predefined questions; nevertheless still leave enough room for further probe questions within related topic. A series of questions were asked, as outlined in Table 25.1. To guide the question, the study used Carù and Cova's (2003) consumption experience stages, which perfectly captured moments of before, during, and after vacation. The questions are arranged according to experience phases (before, during, and after the vacation). The scope of questions includes the budgeting, management, and spending behavior. In addition, the study also explores the possibility of introducing cashless payment system for island tourism and preferred form of cashless payment system.

The participants were selected on two criteria, (a) a smart phone owner and (b) have experience with using noncash method in various setting (e.g., food court, transportation, game center, and event). The criteria are based on Pihlström (2008), which stated that only technology users are able to report experiences related to technology usage. Participants were six students of a university in Jakarta with equal number of female versus male. The age of the participants ranges from 18–22 years old. The session was tape-recorded and transcribed (in Bahasa Indonesia). The length of the session is about 1.5 h.

**Table 25.1** FGD questions

| Phases of experience | Phases of experience |
|---|---|
| Before departure (pre-consumption experience stage) | What information you search to determine the vacation budget?<br>How do you usually fund your vacation?<br>How do you determine your vacation budget? |
| During vacation (consumption of experience stage) | What are the items to carry while exploring the destination? (List and rank)<br>How is the money management during the trip?<br>How do you pay for things in the destination? (Cash vs. card)<br>What do you like/dislike about using cash/card method? |
| After vacation (remembered consumption experience) | What makes you want to re-visit a destination?<br>What makes a memorable vacation experience?<br>What happened to the money leftover after vacation? |
| Additional questions | Have you had experiences using noncash payment system?<br>How do you feel/think about noncash payment system to be implemented in island destination?<br>Exploring preferences for non-cash payment system related to vacation experience (device forms) |

## Result

The main objective of the study is to learn about payment system as part of tourism experience. The findings of FGD are presented into three general themes. First is the value of cashless payment versus cash payment. Second is the challenge of current cashless payment implementation and lastly, the preferred form and feature of cashless payment.

## The Values of Cashless Payment

The discussion extracted three values of cashless payment compared to cash payment system, namely, safety, different, and convenient.

- Safety
  The findings on the use of cash revealed that carrying cash still results in the feeling of unsafe. One participant admitted that he carries all the money everywhere he goes for security reason and he felt unsafe to leave it behind. The feeling of unsafe also triggered the behavior of checking his pocket often to check whether the wallet is still there. Another participant also expresses the behavior of dividing the money into different places for precautionary reasons. In essence, carrying cash could create feelings of unsecure and anxiety.

  [Randi, Male, 20 yo] I divide (the cash), for example if I have seven hundred thousand (rupiah), two hundred (I keep) in the wallet, two hundred in the bag, the rest I put in my room. In case the wallet is missing, (I still) have automatic teller machine (ATM) (or debit card), and (some money) in the bag.

- Different
  This particular value is related with the question of using cashless payment system in tourism destination. Although all participants admitted that they have used cashless payment system in various forms, no participants have ever used it in a destination. Thus, the idea to use cashless payment system intrigues their curiosity. Thus, the discussion findings suggested that the cashless payment creates the sense of new experience. It also made the destination "different" from other destination in the eye of tourists.
- Convenient
  The reason for participants that favor cashless payment over cash is convenience. He stated that a debit card is simpler and he is lazy to go back and forth to the ATM to get cash for payment. On the contrary, handling coins and small changes is considered a hassle.

  [Rizki, Male,20 yo] On coins and cash —It is a hassle, that's why I prefer (to use) card, I lost small money and coins easily, and otherwise it is somehow spent

## Challenges and Criteria of Implementation

All of the FGD participants admitted that they have experience using noncash payment system in various settings, for example using coupons in music festival, prepaid card in food court, and game/amusement park. Based on the experience, it is found that there are challenges with the current noncash payment method used that providers need to address for improvement of future system. Below are the important findings emerged from the discussions:

- The study found that the process of transferring and topping up the money into the cards is stressful for the customers. One of participant felt the process of using the card is a bit of a hassle since she is required to put some cash into the food court's card first and irritated with the need to top up the money in the card every time the money runs out.

  [Amanda, Female, 18 yo] … it is a hassle, especially if there is only one cashier, we need to queue to deposit (some money into the card), and if it is not enough (the money), we need to pay again (top up the card) and queue again.

  The future cashless payment system implementation should make the process acquiring the service to be available easily and swiftly. Adding cashiers and reducing queue in any way possible is important in the process of using a cashless payment. Fast is the key criteria for the cashless system implementation.
- The study also found that participants felt lack of fairness in the current cashless system. One participant expressed objection in using cashless (card) system because there is a minimum balance to be retained in the card. The participant often use the card and left a small nominal balance that they do not want to spend anymore or somehow cannot spend since the cheapest item is more than the nominal, but cannot cash in due to the rule of minimum balance. In this situation, the participant felt the company is taking advantage over the small change and he feels cheated.

  [Rafi, Male, 22] I refused (of using the card as a noncash payment system), if there is some money left and we don't want to spend it no more, we can only keep the card (it will be a waste) and there is small nominal that is becoming the company's advantage. I felt at loss as a customers, because of the money (that is still) inside the card.

  There should be a mechanism to extract money left in the card, no matter how small, to avoid the feeling of being cheated or unfair for the user. Thus, fairness is one of the criteria to consider when designing the system.
- Other participants felt that using cashless payment system would bring difficulties to travel experience as not all service provider on the island might have (and provide) debit card/cashless payment facilities. One participant further added that she agreed to try the cashless payment system on her vacation as long as it is integrated with all the facilities (all-inclusive) in the whole island. The commitment and full participation from all destination members is important for

the participation of using cashless payment system. Thus, integration is another criterion for the cashless implementation.

- One particular reason that is often stated by participants who are reluctant to use the cashless payment system is they found it to be a hassle.

> It is a hassle … if it is not enough (the money), we need to pay again (top up the card) and queue again.

Moreover, another participant felt that the reason she has lack of interest of using the card system because of inflexibility due to extra charges. For example, when the card is missing, she has to buy a new one with extra charge. The charge is considered a burden for the participant. Thus, it is crucial that the future implementation of cashless payment system has flexibility and simplicity to use.

## Preferred Form and Features

From the statements given in the discussion, it can be inferred that there are several features that can be considered in developing the future cashless system. First is on the physical form of the cashless device and second on the usability of the payment system.

In previous findings, participants preferred card and bracelet form as the vehicle of cashless payment system. However, should other forms of devices are introduced, the findings suggested several physical features.

1. It is suggested that the cashless payment device is wallet-sized, as wallet is the number one item to be carried when traveling within the destination. That said, since the device is supposedly used to replace cash being carried in the wallet, it can take various forms, as long as it fits in a wallet.
2. It is suggested to personalize the device with pictures of the island, since photos are the triggers for the tourists to return to destination. Photographs and pictures also channels self-expression.
3. In addition, the physical form of the device should be easily in sight. Since one of tourists behavior is to check the physics/presence of the device.

On the usage of the device, there are two additional suggestions:

4. It is suggested that the destination also provide application in mobile-based platform. It can give information on attraction, routes, and destination-related information. Moreover, it can also be used for budget planner with the ability to be accessed pre-departure and checked during the vacation.
5. One of the participant suggested that the cashless system device have extended usage and benefit to use after the vacation is over.

**Table 25.2** Summary of FGD responses

| | Responses |
|---|---|
| Not interested | • One participant is not interested. He felt that it would bring difficulties to his travel as he insisted that not all tourist and service provider in the island might have (and provide) debit card facilities |
| Interested | • Two participants express their interest instantly on the idea of a cashless payment on island destination<br>• One participant felt that it would be easier for him just to swipe, since he categorizes himself as a practical guy<br>• Another said that she is interested and she will put more money in the card deposit from her regular vacation budget |
| Interested, but with condition | • Other participants seemed to have interest with the idea however it comes with condition that addresses their concern on the cashless payment system. The conditions are:<br>• As long as it is integrated with all the facilities in the whole island. An all-inclusive destination<br>• As long as he can cash out his money back upon finishing his vacation in the island |

The last finding from the discussion is on the preferred form of cashless payment. The participants were asked what if a cashless payment system is implemented in a tourism island destination. Although most of them gave positive response, there are several conditions to be met in order to push for the idea's acceptability. Table 25.2 summarizes the responses.

The discussion also asked about the most interesting form of noncash payment for island vacation. The options given are between coupons, card, and RFID bracelet. The study found that most of participants preferred RFID bracelet. The reasons are because a bracelet is different from the others (not a mainstream option); easily in sight thus could be detected more quickly when it is lost; wearable so it will be less likely to get lost or stolen; and avoid frequent use of the wallet (as in the use of card system). However, there is one participant that prefers card, because he found it more advantageous for example perhaps he can use it not only on the island but extended possible use after the holiday.

Overall, most of the participants express curiosity because noncash vacation is a very new concept to them. Some expressed interest on trying out the cashless system. The participants suggested that using cashless payment system will affect their vacation experience. In general, five out of six participants said "Yes" to whether the cashless system will make it easier for them to enjoy the vacation. However, it still depends on the initial perspective of the person, whether they perceive it as a hassle-free system (this means affecting the vacation in a good way) or feel that it will make payment more difficult. Surprisingly, although they express curiosity on the cashless payment, they also indicated that cashless payment system is the main reason for them to visit or not to visit an island destination.

[Safira, Female, 19] It is more important to enjoy the vacation itself (rather than the payment system the destination use).

Although the participants have several concerns about the card system and it is a new system for them, they realize that nowadays businesses going towards the cashless method thus they have no choice but to follow through with the cashless plan.

[Randi, Male, 20] Maybe because it is still new, (I am) not yet comfortable, but now even when you park in the station, you cannot do it unless you use XXX (debit card brand).

The above statement confirmed that cashless system is considerably a new concept thus have the tendency to make participants uncomfortable at first. However, the implementation of the system will eventually push the society to adapt.

## Conclusion

The preliminary study aims to learn about payment system as part of the tourism experience and how consumers respond to the cashless payment system in the tourism context. Although it is too early to conclude, the findings can give indication for the next study.

The study suggests that most of participants have positive response toward the concept of cashless payment system; however, there are still challenges with the implementation of the system. In order to develop future cashless payment system in tourism destination context, the study learns that certain criteria have to be met, namely, fast, fair, integrated, simple, and flexible.

The study also found three values of cashless payment to pose against the limitation of using cash payment. These values are safe, different, and convenient. The three values can be used as marketing values to be communicated to promote the cashless payment system. For example, the cashless payment can be communicated as a safer alternative in comparison to the cash payment system. While convenience translates to hassle free and different can be defined as anti-mainstream and self-expression.

The study also contributes in identifying preferred form and features related to physical and usage of the cashless payment system. The physical aspect suggestions are to have wallet-sized device, personalization, and easily in sight. The features suggestion includes the link with mobile-based application to share relevant information and perform budgeting function for the trip.

Finally, the study has not found clear evidence linking cashless payment implementation to experience of tourists. There is difficulty in examining this link since all participants have never been exposed fully to the situation before the discussion and the idea is very new to them. Thus, more studies with different methods could be conducted to tackle the issue.

## Limitation

As the first exploratory study on cashless payment system in tourism context, this preliminary study has many limitations. First limitation lies with the number of focus group conducted and the single-perspective of tourists, which limits the generalization of the findings. Further research should conduct more focus group to increase the quality of the findings. The next study can recruit participants from different demographic and travel styles. Second, the study can proceed by investigating provider perspective on cashless tourism system to give all-round perspective on the subject and criteria. Although limitation exists, this study has given a pathway and base for future cashless payment implementation in tourism industry and contribution for experience studies in the same context.

## References

Buhalis D, Law R (2008) Twenty years on and 10 years after the Internet: the state of eTourism research. Tourism Manag 29:609–623

Carù A, Cova B (2003) Revisiting consumption experience a more humble but complete view of the concept. Mark Theory 3(2):267–286

Dan W, Xiang Z, Fesenmeier DR (2014) Smartphone use in everyday life and travel. J Travel Res 55:52–63

Dube L, Le Bel J, Sears D (2003) From customer value to engineering pleasurable experiences. Cornell Hotel Restaurant Adm Q 44:124–130

Johns N (1999) What is this thing called service? Eur J Mark 33:958–973

Kim J, Tussyadiah IP (2013) Social networking and social support in tourism experience: the moderating role of online self-presentation strategies. J Travel Tourism Mark 30(1):78–92

McCabe S, Sharples M, Foster C (2012) Stakeholder engagement in the design of scenarios of technology-enhanced tourism service. Tourism Manag Perspect 4:36–44

Neuhofer B, Buhalis D, Ladkin A (2013) A typology of technology-enhanced tourism experiences. Int J Tourism Res 16:340–350

Pihlström M (2008) Perceived value of mobile service use and its consequences. Doctoral Thesis. Finland: Swedish School of Economics and Business Administration, Helsinki

Pine JB, Gilmore JH (1999) The experience economy. Harvard Business Review Press, Boston

Schmitt B (1999) Experiential marketing. J Mark Manag 15:53–67

Sfandla C, Björk P (2013) Tourism experience network: co-creation of experiences in interactive processes. Int J Tourism Res 15(5):495–506

Sigala M (2012) Social networks and customer involvement in new service development (NSD): the case of www.mystarbucksidea.com. Int J Contemp Hospitality Manag 24(7):966–990

Tussadiyah IP, Fesenmaier DR (2009) Mediating tourist experiences: access to places via shared videos. Ann Tourism Res 36:24–40

UNWTO. (2011). UNWTO tourism highlights. UNWTO, Madrid

Wang D, Park S, Fesenmaier DR (2012) The role of smartphones in mediating the touristic experience. J Travel Res 51(4):371–387

Werthner H, Klein S (1999) ICT and the changing landscape of global tourism distribution. Electron Markets 9:256–262

# Chapter 26
# Monitoring Growths and Product Quality of Local Souvenirs to Enhance Tourist Destination Image of North Sulawesi, Indonesia

**Agustinus Walansendow and Bet El Silisna Lagarense**

**Abstract** Tourism development in North Sulawesi grows with a variety of industry sectors including souvenir industry. Souvenir representing local tourism products is an important component of the tourist experience visiting North Sulawesi. However, the souvenir industries in North Sulawesi have not yet been produced distinctively as local souvenirs and are still limited in both quality and quantity. This study is aimed at identifying and developing souvenir model and design as a supporting element of tourism product in North Sulawesi in terms of uniqueness, marketability and competitiveness with its local characteristic of North Sulawesi. Data and information are collected from home industries in North Sulawesi which produced souvenirs made from wood and coconut shell materials. This study employed filed observation, interviews and FGD (Focus Group Discussions). The informants are craftsmen and the souvenir sellers in North Sulawesi. The interview was also conducted to tourists visiting North Sulawesi to get their opinion and recommendations. The results of the study identified several models and designs for coconut souvenirs related such as the miniature related to local heroes for examples Sam Ratulangi, Toar-Lumimuut and Dotulolong Lasut. The miniatures related to local flora and fauna such as the ancient fish (Coelacanth), *manguni* bird and tarsius spectrum with its uniqueness and speciality of North Sulawesi souvenirs.

**Keywords** Monitoring · Growth · Product quality · Souvenirs · Tourist · Destination image

A. Walansendow (✉)
Business Administration Department, Manado State Polytechnic, Manado, Indonesia
e-mail: walansendowagus@gmail.com

B.E.S. Lagarense
Tourism Department, Manado State Polytechnic, Manado, Indonesia
e-mail: bet_lagarense@yahoo.com



A. Saufi et al. (eds.), *Balancing Development and Sustainability in Tourism Destinations*, DOI 10.1007/978-981-10-1718-6_26

## Introduction

Utilization of North Sulawesi's natural potential namely coconut which has derivative products such as CW and coconut shell can be developed to be souvenir products by the local community to support North Sulawesi tourism development. Forum of community plays role in discovering the natural potential of coconut farmers through training for the development of handicraft and small-scale businesses in the form of home industry. From the result of the preliminary research, collecting information about the existing souvenir products and the better development model, there will be training for small-scale business development and training for handicraft using CW and nutshell as the basic materials for the community or villagers who have talents or interest in these activities. Hence, these souvenir-making activities apply community-based development theories that can enhance the local community empowerment through direct participant.

The objectives of this research are (a) to describe the characteristic of souvenirs with CW and coconut shell as their basic materials having a distinctive feature of North Sulawesi, (b) to map the groups of industries as well as to determine the zonation of craft makers for easier control and better quality of products, (c) to improve the competitiveness of souvenir business with CW and coconut shells as the basic materials having a distinctive feature of North Sulawesi. Souvenir product development in North Sulawesi will be sustainable and have a good quality if it involves the local community to participate directly. This must be done starting from the model development planning stage up to the program implementation or the income raising activities and the expansion of job opportunities for the local community in North Sulawesi that will finally improve the community's quality of life.

Participation is a process of empowerment that helps to involve local people in problem identification, decision-making and implementation processes, thereby contributing to sustainable development (France 1998: 224). This is strengthened by the delegation to community (community empowerment) that is the level of community involvement in the souvenir product management and development in North Sulawesi that be more profitable for the local community and will have positive influence on tourism development. Wearing and McLean (1998) put forward that during the planning process it could be possible for the community to realize the potential benefits of such tourism without disadvantaging some sectors of the community. Therefore, community development or delegation to the local community requires community consultation from the planning stage, development up to the implementation of development activities, monitoring and growth of tourism souvenir products of North Sulawesi. In the development of souvenir product for having a strong competitiveness, there is a need for support from all the community, government and private sector.

Some circles of society have enjoyed the direct benefits of tourism sector, but some are not yet aware of the importance of tourism development, for instance, the farmers and fisherman who usually oppose the development of tourism in their regions because they opine that tourism activities hinder their means of livelihood everyday. Hence, there is a need for maximal efforts to internalize the concepts of community-based tourism product development correctly and precisely and to continuously monitor its growth in order to be able to maintain the handicraft product quality. The local community are not only given the freedom to interact with the ecosystem around them but also given the opportunities to utilize and manage the natural resources in the region that will positively influence the life quality and the sense of responsibility of the management and the feeling of pride of having the natural resources. The development and growth of handicraft product revenue of Indonesia in the international market in 2014 can be seen from graph 1 below where the United States has 67 % and Japan has 17 % showing that these two countries are potential market targets of handicraft. The other countries are England with 6 %, Germany and Netherlands with 5 % each which means that these countries can also be the market target for the Indonesian handicraft. This shows that handicraft products have given significant contribution to the country's income (Badan Pusat Statisik 2014). Figure 26.1: Indonesian handicraft product revenues in international market in 2014.

Further, revenues from the Indonesian handicraft products have given a significant contribution to the national foreign exchange since 2012 as great as 695,000 million American dollars showing that the Indonesian handicraft products have a quite good market segment that the country can rely on and therefore the government should maximize the foreign exchange from informal sector. The following table presents the revenue from the Indonesian handicraft products in American dollars (BPS 2014). Figure 26.2: Revenue from the Indonesian handicraft products in $US

The competitiveness of coconut products lies on downstream industry. The end products which have been well developed are coconut oil (CO) crude coconut oil (CCO), DC, coconut milk (CM), coconut charcoal (CCL), activated carbon (AC), brown sugar (BS), nata de coco (ND) and coconut fiber (CF). Two new products are

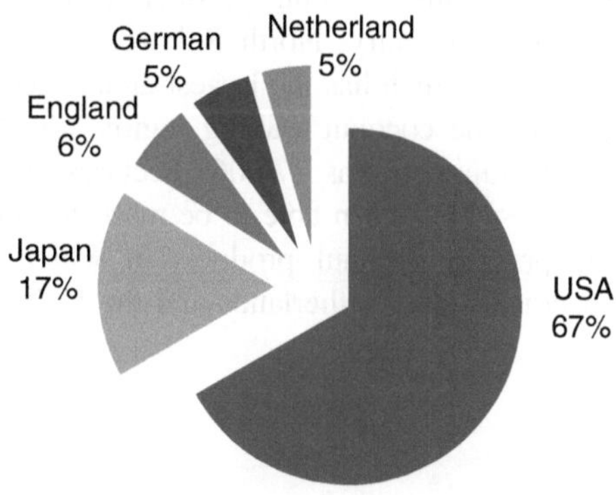


**Fig. 26.1** Indonesian handicraft product revenues in international market in 2014 (BPS 2014)

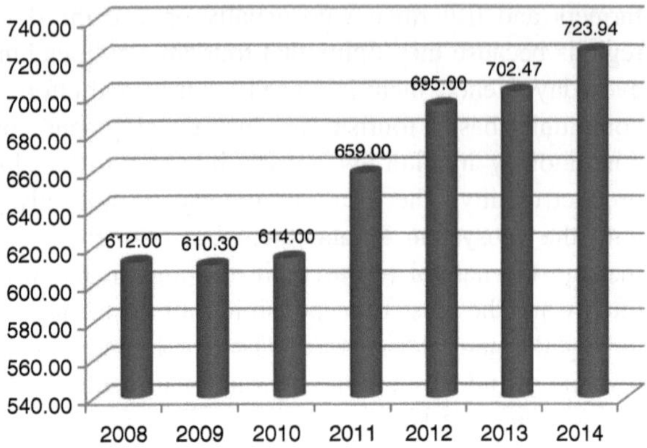


**Fig. 26.2** Revenue from the Indonesian handicraft products in $ US (BPS 2014)

**Table 26.1** Area and coconut production of North Sulawesi 2007–2010

| | 2007 | 2008 | 2009 | 2010 |
|---|---|---|---|---|
| Area (ha) | 267.625 | 272.137 | 274.917 | 276.069 |
| Production (millions) | 229.613 | 209.994 | 265.451 | 273.234 |

*Source* Biro Pusat Statistik of North Sulawesi (2011)

virgin coconut oil (VCO) and coconut wood (CW). CO, CCO, DC, CM, CCL, AC, BS and CF products have entered the export market, although the derivative coconut products are highly demanded by export market, only very few Indonesian producers have gained profits from the coconut's value-added process. According to the Indonesian Coconut Forum (FOKPI), the Philippines is still the leader of export market with its 125 kinds of coconut processed products. Table 26.1: Area and coconut production of North Sulawesi 2007–2010

In North Sulawesi the development of agricultural sector is very important considering that North Sulawesi has a great potential of natural resources in agricultural sector. Therefore, changes in price of derivative coconut products will directly influence the life standard of the farmers and their families. The revenue from derivative coconut products can contribute to the foreign exchange for the producing country. North Sulawesi is one of the coconut producing regions in Indonesia which has the largest areal of coconut plantation that this region is often called 'The coconut waving region'. In 2010, the area of coconut plantation in North Sulawesi was 276.069 hectares. The derivative coconut products of North Sulawesi have been able to be marketed internationally due to the good quality of the pressed coconut products in which the international market is interested, specifically the Netherlands and the United states (Tatengkeng 2011).

## Kinds of Derivative Coconut Product Export of North Sulawesi

Coconut (*cocos nucifera L*) is a strategic commodity which has social, cultural and economic roles in the life of Indonesian people. The part of coconut that is beneficial is not only its meat that can be processed to be CM, copra and CO but also its other parts. So great is the benefit of coconut that some people call it ‘the tree of life’ or ‘the heaven tree’ (Asnawi and Darwis 1985). Coconut plant is a tropical plant which has been long known by the Indonesian people. This can be seen from the spread of coconut plants in almost all plants of Indonesia, namely Sumatra with an area of 1.20 million hectares (24.30 %), Sulawesi with 0.716 million hectares (19.305), Bali, NTB and NTT with 0.305 million hectares (8.20 %), Maluku and Papua with 0.289 million hectares (7.80 %), and Kalimantan with 0.22 million hectares (7.50 %).

Coconut plants are grown by the farmers in both plantations and yards. Basically, all parts of coconut can be processed to be various products for various needs. Its processing technology and quality standard as well as certification have been mastered by the Indonesian experts. How ever, various weaknesses are still adhering to out coconut processing industry such as raw materials supply because the industry does not have coconut plantation and relatively large investment and therefore does not interest investors (FOKPI 2006). Most of the coconuts are processed to be copra that is further processed to be cooking oil. But this business is getting weaker in both domestic and foreign trades in competition with palm oil.

Besides being processed as cooking oil, diversified coconut products have been developed such as desiccated coconut (DC), coconut sugar, ND, various products of coconut meat, dry fibrous shell, coconut wood furniture and the latest development is ready to drink CM with various packages (Allorerung and dan Lay 1998). There are wider opportunities for developing coconut agribusiness with products of high economic value. Alternative products that can be developed among which are VCO, oleochemical (OC), DC, coconut milk/cream (CM/CC), coconut charcoal (CCL), AC, BS, CF and coconut wood (CW) which are produced partially and integrally. These product business doers can increase their income 5–10 times high than what they get just by selling copra. Based on the fact that there is a great potential for product development, coconut industry is economic advanced at micro level (farmer’s income, domestic value added and important substitution) and seems to require a more serious cluster coconut industry development support as a prerequisite (Allorerung et al. 2005).

The derivative coconut products which are exported all over the world, which includes CO, coconut flour, copra, copra residue, CCL and AC. Most of the derivative coconut products in North Sulawesi are likely to be exported to some destinations. Some countries of destinations and the kinds of exported products are presented in Table 26.2.

Table 26.2 shows that the derivative coconut products of North Sulawesi are exported to many countries of export destinations. Various kinds of derivative coconut products become main export commodities of North Sulawesi that give

**Table 26.2** Kinds of the exported derivative coconut products and countries of destination

| Kinds of commodities | Countries of destination |
|---|---|
| Coconut oil | Singapore, USA, the Netherlands, Germany, Saudi Arabia |
| Coconut charcoal | Japan, Korea, Malaysia, Thailand |
| Coconut flour | Singapore, USA, the Netherlands, Africa |
| Copra | The Netherlands, Belgium, Malaysia |
| Copra residue | Singapore, India |
| Activated carbon | USA, Thailand |

*Source* Trade and Industry Office of North Sulawesi, 2010

contribution to the export activities of North Sulawesi. These kinds of derivative coconut export commodities play a vital role in economy, especially in North Sulawesi.

## Tourist Destination Image

Kotler and Gertner define a destination image as 'the sum of beliefs and impressions people hold about a place. Images represent a simplification of a larger number of associations and pieces of information connected to a place. They are a product of the mind trying to process and pick out essential information from huge amounts of data about a place' (2004: 42). The main idea of this definition is that the evaluation of a destination is based on quality, value and satisfaction. Destination image can influence a tourist to decide future intention to visit and their recommendation to others on places to visit. Kavaratzis and Ashworth (2005) explain that destination image influences tourists' decision-making in regard to where they will spend their vacation and money. The importance of the destination image for the consumers allows us to think of any destination from the consumer's perspective in terms of how they sense, understand, use and connect to the place. The main thrust of these arguments is that image should be able to increase the benefits to the destination, including the local community.

## Product Quality

Product is the key element of the whole market offer. Also, product can be defined as the perception of consumers that is spelled out by the producers through their products (Tjiptono 2008). Based on the definitions above, product can then be defined as the collection of tangible and intangible attributes including packing, colors, prices, qualities and marks as well as services and reputation of selling. Quality has many senses such as: (1) Compatibility with the requirement;

(2) Appropriateness for use; (3) Sustainable improvement; (4) Free from damage/defect; (5) Fulfillment of the consumer's needs from the beginning and every time; (6) Doing everything correctly; (7) Something which can satisfy the consumers those given. Thus, quality can be defined as the totality of product characteristic (goods and services) that supports its capability to satisfy needs.

In this globalization era, it seems that the community or consumers are increasingly becoming critical in assessing a product. Product quality as an assurance in satisfying the customers' need in choosing a product and in this case the personal taste plays a vital role (Stanton 2006). In this context, the researcher can quote several things in discussing the product quality i.e.: (1) The word 'quaranty' means that the product offered to the consumers has gone through a process of careful and rational process of measuring and thus it is feasible to have guaranty; (2) The word 'taste', which becomes the consumers' motivation in choosing a product, is a factor that becomes the focus of attention of the producers or marketers. Thus who is the customer or buyer is worth knowing by the producers or marketers; (3) Between the guaranty and the need factor, there are rationalizing and relevance that must be accurately translated by the producers or marketers.

## Souvenir as Tourism Product

From a souvenir product we can translate the meaning and value of a culture and region from where the souvenir comes as pointed out that a creative design for a tourist souvenir is an effective way to promote its sales. As a souvenir designer, you are not only to design its beautiful shape and outside looking, but also to give it a clear regional characteristic and rich culture meaning (Xin-ting 2004). Thus, the shape and design of a souvenir show the characteristic of a certain culture or region. So is the development of souvenir product with coconut wood and nutshell as their basic material in North Sulawesi. It becomes very important as a supporting factor for tourism in North Sulawesi because the qualified value and uniqueness contained in the souvenir reflect the uniqueness of the region of tourism destination as explained that the core of tourism merchandise is the souvenir and the uniqueness is the intrinsic quality of souvenir (Xue-Ling 2004).

Souvenir is a handicraft which is a result of creativity of a craftsman that can turn the wasted and worthless things into interesting handicraft products that many people, especially tourists, are interested in. Souvenir is usually small and relatively inexpensive article given, kept or purchased as a reminder of a place visited and an occasion (The Collins cobuild Dictionary 2009). Souvenir is an object a traveler brings home for the memories associated with it. The meaning of souvenir in Indonesia is very similar to the meaning given by the Collins cobuild Dictionary or Webster English Dictionary, which is a small and relatively inexpensive article as a reminder of a place visited. Hence, souvenir is closely related to someone's traveling activities, therefore, it is not surprising that the term 'souvenir' adheres to tourism activities and it even becomes a part of tourism product. Thus, there is a positive influence of souvenir on someone's motivation to travel to a region of travel destination.

## Characteristics of Souvenir Materials from Coconut

**Nutshell**: Physiologically, nutshell is the hardest part of a coconut compared to other parts. The hard structure is caused by a relatively high content of silicate (SiO2) in nutshell. The weight of a nutshell is about 15–19 % of the weight of the whole coconut while its thickness is about 3–5 mm. Front the point of view of quality, a nutshell which is qualified to be activated charcoal must be really old, hard, intact and dry. To produce a high quality activated charcoal, the nutshell must be clean and be separated from its fiber. And to know the good quality of nutshell: when burned the charcoal produced must look black, shiny, intact and easily broken (Mecoho 2009). Nutshell can be used as firewood or be processed to be activated charcoal than can be used by various processing industries. Activated charcoal from nutshell has a high competitiveness because it has a high quality and categorized as renewable resources. This nutshell is a plantation waste which has a considerable potential and can further be used as activated charcoal.

**Coconut trees**: Coconut consists of outer skin, fiber, nutshell, meat skin, meat coconut water and sprout. An old coconut has a fiber whose weight is (35 %), a nutshell whose weight is (12 %), endosperm (28 %) and water (25 %) (Setyamidjaja 1995). Nutshell is one of the AC materials whose quality is good enough for making activated charcoal. The quality of nutshell and the burning process highly determine the sucrose content of the AC produced (Mecoho 2009)

**Coconut wood**: The processing of coconut wood for industrial scale follows the following stages: (a) Tree selection: A physically old tree which bears only few coconuts (unproductive), minimally 60 years old and is physically straight; (b) coconut tree felling and cutting. After being felled and sawn with a 12 m long size, it is cut into desired sizes. These pieces of wood are processed in accordance with their qualities; (c) Wood drying: Air or natural drying under roof, drying with an oven (oven drying) and drying with a kiln (kiln drying). The level of dryness of wood for furniture materials, doors and windows is 10–12 % and for the house building, the water content is about 18 %; (d) selection of coconut wood and its use for making various kinds of furniture, carving and ornaments from first-class (quality) coconut wood on designs. The coconut wood which is categorized as the second-class (quality) is used as building materials or the props for furniture; (e) Finishing: it can be done by using scouring machine or manually. The result of observation shows that the price of 1 cubic meter of first-class (quality) coconut wood (half-finished materials) is IDR 750,000. The coconut wood which is used for furniture making and various kinds or ornaments is IDR 7–8 million. The value added of industry-scaled coconut wood processing is high enough but in handling the process of processing, design and product marketing requires large investment and support of expertise (Gozal 1997).

## Research Methodology

The research method used is descriptive qualitative method. The data used are primary and secondary data. The data are collected through interview, field observation and Focus Group Discussion through information analysis. The researcher uses the theories of quality development related to destination image and competitive products. Data of research are collected from observation, recording, picture-taking and analysis of various document of souvenir product quality development. Research using field observation is done in North Sulawesi region at handicraft industrial centers and personal home industry as well as souvenir shops. The data of research are analyzed qualitatively accentuating the explanation of the data interpretation result through some stages: Identification, classification categorization and analysis to lead to conclusions and recommendations.

## Result and Discussion

The crafts makers of North Sulawesi have important roles in the survival of souvenir business because souvenir business will exist and develop without them. The same are the quality of souvenir and the diversity of products. If they are not creative, innovative and skilled, this will be impossible. The results of interview show that the craft makers of North Sulawesi consist of:

1. Those who depend on their handicraft business for life. They produce handicrafts from rattan, bamboo, cloth and wood.
2. Those who have double profession which means that they do not depend their lives only on their handicraft business. They are housewives, teenage girls who dropped out from school and senior high school graduates who do not continue their education to higher levels. They usually produce embroidered objects both manually and with a machine, knit and sew garments such as traditional costumes, various kinds of bags, wallets, wall accessories, caps, nutshell and coconut wood, etc.
3. Seasonal craft makers exist when there are MICE events in North Sulawesi such as conference exhibitions, visits, meetings and usually come for designing convection (embroidery, T-shirt, silk screening, fabric, jacket, key hooks and hats.

Thus, the resource potential which can enhance community economic empowerment in North Sulawesi is considerable, yet it has not been fully utilized. The community still does not consider such things as potential. However, when it is seriously done, it can overcome big social problems. The resources consisting of goods and service need to be processed to be commodities such as handicrafts, home industry products, traditional medicines including medical herbs, cosmetic product, traditional food and drinks, etc., in to be sold to consumers. In the form of services, these resources consist of traditional medicinal treatment, steam bath and spa.

## Results of Interview

From the results of interview with some craft makers in Bahoi, conclusion can be drawn that souvenirs produced are still mainly based on orders such as material from nutshell which are made kitchen-utensils, key hooks, tissue containers and also animal-shaped objects like decorative fish and tarsius made of coconut wood and nutshells (informant 1: senior High school graduate who used to work as a fisherman and coconut farmer but with innovative creativity he finally started a handicraft business. Now he has a son and has a monthly average income of IRD 6.000.000. In doing this business, the informant employs six permanent workers. She has created handicrafts made forWOC (World Ocean Conference) event in 2009. The handicrafts were made from nutshells whose material were selected coconuts from Tondano village in the form of coelacanth fish. However, this craftsman was still less creative in producing souvenirs with cultural values that it is necessary to socialize again the meaning of souvenirs that have regional characteristics (informant 2 in Teling Atas village). This is supported by the result of interview with another informant from Kalawat village that the handicraft she makes are ordered by souvenir shops located on Jl.B.W Lapian while another informant from Teling Atas said he prefers to open his own outlet when there is a national event in a hotel or when there is a handicrafts exhibition both conducted in North Sulawesi or in other regions. Another informant from Bahoi village often sells through outlets when there ore exhibitions in town.

## Result of Field Observation

From the result of on site observation, it is found that:

1. The souvenirs available do not yet reflect the identity of tourism object.
2. The souvenir shells have low quality and uninteresting
3. The prices of souvenirs are too expensive/imbalanced with their quality
4. The souvenirs are too large-sized, unpacked and hard to bring home
5. The souvenirs offered are not in the form of simple objects.
6. Souvenirs sold are in the form of objects taken from the nature that result in environmental damage such as corals, sea animals, sea shells and snails. Now many people are environmentally aware and hate the actions of taking the natural objects that in turn will result in degradation of natural values.

It is also found that the zonation of craft makers in North Sulawesi covers 11 regencies and four cities, namely: South Minahasa Regency, North Minahasa Regency, Central Minahasa Regency, South East Minahasa Regency, Sitaro Regency, Sangihe Regency, Talaud Regency, South Bolmong Regency, North Bolaang Mongondow Regency, East Bolaang Mongondow Regency, Manado City, Bitung City, Tomohon City and Kotamobagu City. Each of these regencies and cities has its own distinctive motive that becomes the national wealth with uniqueness for the tourists.

Based on the field observation and data from the industry and trade services office of North Sulawesi Province, 2015, it is known that many craft makers stay in Manado because their handicrafts are mainly marketed in Manado or outlets which are mostly found in handicraft shop in Manado. Craft makers are spread in some subdistricts such as Wanea, Wenang, Wonasa, Winangun, Tuminting, Titiwungen, Pinaesaan, Malalayang and Mapanget. Meanwhile handicrafts from earth, wood and bamboo are centered at places where there are raw materials such as Pulutan village in Tondano, bamboo in Kinilow, wood in Woloan village and Tomohon.

## Result of Focus Group Discussion/FGD

Result of focus group discussion among craft makers, tourists and souvenir shop owners is that they agree to socialize the result of discussion to the craft makers.

1. The craft makers should think about the souvenir selling price. Souvenirs which have no selling price should not be produced.
2. Another important consideration is that a souvenir should be rooted in the local culture of the community. This can be done if the souvenir is made by the local craft makers because their spirit of arts is influenced by their culture that is poured into the souvenir. It is noteworthy that a souvenir that has no culture root will not sell well.
3. If the production of souvenir is large enough and varied, there is a need of outlets for selling such as terminals, stations, airports, bus halts and shopping centers. Outlets for selling the souvenir products should be planned well and carefully starting from the location, shape of building, layout, etc. The next step is to fill the outlets with the souvenirs to be sold. The strategic location of outlet with a good attractive building and good layout as well as neat arrangement will make the souvenir products in the outlet more attractive.

## Conclusions

This research concludes that in order to enhance tourist destination image and increase the souvenir sale volume in North Sulawesi, the crafts makers are required to consider some main things in souvenir business as follows: (1) Seriousness of the crafts makers in making the souvenir products; (2) Improvement of the craft makers' skill in making the souvenir products can be done; (3) The craft makers need to know the consumers' taste in order that souvenirs are produced accordingly; (4) Development of outlets or places where the consumers can find the souvenir products; (5) In creating the souvenir products, the craft makers are required to consider making small-size souvenirs with competitive prices because small size and inexpensive price are important considerations when tourists buy souvenir products.

The result of monitoring the growths of souvenir industries shows that crafts makers who fulfill the qualification mentioned above need to do some efforts through: (1) Well planned and directed training and improvement of skills of craft makers in North Sulawesi in order that they can produce souvenirs to compete with other products; (2) Business Capital aid and sustainable guidance are needed to enhance creativity, desire and willingness to develop souvenir products that in turn will result in the growth of craft makers who have sense of belonging of the souvenir business. The existing coconut industry needs to be maintained and developed while the coconut industry which is not yet optimal; (3) Souvenir industries should get serious attention from all the related parties for enhancing the processing capacity and market expansion in order that the potential of the existing raw materials can be optimally utilized for increasing the commodity value added, export value and income of the coconut farmers as well as the craft makers. The following are suggested: (1) Develop the derivative coconut products due to their great contribution to total export of North Sulawesi; (2) The community and

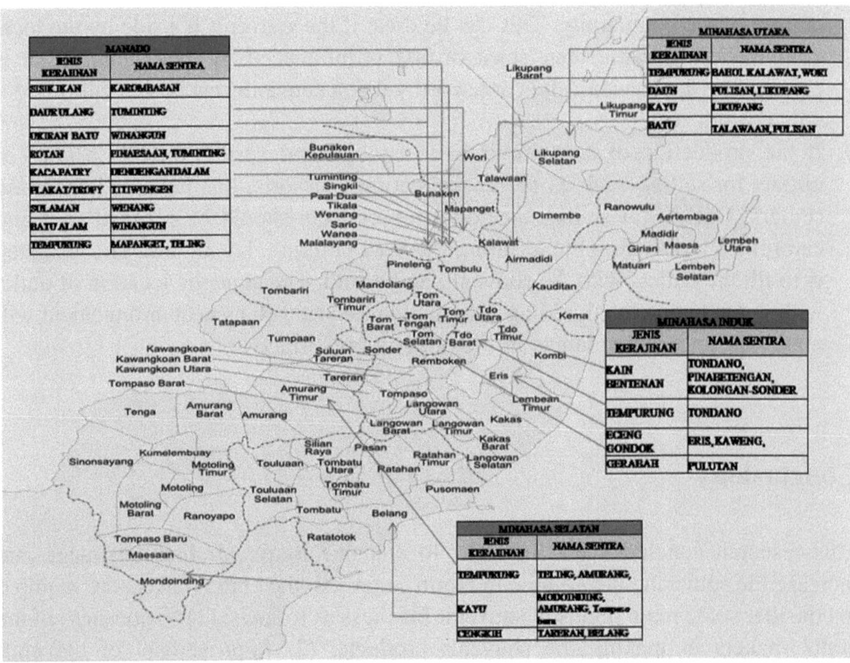


**Fig. 26.3** Map of location and zonation of craft makers in North Sulawesi (Field observation, 2015)

government should pay attention to the plantation area for agriculture commodities, especially coconut that is now getting smaller and smaller dominated by unproductive coconut trees; (3) Processing of coconut products should be continuously increased in order to produce new derivative products that can give considerable value added to derivative coconut products (Fig. 26.3; Table 26.3).

**Table 26.3** Zonation of handicraft industries in North Sulawesi

| Category | Center names | Location | Handicrafts |
|---|---|---|---|
| Fish scales | Manado | Karombasan | Clothes, brooches, earnings, bracelets, key hooks, bags |
| Recycle | Manado | Tuminting | Flower, masks |
| Carved stones | Manado Minahasa | Winangun | Statues |
| Rattan | Manado, Minahasa | Pinaesaan Tuminting | Furniture, baskets |
| Soldered glass | Manado | Dendengan dalam | Modern glass accessories |
| Posters/trophies | Manado | Titiwungen | Various shapes and kinds of trophies and posters |
| Embroidered product | Manado Bolmong | Wenang | Cloth |
| Natural stones | Manado, Minahasa | Pineleng, Winangun, Sonder | Tiles, statues |
| Nutshells | Manado, Minut Minsel | Teling, Wori, Amurang, Bahoi | Table ware, bags, key hooks, decorative fish, women's bag, wall accessories |
| Earthenware vessels | Minahasa | Pulutan, Sonder | Table accessories, flower vases, hearts, clay pats, furniture |
| Bentenan cloth | Minahasa | Tondano, Pinabetengan Kolongan, Sonder | Bentenan cloth |
| Sea shells | Talaud, Minut Bitung | Melonguane, Likupang Batu putih | Decorative lamps, baskets, wall/room dividers, musical instruments |
| Bamboo | Tomohon | Tinoor, Kinilow | Decorative lamps, baskets, hats, plaited bags, flower pots |
| Wood | Tomohon | Woloan | Mini house, table wares, wall accessories |
| Dry flower | Tomohon, Manado | Kakaskasen, wenang | Preserved accessories |
| Water hyacinth | Tondano | Eris, Kaweng | Bags, mats, shoes |

*Source* Field Observation, 2015

## References

Allorerung D, dan Lay A (1998) Kemungkinan Pengembangan Pengolahan Buah Kelapa Secara Terpadu Skala Pedesaaan. Prosiding Konferensi Nasional Kelapa IV. Bandar Lampung. 21–23 April 1998

Allorerung D, Mahmud Z, Wahyudi, Novarianto H, Luntungan HT (2005) Prospek dan Arab Pengembangan Agribisnis Kelapa. Badan Penelitian dan pengembangan Pertanian Departemen Pertanian

Asnawi S, dan Darwis SN (1985) Prospek Ekonomi Tanaman Kelapa dan Masalahnya di Indonesia. Balai Penelitian Kelapa, Manado. Terbitan Khusus 1985. hal 6

Badan Pusat Statisik (2014)

Biro Pusat Statistik of North Sulawesi (2011)

Forum Kelapa Indonesia (FOKPI) (2006) Membangun Kemakmuran Bebasis Kelapa di Berbagai Pulau indonesia. Jakarta:FOKPI. hal 3

France L (1998) Local participation in tourism in the West Indian Islands. In Laws E, Faulkner B, Moscardo G (eds) Embracing and managing change in tourism; international case studies. Routledge, London, pp 222–234

Gozal SM (1997) Pengalaman mengolah dan memanfaatkan kayu kelapa untuk meubeler dan bangunan rumah. Prosiding Temu Usaha Perkelapaan Nasional. Manado 6–8 Januari 1997. Buku II Agroindustri

Kavaratzis M, Ashworth GJ (2005) City branding: an effective assertion of identity or a transitory marketing trick? Tijdschrift voor Economische en Sociale Geg afie 96(5):506–514

Kotler P, Gertner D (2004) Country as brand, product and beyond: a place marketing and brand management perspective. In: Morgan N, Pritchard A, Pride R (2002) Destination branding: creating the unique destination proposition. Elsevier, Burlington

Mecoho (2009) Arang Batok Kelapa, Jakarta: www.indonetwork.co.id/mecoho_cv/117933/Arang_Batok_Kelapa/html

Setyamidjaja D (1995) Bertanam kelapa. Yogyakarta, Penerbit Kanisius

Stanton WJ (2006) Fundamental of marketing. International Book Company, Singapura

Tatengkeng H (2011) Produk Turunan Kepala Sulut Masih Terbatas. http://www.palakat.com/news/read/6768-produk-turunan-kelapa-sulut-masih-terbatas.html, diakses 31 Maret 2015

The Collins Cobuild Dictionary (2009)

Tjiptono F (2008) Pemasaran Jasa. Edisi 1. Bayumedia Publishing, Malang

Wearing S, McLean J (1998) Developing ecotourism; a community based approach. HM Leisure Planning Pty Ltd, Williamstown Victoria

Xin-ting W (2004) A study of design for enhancing the value of tourism. J Nanchang University (Social Science) 3

Xue-ling M (2004) Tourism merchandise and souvenirs: discussion on the conceptual definitions and local features. Tourism Tribune 1

# Chapter 27
# Tourism Supply Chain Perspectives on Border Shopping Development at Padang Besar, Malaysia

**Azila Azmi, Ibrahim Abdul Hamid, Johanna Adlin Ahmad and Rafidah Aida Ramli**

**Abstract** Padang Besar is a border town situated between northern Malaysia and southern Thailand, and has been well known for its border shopping activities since the early 1950s. There are many shopping locations to be found here where visitors can find a variety of goods at bargain prices. Though historically it is a shopping attraction for domestic visitors, early observation by researchers showed that there was a lack of adequate facilities for the development of tourism in Padang Besar. Therefore, this research was conducted to try to understand border shopping development from the perspective of tourism supply chains. Grounded theory qualitative data analysis strategy was employed through personal interviews along with observations of, tourism supply chains. The findings indicated that the tourism supply chains, namely the traders, government organizations, and tourism distribution channels, failed to collaborate and this resulted in poor planning and development of the town. The concluding section of this paper suggests that Padang Besar would benefit from concrete support from all levels of supply chains, including government and private investors. This cooperation would undoubtedly help to improve the shopping zone by facilitating the efficient development of shopping infrastructures, facilities, and amenities, not only for tourists but also for local communities as a whole.

**Keywords** Border shopping · Tourism supply chain · Perceptions

A. Azmi (✉) · J.A. Ahmad · R.A. Ramli
Faculty of Hotel and Tourism Management,
Universiti Teknologi MARA (Pulau Pinang), Shah Alam, Malaysia
e-mail: azila.azmi@ppinang.uitm.edu.my

J.A. Ahmad
e-mail: johan921@ppinang.uitm.edu.my

R.A. Ramli
e-mail: wanfidah@ppinang.uitm.edu.my

I.A. Hamid
Kulliyyah of Hospitality and Halal Services, Insaniah University College,
Alor Setar, Kedah, Malaysia
e-mail: ibrahim@insaniah.edu.my



A. Saufi et al. (eds.), *Balancing Development and Sustainability in Tourism Destinations*, DOI 10.1007/978-981-10-1718-6_27

## Introduction

Shopping activities have been the main economic activities in popular tourist destinations and the success of this economic model has attracted several agencies to develop their destinations to lure foreign tourists (Filipović et al. 2013). Projects highlighting shopping destinations have thus become one of the impetuses for national economic development besides becoming a catalyst to attract foreign investors and to enhance the tourism sector. Since then, countries worldwide have developed shopping activities as important agents for their economic growth. Numerous cities around the world have promoted themselves as attractive tourist shopping destinations, with the sole purpose of attracting visitors from near and far, mainly for the purpose of shopping. Among these destinations, Dubai, Hong Kong, and Singapore are both well known and popular in the Middle East and Asia. The media contribute by providing descriptions of places to shop, and even by providing ranking for these venues according to the prices, availability of goods, services, and hospitality. In 2012, a global news network, Cable News Network (CNN) ranked Kuala Lumpur, in Malaysia, as the fourth best shopping city in the world based on a survey of the "10 top best shopping cities in the world". New York was ranked as being the first, followed by Tokyo and London, while Paris was only ranked in fifth place. Hong Kong, Buenos Aires, Vienna, Dubai, and Madrid completed the list (Bernama 2012). It should be emphasized that these listed shopping destinations are all located in city centers or in downtown areas. However, another lesser known area, which has the potential of development as a shopping attraction, is the border town between countries.

Shopping at international border areas has become a common activity in many parts of the world, and border shopping generally attracts both foreign tourists and domestic visitors. These shoppers are prepared to travel outside their home environment for the purpose of shopping. They are sometimes referred to as "out-shoppers." Economics and leisure opportunities have been the two motivating factors that attract the out-shoppers to travel to the border towns to shop (Timothy 1995). Shopping attractions at borderlands are located adjacent to, or directly on political boundaries. Besides shopping, other touristic activities at borderlands usually include gambling, prostitution, alcohol tourism, enclaves and international parks (Timothy 2000). The term "cross-border shopping" refers to the activity where local shoppers cross the national boundary to shop at the neighboring sites, while the aforementioned border shopping includes shopping activities that occur within the national boundary.

This study is an attempt to explore the latitude of development of border shopping in relation to the development of tourism at the border town of Padang Besar, in the state of Perlis, Malaysia. Padang Besarlies adjacent to the national boundary between Malaysia and Thailand. Many Thailand-made products are sold in Padang Besar, particularly on the Malaysian side of the border. This trading arrangement does not require local visitors to produce a travel document when crossing the border. All trading activities are conducted within the Malaysian boundary. Adult entertainments are only available across the national boundary as such activities are illegal in Malaysia.

Most of the products available for sale in Padang Besar consist of foodstuffs, apparels, cooking utensils, imitation costume jewelry, and cosmetics. These products are produced mainly in Thailand, and consequently, the majority of the traders are Thai nationals, although some may hold dual citizenship. The magnetic attraction that draws domestic tourists to Padang Besar is the price of the products. The low-price strategy has made shopping financially worthwhile for domestic tourists from the southern part of the Malaysian peninsula to travel to the northernmost point of the peninsula. Over the course of time, Padang Besar has become well known for its shopping activities.

The development of the infrastructure and shopping facilities available for the visitors to Padang Besar are unsatisfactory, inadequate, and not well maintained. The cleanliness of public amenities such as the food courts, public toilets, and prayer rooms are in a deplorable condition (Amir Salim 2002; Azmi 2013). Concerns have been raised in the past regarding the border shopping facilities offered to visitors, and proposals for improvements from the tourism suppliers for economic developments in the areas have been forwarded. Unfortunately, the lack of attention regarding these matters has been an issue in the well-being of this shopping destination. This study hopes to bridge the research gap by investigating the perceptions of the tourism supply chain towards the border shopping development in Padang Besar, Perlis. The study employed a qualitative investigation of the border shopping scenario in Padang Besar. The aims of this study are to enhance the information and to arrive at a better understanding of the overall border shopping development in Padang Besar, Perlis from the perspective of the tourism supply chain.

## Border Shopping

According to Ju (2010), the political boundary set to separate the different regions (known as the border) has established a different set of political, ideological systems, economic and social structures, and ways of life. These factors have encouraged tourists to travel and experience those differences. International borders can be considered as barriers, inhospitable fortifications, and demarcation markers that are erected for political purposes. Many people are more inclined to cross the border if the language used at the neighboring country is the same as their own. Sommers and Lounsbury (1991) and Timothy and Butler (1995) stated that elements such as the unique landscape of the location, together with its heritage, shopping opportunities, gambling, and sex tourism all contribute to attract tourists. Reynolds (1997) noted that there are barriers to travel to the border, such as visa requirements and psychological barriers, but the genuine needs to travel may overcome these barriers.

Shopping is one of the enjoyable activities carried out by visitors on holiday, and it provides a major attraction and motivation to travel. Many studies have highlighted that new-found levels of mobility have resulted in increased shopping in places far away from people's home communities. Butler (1991) pointed out that the relationships between shopping and tourism could be divided into two categories. The first category can be seen when shopping is the primary purpose of the tourist's trip, while the second category is when shopping is carried out as a secondary activity during a trip, and it might be motivated primarily by something other than shopping. The role of shopping as a form of tourism is examined within the framework or products purchased, together with the selected destination, and the price advantages or value for money.

In travel literature, a limited number of border shopping studies have been carried out, and these are primarily examples of border shopping between the United States and Canada, and between the United States and Mexico. On the other hand, from the Asian perspective, there are limited studies on border shopping due to the limited availability of border towns. Interestingly, however, the border towns along the Malaysian and Thailand border have long shown significant growth in tourism activities. Border towns such as Padang Besar, Wang Kelian, Bukit Kayu Hitam, and Rantau Panjang are examples of border towns in the north of the Malaysian peninsular that have established a landmark for shopping activities. These border towns are located within the Malaysian states that share borders with Thailand. Most Malaysians visit border towns to shop, and are known as border shoppers. Cross-border shopping occurs when domestic shoppers cross the national boundary, where, in addition to shopping, the visitors are known to indulge in leisure activities and adult entertainments.

## Border Shopping at Padang Besar

Padang Besar is a town that borders Songkhla province in southern Thailand and Perlis, the northernmost state in Peninsular Malaysia. This town serves as a major stopover along the rail routes that stretch from Singapore to Thailand. The highway which links Padang Besar to Thailand is considered less busy compared with the North–South Highway that links Malaysia and Thailand. The border town on the Malaysian side is known as Padang Besar, while the border town in Thailand is known as Pekan Siam or Padang Basar, and both these border towns are synonymous with shopping. Tourists visiting Padang Besar would normally cross the border to make their way to Thailand, in order to enjoy the adult entertainments which Thai border towns are famous for. Meanwhile, the tourists' shopping activities mostly take place in Malaysian territory.

There are several marketplaces or complexes that house retail stalls. An example of one of these marketplaces is Padang Besar Business Arcade Complex (PBAC). This is the largest business complex in Padang Besar. Other shopping complexes can be found in Padang Besar, namely The Market (TM), Gapura Square (GS),

Warisan Limpahan Duty Free Shops (WL), and Emas Kerajang Duty Free Complex (EK). The duty-free shopping complexes are situated next to the Malaysian Immigration and Customs Complex. Malaysian travelers, who intend to visit the duty-free complexes, are obliged to declare their intention at the Malaysian checkpoint. The other shopping complexes, namely TM and GS, are essentially warehouses for traders.

From the observation and fieldwork studies conducted, PBAC attract the majority of visitors and shoppers. Domestic tour buses would drop off their entourage at this particular business arcade rather than at other shopping venues. Taxi stands and large public parking areas are available next to this building, which are not found at other business locations. However, automated teller machines (ATMs) are not available at any shopping complexes, and shoppers have to walk a considerable distance to the nearest bank or ATMs available at selected convenience stores along the main streets. The unavailability of ATMs at the business arcade denotes the lack of confidence and trust regarding security issues at the premises, even though there has not been any report of burglary or robbery in the arcade. Most importantly, the use of credit cards or debit cards is nonexistent and in fact, all business transactions are conducted on a cash-only basis.

The main economic activity found in Padang Besar is trading, both for free trade and industrial zone development. There have been in the past, plans and policies drawn up to develop Padang Besar into a logistics and halal hub, while maintaining its status as a tourist attraction. Initially, the idea of turning Padang Besar into a shopping destination was mooted many years ago but Malaysians have always associated Padang Besar as a shopping destination. In 1980s, a proposal was submitted by the Bukhory Foundation to develop Padang Besar as a Satellite Town. The idea was initiated by Malaysia's Prime Minister at the time, Tun Dr. Mahathir Mohamad. Hotels and shopping malls were planned but to date, the plan has not materialized. While holding on to the aspiration to develop Padang Besar as a shopping destination, the Perlis State Economy Development Corporation built PBAC in 1990, in order to strengthen and enhance the shopping facilities. Since then, this business arcade has become the anchor of attraction for tourists visiting Padang Besar.

However, in terms of touristic terminology, the shopping activities surrounding PBAC are not exclusively for cross-border shopping, as the arcade poses as a domestic out-shopping destination. This situation can be explained by the fact that most of the shoppers are Malaysians from the other states in Malaysia. Padang Besar is also considered to be a transit point for international travelers, particularly train passengers stopping by for immigration formalities.

It is rather unfortunate that Padang Besar lacks hotels and touristic amenities. It is noticeable that along the Malaysia–Thailand border, most of the sundry trading is contained within the Malaysian side of the border. The case in point can be observed in Padang Besar in Perlis, Bukit Kayu Hitam in Kedah, Pengkalan Hulu in Perak, and Rantau Panjang in Kelantan. Unlike the border towns of Malaysia, the border towns of Thailand have high economic activities of infrastructural development such as hotels and other tourist ventures, such as those prevalent in Sadao and Bentong.

## Research Methodology

The study began with an intensive literature review, in the attempt to understand the present development at Padang Besar. Following the review, an initial study was conducted to gain information about the development of tourism in Padang Besar, which would further lead to the understanding of the location, and act as a catalyst for border shopping tourism between Malaysia and Thailand. The discussion from the initial study led to the design of the main study, including the context, the participants, the data collection procedures, and the data analysis methods in analyzing and interpreting the qualitative data. Personal (one to one) in-depth semi-structured interviews with the tourism supply chains were conducted for 6 months, in order to understand their perception regarding the border shopping development at Padang Besar. These conversations have been precisely and carefully recorded. This study captured 36 respondents. The respondents were grouped as traders, government organizations, and tourism distribution channels. They were identified as the tourism supply chain at Padang Besar. The data were analyzed using constant comparative analysis with open, selective, and theoretical coding.

## Results and Discussion

Reflecting on shopping development in Padang Besar, the majority of the respondents in this study noted the fact that there have been very few changes over the last 20 years. This perception may be due to the fact that some government officials, both at the state and the federal levels, were considered as not being appreciative of the potential of Padang Besar to be developed as a domestic tourist destination. The following comments were noted:

> Development in Padang Besar is very slow; it can be said that there isn't any change at all. Buildings 20 years ago are still the same until now; maybe the state government does not see the potential here; maybe if they see the potential, they would have developed the place years ago, maybe the North-South Highway would have been built as in Bukit Kayu Hitam so that it is much easier for people to get here. Maybe if they had built the highway 20 years ago, this town of Padang Besar would be better developed than it is today. (View of a respondent)

A perception was mooted that the development of the North–South Highway had bypassed the importance of Padang Besar as a township, which has prompted it to search for its own direction in its development.

Regarding the status of the border shopping development at Padang Besar, most respondents seem to be dissatisfied with the present development. They expressed their optimism that the development at Padang Besar will be revived as planned, and that some improvements concerning the infrastructure, especially the shopping facilities around Padang Besar, are upheld. Proposal of plans to boost the shopping amenities at Padang Besar was discussed, and they had foreseen that with proper planning and development, the town could lighten up the city center of Kangar (the capital of Perlis) and the border towns, such as those already in existence on the Singapore–Johor border. Many respondents visualized Padang Besar becoming a vibrant border town, similar to those towns along the Thai–Laos border. They also had a strong belief that Padang Besar had a great potential to be developed as a border shopping destination. This perception can be summarized from some of the respondents' remarks as quoted below:

> The tourism sector in Perlis has the potential to expand, especially the domestic tourism. Padang Besar is like a cowboy town. The infrastructure in Padang Besar has to be improved. The shops are in unsatisfactory condition, not up to the standard plus the cleanliness of public facilities such as the toilets and prayer rooms is not well maintained. In future I would like to see Padang Besar town becomes just like the towns at the Singapore–Johor border, with facilities up to that standard. (View of a respondent)

> I am not satisfied with the development here, because it is rather slow; maybe because of its location being at the tip of our country; even if there is going to be any development, nothing more can be developed except for the business arcade which is the tourist attraction in Padang Besar. Nevertheless, compared to other border towns in Malaysia, Padang Besar is the best and most suitable town for further development. Padang Besar is more comfortable and is developing compared to others. I don't travel a lot but I like to see Padang Besar becoming like the Thailand and Laos border. Laos is a poor country but the border is ten times better than Padang Besar. The buildings are big and they have good facilities; nicely developed, economic activities; it becomes like a big town and is very much different than Malaysian border town; I wish that Padang Besar can complement Kangar one day; well, it is just my dreams. (View of a respondent)

> the development in Padang Besar is okay; I am satisfied with the planning being carried out at this moment. However, the development project here is a bit slow. We have to understand this; it's a very big project which involves law and red tapes that we have to go through. Most of the development projects in Padang Besar are for industrial purpose, it's just to provide job opportunities for the increasing population, but not for tourism; Anyway, I just hope that one day the town can complement Kangar, the capital of Perlis. (View of a respondent)

> Padang Besar is good and has the potential of becoming a shopping destination because there are many tourism activities that can be done here. We already have the business arcade but the town lacks certain infrastructures like hotels, proper food courts and prayer rooms. In Rantau Panjang, there are about 8 budget hotels and at Pengkalan Hulu there are 6 hotels. Although there isn't much shopping, it has historical attractions like the communist tunnel on the Thailand side in Betong, Perak. At Padang Besar we only have shopping. (View of a respondent)

There are segments of the respondents who seemed to be disappointed with the current developmental situation in Padang Besar. According to them, the majority of the retailers operating in Malaysia are from Thailand. Most local retailers have had to stop from operating at the PBAC. As a result, these respondents felt that the development in the infrastructure had so far only benefited the foreign retailers. Even though many respondents are looking forward to seeing developmental changes in Padang Besar, particularly around the shopping areas, they were also concerned with the probability that the Thai traders might populate the Malaysian business area. Thai traders are found to be more aggressive in exploring business opportunities the moment they see the on-going economic development. They are more attuned to form comradeship with their fellow countrymen in forming the business supply chain. Malaysian traders found that it was almost impossible to break this supply chain, which resulted in losing the business competitive advantages. An in-depth interview with some respondents revealed the profoundly dire situation:

> It is not necessary to make any more improvement to attract people to Padang Besar. Those who come to buy things have only helped in the economy of foreigners. Why would the government want to develop Padang Besar and make it into a popular tourist spot? I do not see any point in doing so; this will only benefit the Siamese. The product and the traders are mostly from Thailand. Every day the Siamese take back the profit they have made here to their own country. What do we local Perlis residents get in return? This has been taking place for a long time. (View of a respondent)

> The economy of Padang Besar is being monopolized by the Thai traders and in some ways this jeopardizes the economy of Perlis. Therefore what is the point of developing the town? Development in the shopping tourism areas at Padang Besar will be beneficial to the Thais if the situations remain the same; they monopolize the economy. (View of a respondent)

This poses the question of how could a shop or a business lot built by the local agency fall into the hands of foreign traders. A possible answer could be that local tenants would have rented the premises to the foreign traders. Therefore, many aspiring local entrepreneurs were left in economic despair.

There are also respondents who are dependent on outside aid for border shopping development. They believe that any development should be borne by the government. This outlook was forwarded by some of the respondents as follows:

> I do not see any shopping development in Padang Besar; but I do agree that Padang Besar is to be made into a shopping centre, but it must be done well. More tourists will come; but who am I to talk about development; shopping development. The government should think of something to increase the number of visitors to the town; yes, it's the government who should do this. (View of a respondent)

> I see the development in Padang Besar as rather slow compared to other places. The development here can be said to be not in proper order. At Padang Besar town the development planning is conducted in a rather disorderly manner; I don't know what the government is doing; it's their responsibility anyway. (View of a respondent)

Personal contributions toward developing and enhancing the border shopping image at Padang Besar are less than enthusiastic. While most of the respondents are aware of their responsibilities towards promoting the shopping destination, they did not feel accountable in developing and promoting tourism in Padang Besar. There was no synergy among all the respondents in this job scope. Apparently, there was no one who volunteered to lead this group of people towards understanding the importance of border shopping development in Padang Besar. Most of them were of the view that Padang Besar is only a trading town and not seen as a tourist destination. They tried to relate to their daily efforts towards solving problems in their routine. They did not show much interest in tourism development per se that could contribute to the well-being of Padang Besar. This apathy and lack of local support could explain the reasons for slow development of and lukewarm support from the government agencies.

Any future economic development on a large scale within the area needs to have some potential economic opportunities that can substantiate future investment decision-making. The shelving of plan to develop Padang Besar as a halal hub was an indication that the economic activities in the area were not feasible in order to provide the catalyst for development. Pearce (1991) pointed out that any involvement by the authority will depend on the size and types of attractions in the destinations. Most of the respondents in Padang Besar were hoping that the state government would develop Padang Besar as the industrial zone, in order for it to become an impetus to generate a high level of economic activities, employment, and development.

## Implication and Conclusion

Physical amenities, attractions, facilities and tourists are all contributory factors, and are the visual signs of the development of the tourist industry (Gunn 1994; Hall and Page 2014). Attractions such as shopping facilities were developed by the local authorities; however, the development in Padang Besar was not only well organized and ill-planned, it also had inadequate facility capacity for the visitors' and communities' well-being.

It is befitting that Padang Besar develops as a domestic border shopping destination, rather than an international tourist destination. While Padang Besar is situated on the border between Malaysia and Thailand, the landscaping of the environment, coupled with the economic environment, did not project the international flavor of a tourist destination. The presence of a duty-free shop and business complexes at Padang Besar are not as presentable as those found in Bukit Kayu Hitam. The location of a duty-free zone and the business complexes are located far apart and are not easily accessible. As a result, the duty-free shop is only visited by a few Malaysians. Most Malaysians frequented the duty-free complex merely for window shopping, or possibly to buy chocolates, perfume and cigarettes. They prefer, instead, to shop at the PBAC. Most tour groups also prefer to shop at PBAC rather than at the duty-free shop.

In order for the government to develop Padang Besar as a border shopping destination, several concerted efforts need to be taken by every stakeholder with an interest in Padang Besar as a tourist shopping destination. The stakeholders include the traders, government agencies, the Chamber of Commerce, and the tourism distribution channels. Currently, these stakeholders do not have the resources or energy needed to develop Padang Besar. Most of them are focusing more on carrying out their own individual activities.

The interesting revelation in this study is the status quo of the players in the trading activities at border shopping development in Padang Besar. The findings show that the Thai traders have received more economic benefits compared with the local entrepreneurs. They form the majority of traders venturing in retailing business at the business complex. Their business activities have somewhat resulted in leakages in the national economy. These activities should prompt the government agencies to address the imbalance in the economic proportions in the area. For example, tourism policy makers within the national tourism organization, both at the state and federal levels, should re-examine the current policy and plan towards developing Padang Besar as a tourist shopping destination. The study has so far highlighted the roles of industry players such as the tourism organization, hoteliers, and service organizations related to shopping tourism activities in Padang Besar. The results could provide the feasibility for providing related attractions, facilities and/or services to the visitors.

## References

Amir Salim AS (2002) Business potential at border, a case of Padang Besar Business Arcade Complex. Unpublished Master Dissertation, Universiti Sains Malaysia

Azmi A (2013) An investigation of border shopping development, activities and facilities from the tourism supply chain perspectives. Unpublished Ph.D Dissertation, Universiti Utara Malaysia

Bernama (2012) Kuala Lumpur ranked by CNN as World's 4th best shopping city. Retrieved 30 Nov 2012 from http://www.theedgemalaysia.com/business-news

Butler RW (1991) West Edmonton Mall as a tourist attraction. Can Geogr 35(3):287–295

Filipović V, Jovanović D, CicvarićKostić S (2013) Attracting tourists to a shopping centre. J Theory Pract Manag 66:15–18

Gunn CA (1994) Tourism planning basics, concepts, cases, 3rd edn. Taylor and Francis, Washington

Hall MC, Page SJ (2014) The geography of tourism and recreation: environment, place and space. Routledge, Abingdon

Ju P (2010) Research on characteristics of spatial behavior of Hong Kong Cross-border consumer. 2010 International Conference on Ebusiness,Management and Economics, v. 3(2011). IACSIT Press, Hong Kong

Pearce PL (1991) Analyzing tourist attractions. J Tourism Stud 2(1):46–55

Reynolds C (1997) U. S. ban aside, Americans visit Cuba easily. Los Angeles Times, pp 1–2

Sommers LM, Lounsbury JF (1991) Border boom towns of Nevada. Focus 41(4):12–18

Timothy DJ (1995) Political boundaries and tourism: borders as tourist attractions. Tour Manag 16:525–532

Timothy DJ (2000) Borderlands: An Unlikely Tourist Destination?. IBRU Boundary and Security Bulletin, pp 57–65

Timothy DJ, Butler RW (1995) Cross-border shopping: a North American perspective. Ann Tourism Res 22(1):16–34

# Chapter 28
# Tour Operators Contribution Towards Sustainable Tourism: A Review from the Literature

**Maisarah Abd Hamid and Salmi Mohd Isa**

**Abstract** Sustainable tourism is growing nowadays and it involves destination, product, process and place (3Ps) of tourism. These are areas of importance for current practices of tour operators. However, current practices by tour operator are lacking in fostering sustainability. This paper illustrates some literatures and the linking of terminology namely supply chain management (SCM), corporate social responsibility (CSR) and ecotourism towards sustainable tourism. The review of the literature is useful for better understanding of the significant contributions of tour operators in implementing sustainable practices in tourism industry.

**Keywords** Sustainable tourism · Tour operator · Ecotourism · Corporate social responsibilities · Supply chain management

## Introduction

Sustainable tourism was introduced in middle of 1990s and United Nation World Tourism Organization (UNWTO) has actively promoted this concept to the world. Even though this concept was used in many areas, the definition is still vague. Many destinations have used the definition by UNEP & UNWTO (2005), which is development that considers its current and future economic, social and environmental impacts, addressing the needs of visitors, the industry, and the environment and host communities. By looking at this definition, three pillars in sustainability are the guidelines for this concept and four important stakeholders in tourism

M.A. Hamid (✉)
Faculty Hotel and Tourism Management, Universiti Teknologi MARA Terengganu, 23000 Dungun, Terengganu, Malaysia
e-mail: Maisarah412@gmail.com

S.M. Isa
Graduate School of Business, Universiti Sains Malaysia, 11800 Gelugor, Penang, Malaysia
e-mail: Salmi.mohd.isa@usm.my



A. Saufi et al. (eds.), *Balancing Development and Sustainability in Tourism Destinations*, DOI 10.1007/978-981-10-1718-6_28

industry are the main concerns of this concept. The implementation of sustainable tourism is required to all forms of tourism regardless in mass tourism or in the various niches of tourism segments.

As stated earlier, an important element in sustainable tourism is the stakeholders and one of the stakeholders in tourism industry is tour operators. They are the intermediaries in tourism industry where they link customers to suppliers and vice versa. In addition, they are the company that plans, markets and organizes tours, either directly to customers or through travel agencies and their involvements in sustainable tourism is very important as they can respond to clients and demands and they can provide the information to influence travel choice (Goeldner and Ritchie 2006). The involvements by tour operators can be done by encouraging their clientele to respect the sociocultural, economic and environmental needs of destination areas (Mbaiwa 2005), but they need to be supported by appropriate national guidelines and regulations to facilitate their work towards creating a more sustainable tourism industry (UNWTO 2013). The awareness inside the tour operator companies itself is also important towards the implementation of sustainable tourism. Currently, in global perspectives, there are guidelines for tour operators to practice sustainable tourism but not many tour operators realize this, and this leads to failure of promoting these guidelines to some countries.

In Malaysia, tour operators have the opportunities to be involved in sustainable tourism because many of the attractions in Malaysia are nature-based and some tour operators are involved in creating ecotourism packages. However, there is still a lack of practices and guidelines towards sustainable tourism by these tour operators. The awareness of this group towards sustainable impacts by looking at economy, social and environment elements do exist, but the implementation of actions towards improved performance is still in its early stages (Ytterhus 2000). In line with that, some tour operators claim to be involved in this sustainability programmes, but in reality, they are just selling it without actually practicing it (Sood 2011). This shows that it is necessary for tour operators to be involved in sustainable tourism due to their roles in the industry and by that, the proper guidelines need to be developed as a reference to the practice of sustainable tourism development.

This study will review sustainable tourism and the three pillars in sustainable tourism, namely economy, social and environment. Second, this paper will look at the roles of tour operators in implementing sustainable tourism, which will focus on some initiatives and code of conducts that are related to sustainable tourism. Third, this paper will focus on some terminologies related to sustainable tourism by tour operators, namely the linking of ecotourism, CSR, supply chain management (SCM) and green business. Finally, this paper will summarize articles on sustainable tourism development practices by tour operators.

## Review of Literature

This section discusses the literature on sustainable tourism in relation to the three pillars in sustainability. Next, the discussion on application of sustainable tourism by tour operator is highlighted with focus on the initiative towards sustainable tourism, which is the main discussion on global sustainable tourism criteria (GSTC). Finally, this section will highlight sustainable tourism practices from previous studies, which comprise of sustainable tourism practices, sustainable SCM practices and CSR practices.

## Sustainable Tourism

The definition of sustainable tourism is still vague even though it is well established in many tourism areas and it has been debated by many academics and practitioners (Butler 1999; Sharpley 2000; Wall 1997). However, the most widely accepted definition is by UNEP & UNWTO (2005), which states that tourism takes full account of its current and future economic, social and environmental impacts, addressing the needs of visitors, the industry, the environment and host communities". The undefined terms of sustainable tourism have led to varying interpretations and brought about many misperceptions on this concept (Butler 1999; Wall 1997) depending on the nature of the study conducted by researchers. Furthermore, the concept should be defined on a case-by-case basis (Manning 1999). As a result, the definition of sustainable tourism is vague and it is open to researchers to make their interpretation of the concept depending on the nature of their study.

In studying sustainable tourism, scholars attempt to focus on destinations and practices in that area, which makes it difficult to be generalized (Gössling 2000). Sustainable tourism should be attainable at local, regional, national and global scales (Hardy and Beeton 2001). The issues of scale, especially global—local nexus, play an important role in this concept (Duffy 2002; Milne 1998; Saarinen 2006). Previous study mainly focused on local and destination level analysis only (Saarinen 2006). However, an analysis in the Journal of Sustainable Tourism showed that in 1993–1997, the most frequently employed research perspectives include tourism destination or product (23.6 %), management (19.3 %) and visitor (17.4 %). From 1998 to 2007, the focus shifted to the role of stakeholders in sustainable tourism development. In addition, the research focus has moved from public to private and from visitor to local communities (Lu and Nepal 2009). Thus, this transaction showed that sustainable tourism studies should focus on all areas, not only on destinations but also include the transaction process in tourism and the operation behind the tourism industry.

Another important element in the study of sustainable tourism is the traditional three pillars of sustainability, namely economy, social and environment. Tourism firms should adopt the triple bottom line (TBL) approach to ensure that firms integrate social, environmental and economic information into managerial decision-making (Dwyer et al. 2009). A key element of this TBLs requires that corporate performance should be geared not only to benefit its shareholders but all of its stakeholders, including groups such as the local community within which business operations are conducted, and it can be used to enhance the performance of tourism development organization (Stoddard et al. 2012). In a previous study on applying the concept of sustainable tourism in practice, the focus of the study is on the economic and environment impact, while the sociocultural factor is given less attention (Jovicic 2013).

In order to understand these three pillars, it is necessary to understand the concept of each pillar. Economic sustainability is about generating prosperity at different levels of society and addressing the cost effectiveness of all economic activities, while social sustainability refers to respecting human rights and equal opportunities for all in society and environmental sustainability means conserving and managing resources, especially those that are not renewable or are precious in terms of life support (Goeldner and Ritchie 2006). From the organization's perspective and as accepted by many scholars and countries, economic sustainability, or also known as profit, is used to ensure viable, long-term economic operations, providing socioeconomic benefits to all stakeholders that are fairly distributed, including stable employment and income-earning opportunities and social services to host communities, and contributing to poverty alleviation. Meanwhile, environmental sustainability, also known as planet, refers to making optimal use of environmental resources that constitute a key element in tourism development, maintaining essential ecological processes and helping to conserve natural heritage and biodiversity. Finally, social sustainability or people aspect refers to sociocultural authenticity of host communities, conservation of their built and living cultural heritage and traditional values, and contribution to inter-cultural understanding and tolerance (UNEP & UNWTO 2005).

These three dimensions are the guidelines for the tourism enterprise to implement sustainable tourism and it has proven to yield benefits. The economy, social and environment dimensions may provide some useful and dependable indicators of societal and environmental indicators of "value added" in relation to corporate responsibilities and accountabilities. In addition, it can be used to differentiate the company from their competitors to attract investors and customers who are concerned on sustainability (Stoddard et al. 2012). Even though the TBLs are useful, the implementation is harder. The measurement of societal and environment is a difficult task in organizations. Thus, a proper study on guidelines from these three perspectives needs to be conducted so that each of the elements provides the sustainability value.

As a conclusion, the term 'sustainable tourism' is broad and it depends on which aspects of development that needs to be addressed. From the tour operators' perspectives, sustainable tourism should address the three pillars, namely economy (profit), social (people) and environment (planet). Other than that, in identifying the application of sustainable tourism, it must focus on all aspects to ensure the needs of visitors, the industry, the environment and host communities can be fulfilled.

## Tour Operators and Sustainable Tourism Development Practices

Tour operator and travel agencies have direct access to the customers, which makes them the main player in the tourism industry in fostering sustainability. Among the involvements that can be done through the interaction with travellers is by providing information about appropriate behaviours in the green tourism business and products and by giving preferential exposure to the travellers (Hjalager 1999; Miller 2010). However, in order to attract tour operators to be involved with these types of development, many studies have to be conducted on sustainable tourism and tour operators (Budeanu 2005, 2009; Tepelus 2005; Khairat and Maher 2012; Van Wijk and Persoon 2006; Xin and Chan 2014).

Tour operators nowadays have recognized their roles in sustainability. Many tour operators have taken several initiatives to evaluate impacts and improve performance (Budeanu 2005). Furthermore, a large number of tour operators have taken a more proactive attitude and have started to develop environmental policies and plans (Khairat and Maher 2012). By looking at regional perspectives, the European travel agency sector appears to be more proactive than US in adopting sustainability principles and practices (Weaver 2006). Meanwhile, Asean countries are moving towards these actions, which can be seen through some initiatives undertaken by tourism organizations by creating code of conducts for the tour operators.

The initiative to improve the sustainability of the tourism sector has been taken by international organizations, industry associations and government bodies. This is proven with the initiative for tour operators by nongovernmental organizations and tour operators (Van Wijk and Persoon 2006). Some of these organizations create code of conducts for tour operator to implement sustainable tourism. Even though many initiatives have been taken towards sustainable tourism, actions by the tour operators themselves are still limited (Table 28.1).

In studying the sustainable tourism development practices by tour operators, the global sustainable tourism criteria for Hotels and Tour Operators can be used as the basis. The Global Sustainable Tourism Council (GSTC) Criteria and Suggested Indicators for Hotels and Tour Operators (GSTC C-HTO) was created in an effort to come to a common understanding of sustainable tourism, and are the minimum that any tourism business should aspire to reach. They are organized around four main themes

- effective sustainability planning
- maximizing social and economic benefits for the local community
- enhancing cultural heritage
- reducing negative impacts to the environment.

**Table 28.1** The initiative towards sustainable tourism

| The programme/code of conduct | Objective |
|---|---|
| Blue flag—certification programme | To work towards sustainable development at beaches/marinas through strict criteria dealing with water quality, environmental education and information, environmental management, and safety and other services |
| Global sustainable tourism criteria for hotels and tour operators | To create a platform to develop ideas and projects to address the environmental, social, economic and cultural aspects of sustainable development within the tourism sector |
| EUROPARC federation | To unite national parks, regional parks, nature parks and biosphere reserves in 37 countries, with the common aim of protecting Europe's unique variety of wildlife, habitats and landscapes |
| Green Globe 21—certification programme | An Agenda 21-based improvement programme that provides guidance material and a certification process both linked to ISO standards and Agenda 21 principles, throughout the world |
| Better place programme by ABTA | Our better places programme is a step-by-step practical guide that can help all our members develop a sustainability approach that's right for their business |
| Code of practice for ecotourism operator by ecotourism association of Australia | Provide the code of practice for ecotourism operators to applied ecotourism and get involved in eco certification programme at Australia |
| The APEC/PATA code for sustainable tourism | The code for sustainable tourism was adopted by both PATA and APEC as a reflection of their strong commitment to tourism growth across the Asia and Pacific region that is viable and sustainable over a long-term future |

*Source* Author's compilation

This guideline was created in 2007 and the first version of the criteria was released in October 2008 and was publicly available for comment until April 2011. The GSTC is a USA-registered non-profit organization that represents a diverse and global membership including UN agencies, leading travel companies, hotels, country tourism boards, tour operators, individuals and communities—all striving to achieve best practices in sustainable tourism. Previously, the sustainable tourism guideline was created by Tour Operator Initiative (TOI). However, on December 2014, the TOI merged with Global Sustainable Tourism Council (GSTC) and the new guidelines by GSTC was used as guidelines for tour operators. The guidelines consist of 40 criteria and more than 70 indicators for tour operator (Global Sustainable Tourism Council 2013).

## Tour Operator and Linking of Terminology in Sustainable Tourism

Studies on application of sustainable tourism have used other terms. This occurs due to confusion of the term 'sustainable tourism' and scholars have begun to introduce similar terms like ecotourism or alternative tourism, such as green tourism (nature-based), soft tourism, nature tourism, environmental friendly/environmentally sensible tourism, responsible tourism, discreet tourism, appropriate tourism and eco ethno tourism (e.g. Boo 1990; Butler 1992; Mose 1993; Font and Tribe 2001). Even though many terms have been introduced as synonyms to sustainable tourism, not many of it is truly synonymous with this concept (Swarbrooke 1999). In this review, an application of sustainable tourism will be used with other terms, namely ecotourism, CSR, SCM, green tourism and sustainable tourism development practices (STD) (Fig. 28.1).

In studying the roles of tour operators toward sustainability, the terminology 'corporate social responsibility' (CSR) is among the topics that are discussed by researchers. A review of tourism and CSR was conducted to present a critical review of recent progress on CSR in tourism management and provide possible direction for future research (Coles et al. 2013). The term CSR is difficult to define; however, in simplified the terms, it is essentially 'sustainable development for businesses' whereby businesses voluntarily take into account, and act on, their environmental and social gains and losses along with economic ones, or the so-called 'triple bottom line' approach (O'Rourke 2003; Moneva et al. 2006). Studies on CSR and tour operator have gained interest by some researchers (e.g. Miller 2001; Kim 2009; Dodds and Kuehnel 2010; Van De Mosselaer et al. 2012).

In CSR practices, good practices are still indefinable. Some businesses may be practising a more responsible approach to business but do not recognize their actions as constituting a programme of CSR activity (Coles et al. 2013). There is also evidence that there is good awareness of CSR conceptually among Canadian mass tour operators but little action followed, where among the small sample (seven businesses), 80 % had heard of the TBL, all were aware of CSR, but only one had published a CSR policy and one had issued a CSR report (Dodds and Kuehnel 2010). In the same study, the measurements of CSR by tour operators used are

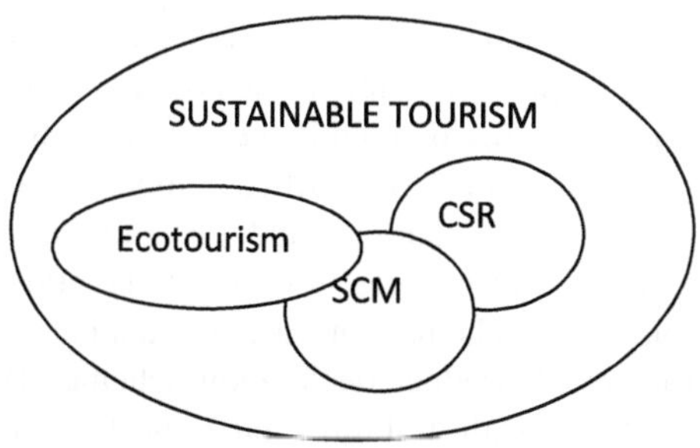


**Fig. 28.1** The link between terminologies in sustainable tourism. *Source* Author's illustration

specific employees, environmental, community, philanthropic and consumer practices and the result showed that majority of the focus is in destination support of local people through building orphanages, schools or water, waste and energy projects and support for employee issues.

Other than CSR, the term SCM has also gained interest among researchers. SCM is also a part of CSR. In the study by Dodds and Kuehnel (2010) in creating the framework to help CSR practices, among the important elements that must be addressed by tour operators to be more responsible in tourism is the supply chain efforts. The term SCM can be defined as 'a philosophy of management that involves the management and integration of a set of selected key business processes from end user through original suppliers, that provides products, services, and information that add value for customers and other stakeholders through the collaborative efforts of supply chain members' (Ho et al. 2002). The elements of sustainability is included so that these also consider the environmental, social and economic impacts of business activities and in tour operations, this relates to the development of operator purchasing policies and practices that positively contribute to supplier sustainability, and at the same time, improve the quality of the products and services offered (Schwartz et al. 2008). Studies on SCM in sustainability by tour operators have been done by several researchers (e.g. Sigala 2008; Schwartz et al. 2008; Budeanu 2009; Khairat and Maher 2012; Morali and Searcy 2013).

In the application of SCM in tourism industry, there is lack of the literature discussed on this topic (Schwartz et al. 2008). In tour operations, the sustainability SCM brings environmental and socioeconomic considerations into suppliers management process, including considerations relating to energy and waste management, conservation of nature and heritage, preservation of cultural identity, respect to local communities and contribution to economic development in destinations (TOI 2004). In good sustainability SCM practices, there is more evidence that tour operators are focusing on environmental rather than socioeconomic issue (Schwartz et al. 2008). Among the practices that reflect sustainable SCM is by TUI as a leading tour operator, which has a strong commitment in sustainability where numerous TUI subsidiaries have also established their own environmental codes of practice, policies or corporate guidelines focusing on the following three areas: being a responsible business (internal sustainable management); being a good neighbour (external network sustainable management and collaboration, e.g. local communities, staff, tourists, authorities); and protecting holiday destinations [destination sustainable management (Sigala 2008)]. In the same study, the four stages in supply chain for TUI were created, namely sustainable product design, sustainable procurement, sustainable production and sustainable delivery. Each of the stages is provided with the activities that relate to sustainability, for example: selection and use of certified companies, TUI Group's wide sustainability activities reflect social dialogue and cooperation with multiple tourism stakeholders and entities, priority selection and contracting with suppliers abiding to TUI's environmental schemes and achieve annual sustainable targets, development of environmental product declaration schemes (EPDS) for the development of tourism products, staff training and education (through seminars, videos and in-house management schemes) as well as exemplary consumer information on environmental aspects.

Another term that is used with the application in sustainable tourism is ecotourism. This sector is one of the fastest growing sectors in tourism industry worldwide (UNWTO 2015). It is a nature-based tourism and it is defined as tourism that allows for the enjoyment and understanding of the nature and culture of a destination while producing economic benefits and actively promoting environmental conservation (Ayala 1996). Sustainability issue is important for this sector as it contributes to the environmentally sensitive tourism development and protection of natural resources from detrimental environment impacts (Hashemkhani Zolfani et al. 2015). A few studies related to ecotourism and sustainable tourism from tour operator perspectives are available (e.g.: Ahmad 2014; Sangpikul 2011; Sirakaya and McLellan 1998; Wood et al. 1996). It is seen that there is limited literature on ecotourism practices by tour operators.

In identifying the application of ecotourism by tour operators, some practices have been used by the researchers. A study conducted in Brunei on ecotourism sector highlighted some of the practices by tour operators, which include recycling paper/glass/can/plastic, using energy-saving light bulbs, buy organically produced food, buy recycled products, ban smoking, cut down on the use of water, cut down on the use of electricity and contribute to wildlife conservation. Furthermore, the results show that 29 % of the tourism industry surveyed adopts recycling practices in their business, whilst 28 % of them ban smoking in their working spaces (Ahmad 2014). Meanwhile, in Thailand, in a study on developing good practice for tour operator, the author has highlighted six elements to create ecotourism practices that include nature element, education/learning element, conservation element, impact element, community element and corporate elements. From these elements, almost 26 indicators have been formulated for tour operators in Thailand (Sangpikul 2011). Additionally, some codes of conduct or guidelines have been introduced by some organizations on ecotourism practices, namely The International Ecotourism Society, who prepared ecotourism guidelines for nature tour operators that consist of seven guidelines to prepare travellers, to minimize visitor impacts, minimize nature tour company impacts, provide training, contribute to conservation, provide competitive local employment and offer site sensitive accommodation (Wood 2002). Even though the practices exist, the application is still limited. In addition, the link between these practices and the TBL needs to be explored so that the goals of sustainable tourism can be achieved.

Finally is the application sustainable tourism development by tour operators from the previous literature. In identifying the sustainable tourism development practices, many scholars attempt to focus on hotels as the main suppliers in the tourism industry (e.g. Enz and Siguaw 1999; Erdogan and Baris 2007; Le et al. 2006; Hamid et al. 2012). However, there are some studies that focus on tour operators but are still limited (e.g. Budeanu 2005; Forsyth 1995; Frederick et al. 2008; Lazarenko 2014; Le and Hollenhorst 2005; Tepelus 2005). A study on business attitudes to sustainable tourism highlights some practices that are related, namely provision of eco-tips in brochures, donating to local charities, sponsoring or supporting research related to sustainability, promoting green holidays, lobbying in destinations to improve infrastructure, recycling brochures, selecting tour guides from local communities and promoting codes of conducts (Forsyth 1995). Moreover, a study on aiming for sustainability in tour operating businesses has identified some guidelines, namely information, education, environmental management work as well as environmental or social investment (Tepelus 2005). In a thesis on UK Tourism Operators and their response to sustainability, the author used two sets of criteria and indicators by United Nations Environmental Programme (UNEP) and the Global Sustainable Tourism Council (GSTC) (Lazarenko 2014). GSTC criteria are organized around four main themes: effective sustainability planning; maximizing social and economic benefits for the local community; enhancing cultural heritage; and reducing negative impacts to the environment (GSTC 2013). Additionally, UNEP's criteria address areas like internal management, product development and management, customer relations, and cooperation with destination (UNEP 2005). The application and practices are available for tour operators. However, the linking of these guidelines with three pillars in sustainability needs further exploration so that the goals of sustainable tourism can be achieved.

## Conclusion

This article reviews previous studies in relation to sustainable tourism development practices by the tour operator. This paper illustrates that the sustainable tourism development practices are a combination of many mechanisms, such as SCM, CSR, ecotourism and sustainable tourism itself. Thus, in order to develop sustainable tourism practices by the tour operator, each of the practices and guidelines should be integrated and linked to sustainability pillars, which are social, economy and environment. Besides that, it is also important to form systematic guidelines and code of conduct for tour operators in creating the practice model for sustainable tourism.

## References

Ahmad A (2014) The disengagement of the tourism businesses in ecotourism and environmental practices in Brunei Darussalam. Tour Manag Perspect 10:1–6. doi:10.1016/j.tmp.2013.12.002

Ayala H (1996) Resort ecotourism: a master plan for experience management. Cornell Hotel Restaurant Adm Q 37(5):54–61

Beugen MV (2005) Sustainability of Dutch Tour Operators (Doctoral dissertation, Doctoral Thesis. Tilburg University, Netherlands, Department of Leisure Sciences)

Boo E (1990) Ecotourism: potentials and Pitfnlls. World Wildlife Fund-U.S, Washington, D.C.

Budeanu A (2005) Impacts and responsibilities for sustainable tourism: a tour operator's perspective. J Clean Prod 13(2):89–97. doi:10.1016/j.jclepro.2003.12.024

Budeanu A (2009) Environmental supply chain management in tourism: the case of large tour operators. J Clean Prod 17(16):1385–1392. doi:10.1016/j.jclepro.2009.06.010

Butler RW (1992) Alternative tourism: the thin edge of the wedge. Tour Altern 31–46

Butler RW (1999) Sustainable tourism: a state-of-the-art review. Tour Geogr. doi:10.1080/14616689908721291

Coles T, Fenclova E, Dinan C (2013) Tourism and corporate social responsibility: a critical review and research agenda. Tour Manag 6:122–141

Dodds R, Kuehnel J (2010) CSR among Canadian mass tour operators: good awareness but little action. Int J Contemp Hospitality Manag 22(2):221–244. doi:10.1108/09596111011018205

Duffy R (2002) A trip too far: ecotourism. Earthscan Publications, Politics and Exploitation

Dwyer L, Edwards D, Mistilis N, Roman C, Scott N (2009) Destination and enterprise management for a tourism future. Tour Manag 30:63–74. doi:10.1016/j.tourman.2008.04.002

Enz CA, Siguaw JA (1999) Best hotel environmental practices. Cornell Hotel Restaurant Adm Q 40(5):72–75

Erdogan N, Baris E (2007) Environmental protection programs and conservation practices of hotels in Ankara, Turkey. Tour Manag 28(2):604–614

Font X, Tribe J (2001) Promoting green tourism: the future of environmental awards. Int J Tour Res 3(1):9

Forsyth T (1995) Business attitudes to sustainable tourism: self-regulation in the UK outgoing tourism industry. J Sustain Tour 3(4):210–231

Fredericks L, Garstea R, Monforte S (2008) Sustainable tourism destinations: A pathway for tour operators. Blekinge Institute of Technology, Karlskrona, Sweden

Global Sustainable Tourism Council (2013) Global sustainable tourism council criteria hotels and tour operators (December), 1–11. Retrieved from http://www.gstcouncil.org/sustainable-tourism-gstc-criteria/criteria-for-hotels-and-tour-operators.html

Goeldner CR, Ritchie JRB (2006) Tourism: principles, practices, philosophies. Wiley, New York City

Gössling S (2000) Tourism—sustainable development option? Environ Conserv. doi:10.1017/S0376892900000242

Hamid MA, Ismail N, Fuza ZIM, Ahmad KN, Awang KW (2012) Sustainable tourism development practices of MICE venue provider in East Coast Region, Peninsula Malaysia. Curr Issues Hospitality Tour: Res Innovations 87

Hardy AL, Beeton RJS (2001) Sustainable tourism or maintainable tourism: managing resources for more than average outcomes. J Sustain Tour. doi:10.1080/09669580108667397

Hashemkhani Zolfani S, Sedaghat M, Maknoon R, Zavadskas EK (2015) Sustainable tourism: a comprehensive literature review on frameworks and applications. Econ Res-Ekonomska Istraživanja 28(1):1–30. doi:10.1080/1331677X.2014.995895

Hjalager A (1999) Consumerism and sustainable tourism. J Travel Tour Mark 8(3):1–20

Ho DC, Au KF, Newton E (2002) Empirical research on supply chain management: a critical review and recommendations. Int J Prod Res 40(17):4415–4430

Jovicic DZ (2013) Key issues in the implementation of sustainable tourism. Curr Issues Tour 17 (4):297–302. doi:10.1080/13683500.2013.797386

Khairat G, Maher A (2012) Integrating sustainability into tour operator business: an innovative approach in sustainable tourism. Tourismos Int Multi J Tour 7(1):213–233

Kim JE (2009) Understanding corporate social responsibility in the tourism. Graduate School of The University of Florida, Gainesville

Lazarenko V (2014) UK Tourism Operators and their response to sustainability: rhetoric or reality? (Master Thesis. Cardiff University)

Le Y, Hollenhorst S (2005) Perceptions of Vietnamese tour companies toward adopting sustainable tourism practices. Anatolia 16(1):79–99. doi:10.1080/13032917.2005.9687166

Le Y, Hollenhorst S, Harris C, McLaughlin W, Shook S (2006) Environmental management: a study of Vietnamese hotels. Ann Tour Res 33(2):545–567

Lu J, Nepal SK (2009) Sustainable tourism research: an analysis of papers published in the journal of sustainable tourism. J Sustain Tour 17:5–16. doi:10.1080/09669580802582480

Manning T (1999) Indicators of tourism sustainability. Tour Manag 20(1):3–6
Mbaiwa JE (2005) The problems and prospects of sustainable tourism development in the Okavango Delta, Botswana. J Sustain Tour 13(3):203–227. doi:10.1080/01434630508668554
Miller G (2001) Corporate responsibility in the UK tourism industry. Tour Manag 22(6):589–598
Miller G (2010) Public understanding. Ann Tour Res. doi:10.1016/j.annals.2009.12.002
Milne S (1998) Tourism and sustainable development: the local-global nexus. In: Hall M, Lew A (eds) Sustainable tourism: a geographical perspective. Longman, New York, pp 35–48
Moneva JM, Archel P, Correa C (2006) GRI and the camouflaging of corporate unsustainability. Acc Forum 30:121–137
Morali O, Searcy C (2013) A review of sustainable supply chain management practices in Canada. J Bus Ethics 117(3):635–658. doi:10.1007/s10551-012-1539-4
Mose I (1993) Hohe Tauern National Park test case for "Soft Tourism" in The Austrian Alps? Experiences with projects in the Upper Pinzgau Region. Tourism Recreation Research 18 (1):11–19
O'Rourke A (2003) A new politics of engagement: shareholder activism for corporate social responsibility. Bus Strategy Environ 12:227–239
Saarinen J (2006) Traditions of sustainability in tourism studies. Ann Tour Res 33(4):1121–1140. doi:10.1016/j.annals.2006.06.007
Sangpikul A (2011) Developing good practices for ecotourism tour operators. Sasin J Manag 17 (1):53–86
Schwartz K, Tapper R, Font X (2008) A sustainable supply chain management framework for tour operators. J Sustain Tour 16(3):298–314. doi:10.2167/jost785a.0
Sharpley R (2000) Tourism and sustainable development: exploring the theoretical divide. J Sustain Tour. doi:10.1080/09669580008667346
Sigala M (2008) A supply chain management approach for investigating the role of tour operators on sustainable tourism: the case of TUI. J Clean Prod 16:1589–1599. doi:10.1016/j.jclepro.2008.04.021
Sirakaya E, McLellan RW (1998) Modeling tour operators' voluntary compliance with ecotourism principles: a behavioral approach. J Travel Res 36(3):42–55. doi:10.1177/004728759803600305
Sood S (2011) Sustainable travel in Malaysia. Retrieved 1 April 2015, from http://www.bbc.com/travel/story/20110112-travelwise-sustainable-travel-in-malaysia
Stoddard JE, Pollard CE, Evans MR (2012) The triple bottom line: a framework for sustainable tourism development. Int J Hospitality Tour Adm 13(3):233–258. doi:10.1080/15256480.2012.698173
Swarbrooke J (1999) Sustainable tourism management. Cabi, Wallingford, Oxfordshire
Tepelus CM (2005) Aiming for sustainability in the tour operating business. J Clean Prod 13 (2):99–107. doi:10.1016/j.jclepro.2003.12.018
TOI (2004) Supply chain engagement for tour operators. Paris: Tour Oper' Initiative Sustain Dev. On WWW at http://www.toinitiative.org. Accessed
UNEP, UNWTO (2005) Making tourism more sustainable: a guide for policy maker
UNWTO (2015) UNWTO tourism highlights 2015 edition
UNWTO: leading the global drive to sustainable tourism (2013) Blue and green tomorrow. Retrieved 19 May 2015, from http://blueandgreentomorrow.com/features/unwto-leading-the-global-drive-to-sustainable-tourism/
Van De Mosselaer F, Van Der Duim R, Van Wijk J (2012) Outbound tour operators. Tour Enterp Sustain Agenda Across Europe 71
Van Wijk J, Persoon W (2006) A Long-haul destination. Euro Manag J 24(6):381–395. doi:10.1016/j.emj.2006.07.001
Wall G (1997) Is ecotourism sustainable? Environ Manage. doi:10.1007/s002679900044
Weaver D (2006) Sustainable tourism. Elsevier, Oxford
Wood ME (2002) Ecotourism: principles, practices and policies for sustainability. Int Ecotourism Soc. doi:10.1079/9781845934002.0000

Wood ME, Norman W, Frauman E, Toepper L (1996) The green evaluation program: preliminary results from a pilot ecotourism monitoring program for nature tour operators. Environ Res Yale Edu 211–231. Retrieved from http://environment.research.yale.edu/documents/downloads/0-9/99WoodNorman.pdf

Xin TK, Chan JKL (2014) Tour operator perspectives on responsible tourism indicators of Kinabalu National Park, Sabah. Procedia—Soc Behav Sci 144:25–34. doi:10.1016/j.sbspro.2014.07.270

Ytterhus BE (2000) Supply chain management. Intermediary firms and sustainable developments: some examples from the tourism and retailing sectors. Norwegian School of Management, Sandvika, Norway

# Chapter 29
# Derawan Islands as a Domestic Destination: Insights of Young Indonesians Living in Singapore

**Jenny H. Panchal, Ugra Prasetyo Bawono and Jordan Napitupulu**

**Abstract** Derawan Island in East Kalimantan, Indonesia has seen growth in recent times. The island is endowed with natural resources, particularly with marine biodiversity, which make them appealing to international and domestic tourists. This paper reports a part of the findings of a larger study that explored young Indonesians' level of awareness and perceptions of domestic attractions in Indonesia. More specifically, this paper presents the insights of Indonesian students in Singapore about Derawan Island. The data collection involved an online survey which yielded 126 responses. The findings suggest that the respondents had low level of awareness of Derawan's features and accolades as a destination. Additionally, the results imply that visiting Derawan with the intention to "see and experience" the island rather than "do to experience things there" is a common rationale for young Indonesians in visiting.

**Keywords** Derawan Islands · Indonesia · Young Indonesians · Destination image · Perceptions

## Introduction

Tourism has become one of the most essential factors around the globe. Tourism development in many destinations has contributed significantly to the economic growth in countries and regions. Indonesia, being the largest archipelago and one of the most populous countries in the world, is enjoying the benefits and equally facing the challenges that tourism brings. Being a large country of more than

J.H. Panchal (✉) · U.P. Bawono · J. Napitupulu
James Cook University, 149 Sims Drive, Singapore 387380, Singapore
e-mail: jenny.panchal@jcu.edu.au

U.P. Bawono
e-mail: ugraprastyo.bawono@my.jcu.edu.au

J. Napitupulu
e-mail: jordansaptagihon.napitupulu@my.jcu.edu.au



A. Saufi et al. (eds.), *Balancing Development and Sustainability in Tourism Destinations*, DOI 10.1007/978-981-10-1718-6_29

17,500 islands with only about 6,000 inhabited, Indonesia is a home of about 253.2 million people and receives several millions of tourists every year.

The geographical context of this paper is focused on the Derawan island chain in East Kalimantan, Indonesia, which is considered to be one of the most biologically rich areas in the country. Kalimantan is the third largest island in the world, and the largest in Asia (reference), which comprises 31 islands. Derawan Island is a peripheral destination, which fits the descriptions of peripheral areas in the literature. As a destination, very little is known about tourists' perceptions of Derawan. This paper takes into consideration the perceptions of young Indonesians, whose travel behavior shapes the future of domestic tourism in Indonesia. The observed dearth of knowledge about young Indonesian travelers as well as perceptions of Indonesians of various domestic destinations in Indonesia provided an opportunity to explore the phenomenon. This also provided the opportunity of identifying prospective ways for the domestic destinations and their marketing organizations to further improve the marketability of the place to domestic and international tourists. This paper presents a part of a larger study that explores how young Indonesians living in Singapore perceive Indonesia and other places in the country as a tourist destination. More specifically, this paper presents how Indonesian students in Singapore perceive Derawan Island based on general information about the destination.

## Review and Clarification of Key Concepts

### *Youth Travel*

Over the last two decades, the role of young travelers has been recognized as an important segment in the travel and tourism industry at a global scale. The United Nations World Tourism Organization (UNWTO) defines youth tourism as a form of tourism that includes all travel by young people aged between 15 and 29 years (du Cros 2014; UNWTO 2008). The UNWTO estimates that around 20 % of the 940 million international travelers in 2010 were young people (UNWTO and WYSE Travel Confederation 2011). It is also estimated that the youth travel market is about 18 % of worldwide international tourism receipts which is an estimated US $ 136 billion per annum (UNWTO 2008). The future of the youth market is perceived to represent majority of future growth in the travel industry, and it is estimated that the youth travel industry will contribute to about 300 million international trips per year by 2020 (UNWTO and WYSE Travel Confederation 2011).

Richards (2011) describes youth markets as resilient as he relates the increasing growth in youth travel despite economic challenges, political unrest, epidemics, and other problems related to travel and tourism. At local level, the contributions of young tourists can be related to greater proportion of travel budget spent and longer stays in the destination. This is underpinned by the World Youth Student and

Educational WYSE Travel Confederation whose research revealed that about 60 % of youth travel budgets are spent in the destination (UNWTO and WYSE Travel Confederation 2011). Further, Frost and Shanka (1999), Theuns (1992) and Richards and Wilson (2003) suggest that young travelers generally travel more frequently and for extended periods than average adult tourists. Richards (2011) adds that one of the biggest factors in the economic impact of tourism for local communities is the scale of leakage from the local economy, especially in smaller economies in developing regions of the world.

## *Young Indonesians as Domestic Tourists*

Domestic tourism is defined as tourism that comprises the activities of a resident visitor within the country of reference, either as part of a domestic tourism trip or part of an outbound tourism trip (UNWTO 2015). In this study, the focus is on the analysis of Singapore-based Indonesian students on their perceptions of Derawan Island. One may argue that these individuals are international students by definition, because they live in Singapore and not in Indonesia at the time when the study was conducted. While this may be true, they are considered domestic students in this study because of the criteria that were set in the recruitment of participants for this study. The criteria included: (a) the students held an Indonesian passport; and (b) they held a valid student pass (or visa) in Singapore at the time of data collection. This rationale is linked with the notion that international students in Singapore are temporary residents in the country, and are assumed to eventually return to their respective countries between or upon completion of their studies.

## *Destination Image and Peripheral Destinations*

Destination image has been widely researched for the past five decades. It is an important aspect in successful destination marketing as well as tourism management (Molina et al. 2010). Crompton (1979) suggests that cognitive images refer to beliefs, impressions, ideas, knowledge, and perceptions that people hold on objects. Prior to Crompton's work, Gunn (1972) posited that destination image has two elements: organic images which are formed through past experiences, and induced images, which are created through information from external sources such as advertising and promotion of a destination. Baloglu and McClearly (1999) upheld this emphasis on the cognitive element of destination image and offered an affective component to the destination image concept. They argued that tourists' emotions are vital in the formation of an image. Additionally, Baloglu and Mangaloglu (2001) illustrated that the relationship between an object or a place (in the case of a destination) against the evaluation of cognitive and affective would form an image. Later, Ekinci (2003) argued that the image of a destination involves the question of the extent to which the overall view of the purpose of subjective and emotional knowledge of a destination. These underpinnings from the literature suggest a holistic view of destination image, because researchers have studied the topic from various perspectives including demand (i.e., tourists) and supply (i.e., marketing of the destination).

A significant but largely specialized subtopic of destination image and marketing are peripheral destinations. Although much of the studies on peripheral areas were in the European context, the understanding of the term and concept of "peripheral areas" provided in the literature are mostly universal and are essentially applicable in other geographical contexts. Hall (2007), for example, argues that peripheral areas are characterized by various interconnected features that impact tourism development and other industry-related sectors. Some of the features lie on definition, economic linkages, as well as political and economic control. Peripheral areas, in terms of tourism, are destinations that are geographically remote from mass markets (Hall 2007). This description implies greater transportation and communication costs for suppliers and tourists at the destination. The remoteness of these areas means that internal economic linkages tend to be weaker. Another feature of peripheral areas is linked with economic and political control over major decisions affecting these places' economic and well-being, which are mostly lacking (Hall 2007; Muller and Jansson 2007; Strauss et al. 1997). Muller and Jansson (2007) suggest that a popular vision for peripheral areas is, through their inherent supply of natural landscape and resources, an enhanced ability to cater for the recreational demand of visitors from all over the world. Derawan Island is considered a peripheral destination because it possesses the features outlined by Hall (2007).

## Study Method

The data was collected using an online survey which consisted of six parts: (1) the individual's travel experience; (2) factors that affect destination choice; (3) perceptions of Indonesia as a tourist destination; (4) perceptions of Derawan Islands as a tourist destination; (5) likelihood of visiting Derawan Islands; and, (6) profiling information. While the survey has yielded a plethora of information, this paper chiefly reports the respondents' level of awareness and perceptions of Derawan Island as a tourist destination. It also briefly discusses the islands' characteristics as factors that affect young Indonesian's decision in visiting the destination. Stratified sampling technique was used in the study whereby Indonesian students James Cook University Singapore aged 15–35 years were recruited to participate in the survey. The sample consisted of mostly undergraduate students, who are categorized according to the level of undergraduate program: (a) bachelors degree (69 %); and, (b) foundation/diploma/pre-bachelor program (11 %). The rest are doing their master degrees (20 %).

**Table 29.1** Travel experience (in the last five years)

| Previous travel | Within Indonesia (domestic) | Outside Indonesia (international) |
|---|---|---|
| Never | 3 % ($N = 4$) | 0 % |
| 1–4 times | 48 % ($N = 70$) | 54 % ($N = 78$) |
| 5–10 times | 31 % ($N = 45$) | 30 % ($N = 44$) |
| >10 times | 18 % ($N = 26$) | 16 % ($N = 23$) |

The age limits used in this study was based on the works of Aramberri (1991) who suggested that the minimum age of young tourists is 15 years old, and of Lilley (1997; *cited in Carr,* 1998) who proposed that is still a "young tourist." While there are questions regarding this aspect of the study, the researchers have considered the gaps and limitations in the literature about youth travel in the Asian context, particularly in Indonesia. The researchers have expanded the age limit to 35 years old to capture various perceptions among age groups. Notwithstanding the age limitations, the sample consisted of mostly 18–20 year olds (44 %), 21–25 year olds (29 %), and 25–30 year olds (17 %). More than half of the sample were females (60 %). The survey yielded a total of 175 responses, but only 145 were usable and hence were included for analysis (about 83 %).

## Findings and Discussion

### *Travel Experience*

The initial descriptive analysis of the data revealed that the young Indonesians in the study were by and large interested in traveling and were generally experienced travelers. More specifically, more than half of the sample like to travel and will do so at every opportunity (53 %), and about 39 % said that traveling is something they would do once in a while. Table 29.1 shows that the participants have traveled within and outside Indonesia at least 1–4 times for holidays in the last five years. It is important to note, however, that the participants in this study were international students in Singapore. Caution is therefore taken in the inferences that are offered in this study. The researchers are thoughtful that the results reported here represent the young Indonesians in Singapore.

### *Pull Factors*

The push-pull model is a foundation approach for understanding tourist motivations (Dann 1977). The pull factors are characteristics of destinations that attract tourists,

**Table 29.2** Importance of destination characteristics

| Destination factors | Mean |
|---|---|
| Safety and security | 5.51 |
| Local cuisine | 5.23 |
| Availability of transportation within destination | 5.2 |
| Price (i.e., amount to be spent here as compared to similar destinations) | 5.18 |
| Natural setting of the place (e.g., beach, mountain) | 5.17 |
| Quality of the accommodation | 5.13 |
| Accessibility of the destination | 5.02 |
| Friendliness of the local people | 4.97 |
| Cleanliness of the place (i.e., general sanitation) | 4.93 |
| Diversity of attraction (cultural, historical, etc.) | 4.76 |
| Season/time of the year (i.e., summer) | 4.63 |
| Cultural significance of the place | 4.55 |
| Popularity of the destination | 4.53 |
| Climate at the destination (e.g., tropical, temperature) | 4.48 |
| Shopping opportunities | 4.47 |
| Historical significance of the place | 4.32 |
| Water-based activities at the destination (e.g., diving, swimming, surfing, scuba diving) | 4.2 |
| Land-based activities at the destination (e.g., hiking, mountain climbing) | 4.13 |
| Spa/wellness activities and/or treatments | 4.1 |
| Nightlife and entertainment | 4.08 |

while the push factors are sociopsychological motivations that prompt individuals to travel. These push and pull factors are the essential components of tourists' travel decisions. In relation to the travel experience, the students were asked to rate the importance of a predetermined list of destination qualities that encompass both physical and social components as prescribed by Pearce (2005), where 1 was not at all important and 7 was extremely important (Table 29.2).

The findings suggest that safety and security, although they are not of extreme importance by mean score ($\bar{x} = 5.51$), are the top consideration for young Indonesians. Weaver and Lawton (2010) suggested that the tourist market, in general, is sensitive to any suggestion of social or political instability within a destination. The importance of destination security stems from the immediacy of tourists' consumption in and interaction with the setting. Safety and security, after all, are considered to be highly important in other forms of tourism and in other geographical contexts, especially in Asia. Panchal (2012), for instance, found out that spa-going tourists in Asian countries such as the Philippines, India, and Thailand regard safety and security as very important factors when choosing and traveling to a destination.

Accessibility to and within the destination are key concerns of young Indonesians within their country. The availability of transportation within the destination ($\bar{x} = 5.2$) was also observed to be of slightly greater importance than the accessibility of the destination itself ($\bar{x} = 5.02$). This can be linked to the remoteness of Derawan Island as a tourist destination. As discussed earlier, Derawan is a peripheral area that is geographically isolated and less developed as a tourist space. The infrastructure and services directed to tourists may not be as prevalent and as urbane compared to those in places like Bali, Komodo, or even Batam and Bintan islands. The other factors that are of relatively higher importance include price ($\bar{x} = 5.18$), the natural setting of the place ($\bar{x} = 5.17$), and quality of accommodation in the island ($\bar{x} = 5.13$).

Overall, the results suggest that visiting Derawan with the intention to "see and experience" the island rather than "do to experience things there" is a common rationale for young Indonesians in visiting. As noted, safety and security, accessibility, and the natural setting of the place were their top concerns. Similarly, price, food, and quality of accommodation in the island were also of fair importance in choosing Derawan as a destination. Conversely, water-based activities, by which Derawan is known for, are not the key attractions for this cohort of the domestic tourists. Spa and wellness activities and nightlife offerings have low relevance to them as pull factors.

## *Information Sources*

When asked about the sources of information about Derawan Island, the respondents reported that word of mouth (WOM) from family and friends (40 %) and social networking sites (32 %) are the most common. In this study, however, it is difficult to identify the means by which WOM information was obtained, because they can be through face-to-face or telephonic conversations, or through digital means specifically through social media (e.g., Facebook posts and private messaging). The role of online interpersonal influence, or eWOM in marketing, and information dissemination are very important (Litvin et al. 2008). The popular use of Facebook, Twitter, and similar platforms among the respondents also underpins Chadee and Cutler's (1996) notion that young tourists are strongly influenced by social networks. The Indonesian tourism website is also a popular source of information, while mobile apps and non-tourism-related events are less popular sources.

## *Perceptions About Derawan Islands*

Information on Derawan and the manner in which they are channeled and received by audience play an important factor in their perceptions of the island as a destination. In knowing how they perceived Derawan, the respondents were provided with specific facts as well as general statements about Derawan. They were asked to rate their level of agreement to the statements, where 1 = strongly disagree, and 5 = strongly agree. As one of the top diving spots in the world, young Indonesian travelers are well aware of Derawan's beaches ($\bar{x} = 3.9$) and marine diversity ($\bar{x} = 3.78$) as the island's key selling points. These perceptions of Derawan may be linked to the respondents' notion that the island is a unique destination ($\bar{x} = 3.76$), because there are no other place in Indonesia that shares the same fame of being the best dive destination in the world. This can also be related to the remoteness of the island, which is linked to exoticness and wilderness. Another observation that can be linked to the island's remoteness lies in respondents' perception of safety and security in Derawan. The students' degree of agreement to Derawan as a safe place that they can explore on their own was relatively lower ($\bar{x} = 3.17$).

Similar to the previous section, the students were asked to rate their level of agreement to the facts which were gathered from news articles and websites (UNESCO 2015), where 1 = strongly disagree, and 5 = strongly agree. Overall, their level of awareness about Derawan's characteristics, resources and accolades as a destination was relatively lower than expected. It was not surprising, however, that the island's reputation as an iconic spot was known amongst most of the students ($\bar{x} = 3.01$). Also, the accessibility of other islands in the archipelago from Derawan ($\bar{x} = 2.83$) was another factor known to students. Interestingly, however, the promotional efforts of Derawan by the Indonesian Embassy in Singapore was a fact that the students knew very little about ($\bar{x} = 2.5$).

Given that the Indonesian government's marketing efforts for Derawan Island is directed toward international tourists, it is not surprising that young Indonesians living in Singapore have very vague intentions of visiting the destination. Sixty percentage of the respondents said that they do not have the intention to visit the islands, but they may consider it as a destination as a result of the survey. About 27 % have expressed their intention to visit in the next 6–18 months after completing the survey. The findings suggest that the most attractive factors for these individuals who intend to travel to Derawan Island are the marine life, nature (flora), and wildlife (fauna). The local culture is also a factor that would make them visit the islands. A follow-up study of these respondents would provide richer insights about their experiences in Derawan, but because of ethical limitations, the responses in the current study cannot be used to identify the participants. Additionally, about 12 % have no intention to visit the islands. The lack of time to visit and the lack of information about the islands were identified as reasons for not visiting Derawan Island. The remoteness of the islands also makes it unappealing to the respondents.

## Conclusions and Future Studies

This study aimed to explore Singapore-based Indonesian students' level of awareness and perceptions of Derawan Island in Indonesia. The results suggest that WOM, the social media, and the Indonesian tourism websites are the most popular sources of information for the young Indonesian travelers. It has also been found that the islands' beaches and marine biodiversity are known to be Derawan's selling points. The key influential factors for young Indonesians to visit the islands include flora and fauna, marine life, local culture and to some extent, the historical value of the place.

The current study merely provides a small but insightful view of Derawan as perceived by young Indonesians. Future studies can further investigate this topic by including respondents who are based not only overseas but also in Indonesia. Their perspectives can be compared with Indonesians of other generations such as baby boomers and generation X members. Comparisons can also be made using demographic variables including gender, income level, and educational attainment. Apart from the tourist's perspectives, the local's perspective of Derawan as a destination will also provide greater understanding of the level of awareness and perceptions about the island. Triangulation as a key methodological approach can be used in future studies. The juxtaposition of various perspectives (i.e., of tourists, of locals, of local businesses, and of the destination marketing organization) can be undertaken to advance the current view of this topic. Future studies can also be undertaken to explore the extent in which both public and private sectors in the local, provincial, and national levels contribute in the promotion of Derawan and/or improve the public awareness of the island as a destination (Tables 29.3 and 29.4).

**Table 29.3** Perceptions about Derawan Islands

| Statements about Derawan Islands | Mean |
|---|---|
| The beaches in Derawan are the islands' main selling point | 3.9 |
| The marine diversity are the islands' main selling point | 3.78 |
| It is a unique destination (within Indonesia) | 3.76 |
| I think most people have a positive opinion about Derawan Islands | 3.66 |
| Derawan Island is a family-friendly destination | 3.41 |
| Derawan Island is a popular destination among both international and Indonesian tourists | 3.39 |
| Infrastructure in Derawan are well planned, prevalent, and support the advancement of tourism | 3.31 |
| It is rich in cultural attractions | 3.26 |
| It is only the beaches that can be enjoyed in Derawan | 3.19 |
| Derawan is a safe destination that I can explore on my own | 3.17 |
| Derawan is popular among honeymooners only | 2.9 |

**Table 29.4** Young Indonesians' awareness of Derawan Island as a destination

| Facts about Derawan and its surrounding islands | Mean |
|---|---|
| Derawan Island is the iconic diving spot of Borneo | 3.01 |
| Derawan Island is a gateway to Maratua, Sangalaki, and Kakaban Islands | 2.83 |
| In Maratua Island, there is "turtle traffic," which means that green turtles around the world go there to reproduction | 2.83 |
| At the Turtle Conservation Centre, tourists can see turtles laying eggs and hatching from eggs | 2.8 |
| The government built a new seaport and an airport to boost tourism in Derawan Island | 2.77 |
| Derawan Island dive spot is better than any other dive spots in Indonesia | 2.75 |
| The turtle conservation area is for two species of turtles, they are Hawkbills, and Green Turtle | 2.65 |
| Derawan Island awarded as third best diving spot in the world by UNESCO | 2.61 |
| UNESCO awards Kakaban lake as World Natural Heritage Area | 2.53 |
| The Indonesian Embassy in Singapore is promoting Derawan Island | 2.5 |

# References


Aramberri J (1991) "The Nature of Youth Tourism: Concepts, Definitions and Evolution." Keynote paper presented at the International Conference on Youth Tourism, New Dehli, India

Baloglu S, Mangaloglu M (2001) Tourism destination image formation. Ann Tour Res 26: 868–897

Baloglu S, McCleary K (1999) A model of destination image formation. Ann Tour Res 26: 868–897

Carr N (1998) "The young tourists: a case of neglected research". Progress in Tourism and Hospitality Research, 4(4):307-318

Chadee DD, Cutler J (1996) Insights into international travel by students. J Travel Res 35(2): 75–80

Crompton J (1979) Motivations for pleasure travel. Ann Tour Res 6:409–424

Dann G (1977) Anomie, ego-enhancement and tourism. Ann Tour Res 4:184–194

du Cros H (2014) New models of travel behavior for independent Asian youth urban cultural tourists. [Working Paper]. ARI Working Paper Series, (217). Asia Research Institute, National University of Singapore, Singapore

Ekinci Y (2003) From destination image to destination branding: an emerging area of research. E-Review Tour Res 1(2):21–24

Frost F, Shanka T (1999) Asian Australian student travel preferences: an empirical study. Asian Pac J Tour Res 42(2):19–26

Gunn C (1972) Vacationscape: destiantion tourist regions. University of Texas, Austin, TX

Hall CM (2007) North-south perspectives on tourism, regional development and peripheral areas. In: Muller DK, Jansson B (eds) Tourism in peripheries: perspectives from the far north and south. CABI, Oxfordshire, UK

Litvin SW, Goldsmith RE, Pan B (2008) Electronic word-of-mouth in hospitality and tourism management. Tour Manag 29(2008):458–468

Molina A, Gomez M, Martin-Consuegra D (2010) Tourism marketing information and destination image management. Afr J Bus Manage 4(5):722–728

Muller DK, Jansson B (2007) The difficult business of making pleasure peripheries prosperous: perspectives on space, place and environment. In: Muller DK, Jansson B (eds) Tourism in peripheries: perspectives from the far north and south. CABI, Oxfordshire, UK, pp 3–18

Panchal JH (2012) The 'positive tourism' linkage: a study of motivations, flow and benefits of spa experiences in Southeast Asia. Paper presented at the 6th world conference for graduate research in tourism, Hospitality and leisure, Fethiye, Turkey

Pearce PL (2005) Tourist behaviour: themes and conceptual schemes. Channel View Publications, Clevedon, UK

Richards G (2011) An economic contribution that matters. The power of youth travel. UNWTO, Madrid, pp 7–8

Richards G, Wilson J (2003) Today's youth tourists: tomorrow's global nomads? New horizons in independent youth and student travel. International Student Travel Confederation & Association of Tourism and Leisure Education, Amsterdam and Arnhem

Strauss CH, Lord BE, Grado SC (1997) Economic impact of tourism within a regional economy: sources, size and direction of gain. Paper presented at the fourth international conference on retailing and services science, Scottsdale, Arizona

Theuns L (1992) The significance of youth tourism: economic dimensions. Acta Turistica 4(2): 165–187

UNWTO (2008) Youth travel matters: understanding the global phenomenon of youth travel. World Tourism Organization, Madrid

UNWTO (2015) Understanding tourism: basic glossary. From http://media.unwto.org/en/content/understanding-tourism-basic-glossary

UNWTO and WYSE Travel Confederation (2011) The power of youth travel, vol 2. UNWTO, Madrid

Weaver D, Lawton L (2010) Tourism management, 4th edn. Wiley, Milton

# Chapter 30
# Responses of Social Media Users to Professional and Tourist Photographs

**Fuadi Afif, Tri Kuntoro Priyambodo and Yulia A. Widyaningsih**

**Abstract** This study examines the responses of social media users to tourist photographs and professional photographs on social media platforms. The research participants were students from the Diploma Program in Tourism, Universitas Gadjah Mada (UGM). The research design was experimental using questionnaires as the data collection instrument. This research used a paired *t*-test to analyze the differences in responses to photographs by tourists and professionals in terms of the photography's component and eWOM's components through AIDA's hierarchy process model. Some different responses were identified.

**Keywords** Professional photograph · Tourist photograph · AIDA · eWOM · Social media

## Introduction

Rapid technological developments in both the Internet and photography have affected tourism activities. The Internet has fundamentally changed the pace of how information related to tourism is disseminated, and the way people plan and conduct travel (Buhalis and Laws 2008). The expressions of the opinions by Internet users through online media is manifested in different forms of consumer generated content (CGC) such as blogs, wikis, virtual communities, social networking, collaborative tagging, and sharing of media files. Sites like YouTube and Flickr, which have gained great popularity, are used online by travelers (Gretzel 2006). Many of

---

F. Afif · T.K. Priyambodo (✉)
Magister Tourism Studies, Universitas Gadjah Mada, Yogyakarta, Indonesia
e-mail: mastri@ugm.ac.id

F. Afif
e-mail: fuadi.afif@mail.ugm.ac.id

Y.A. Widyaningsih
Faculty of Business and Economics, Universitas Gadjah Mada, Yogyakarta, Indonesia
e-mail: yulia.widyaningsih@ugm.ac.id



A. Saufi et al. (eds.), *Balancing Development and Sustainability in Tourism Destinations*, DOI 10.1007/978-981-10-1718-6_30

the social media websites help users with uploading and sharing their comments related to travel, stories, tips, and personal experiences that later serve as information for others. McKean (2013), an expertise in web and SEO copywriting, said that the photograph is able to provide a powerful visual experience. In other words, it is similar to visual of real object which can be sensed by the social media users. They who see it may get real visual experience as feel as if there. Then, images in the social media site can also be an effective supporter of the media, especially in the field of tourism.

With the development of both digital photography and the Internet, it has become more affordable to use photographs as a means of promotion. It has become a feature that is able to attract the interest of social media users and make them to come and enjoy the sights. The combination of photography and the Internet is able to quickly deliver the promotion of tourist destinations, such as those in the Aegean Sea which saw, in a short time frame, its tourist visits increase to 60,000 visits per day. The increase was due to a Taiwanese traveler called Justin who uploaded a photograph reference to one of the famous websites (Lo et al. 2010).

As of January 2015, Facebook had the largest number of active users out of all social media platforms with 1.3 billion. In second place, with 629 million active users, was Qzone (a social media platform in China). The platform owned by Google, which also owns the Internet's most popular search engine, occupied third position with 330 million active users, followed by Instagram with 300 million, and Twitter in fifth with 264 million users (Kemp 2015). The speed of the circulation of the photographs in social media is very fast and is on a massive scale, according to data obtained from infographics platform Gopopcorn (2012) which recorded that in 60 s, on average, Internet users post 136,000 photographs on Facebook, while on Flickr, the famous photograph sharing media, 3500 photographs are uploaded on average every 60 s. This turnover of photographs has increased due to more people using digital cameras and smartphones, and due to Internet packages becoming more affordable.

Photographs on social networking sites can generally be divided into two types, namely tourist or vacation photographs (holiday snaps) on one hand, and professional photographs on the other. The difference is that tourist photographs are usually taken spontaneously and capture experiences or sights during travel (Robinson and Picard 2009), whereas professional travel photography uses techniques such as lighting, compositions arranging backgrounds and foregrounds, and a process in which the texture and color of the photographs are enhanced (Edensor 2005). Professional photographs are subject to a higher level of photographic techniques than tourist photographs.

In the context of tourism promotion, both types of photography can affect the desire of tourists to visit tourist destination that are depicted in different ways (Miller and Stoic 2003). Miller and Stoicare searched the use of photographs and images in media promotion. Their study revealed that increased usage of color can increase the level of attention, but the clarity of the objects contained in the media can generate higher interest. The use of images in social media is a form of media promotion known as Word-of-Mouth (WOM) in tourism. WOM as a form of promotion is inexpensive and effective in spreading information about tourist destinations. WOM has been proven to be able to increase the number of visits to tourist destinations (Lo et al. 2011). It is important to examine the differences between the effects of pictorial stimuli by both types of image through the hierarchy of response model in terms of how tourists are influenced to make a visit so that it can be an input for tourism promotion strategies.

Along with scrutinizing the relationship between photographs, social media and tourism, this research seeks to find the difference between the responses to photographic stimuli on social media platforms in the various phases of the hierarchy of response. Specifically, this study aims to examine how the stimulus in the form of tourist photographs and professional photographs will bring up the different levels of response by social media users in cognitive, affective, and behavioral terms. The third such response can be measured according to Attention, Interest, Desire, and Action. Before the travel activities, a tourist requires references to determine their options. Photographs, as one of these media references, are instrumental in influencing interest in travel.

In social media, there are the two types of photographs that are uploaded by users (i.e., those made by tourists and professionals). Both of these types of photograph can affect the interest in travel of users of social media, so analyzing the influence of both types of photograph is crucial. The research question that is addressed in this research is as follows: Is there a difference in response to the technical components as well as eWOM component between tourist photographs and professional photographs in terms of attention, interest, desire, and action? The second question is: Do the technical and eWOM components become mediators between the stimuli of photographs by professional or photographs by tourists in terms of AIDA?

## Literature Review

In the digital age and Internet era, the spread of information has become very quick. In the past, most visitors relied on travel brochures and TV advertisements. Now, by just looking at photographs and holiday stories uploaded by friends to social media platforms, their interest in visiting a destination can be encouraged. Walters et al. (2012) measured the effects of textual images and stimuli in the dimension of the consumption vision of consumers. The results of this research stated that pictures, along with words, obviously affect the formation of consumption vision; instructions to imagine themselves in the destination depicted in words clearly affect the quality of consumption vision.

A comparison of the media used in tourism advertising has also been made by Miller and Stoica (2003). This study compared the effects, in terms of direct response to two different fictional print advertisements, with one using a photograph of coastal scenery at a Caribbean island resort, and the other using an artistic rendition of the scenery. The results show that the artistic rendition of the photograph is superior in terms of attracting attention. Pictures are better at evoking more, clearer and more affecting mental imagery that is positive. Photographs excel in terms of generating a more favorable attitude towards the advertising for the resort and are strong in terms of intent behaviors, but not better in terms of recall of the name of the resort.

The role of social media in searching for online travel information also has been investigated by Xiang and Gretzel (2010). The purpose of this research is to know the extent to which social media appears in search engine results in the context of searches related to travel. This study uses design studies that simulate the use of a search engine for travel planning using a set of predetermined keywords in combination with nine names of tourist destinations in the United States. Analysis of the search results indicates that social media forms a big part of the search results, indicating that the search engines tend to directly bring users to social media sites. This research confirms the growing importance of social media in the online domain in tourism.

The relationship between social media user behavior in posting photographs and souvenir purchasing behavior is examined by Boley et al. (2012). According to their study, almost half of the tourists have fun posting photographs of their tour on social networking websites (McKercher et al. 2010). Therefore, the study explores whether travelers are posting photographs of their trip on a social media site known as the “trip picture posters” (TPPS) have different souvenir buying behavior than those who did post trip pictures (non-TPPS). The result shows that those who commit to TPPS are more likely to use the TPPS souvenirs as gifts and travel than those with no evidence of TPPS.

Haldrup and Larsen (2003), Jenkins (2003), argue not only that photography inspires travellers to visit destinations but also that photographs of sights are the main focus of activities for the tour.

## Social Media in Tourism

The growth of social networking sites has become a worldwide phenomenon. Sites like Facebook, Twitter, and YouTube dominate the Internet. Social sites thrive because they have several advantages for individuals and other entities. For individuals, these sites offer you the opportunity to engage in better communication with other people around the world and create their social lives.

Rapid growth in Web 2.0 technology has spawned new areas of interest for marketers. Tourism marketers have started to understand the influence of eWOM (electronic Word-of-Mouth) on the tourism industry. Experts have begun to understand how a photograph can be eWOM in social media networks as it offers the opportunity to write, publish, and upload it online to anyone. These activities can then be communicated through their social network of ‘friends’ (Facebook), ‘followers’ (Twitter), or ‘contacts’ (LinkedIn). One of the key benefits of social media is that its users are actually back using one of the oldest forms of communication, i.e., WOM, but the difference is now leveraging the Internet and the latest mobile technologies to connect others together and quickly (Chan and Guillet 2011).

## Methodology

The theoretical framework used in this study is based on the theories that have been described in the previous section. This framework includes photographs by tourist, photographs by professionals, technical components, the photograph component of eWOM, and the hierarchy of response (see Fig. 30.1).

Social media is one of the media for travelers to save their photographs on their way into a form of eWOM for tourist destinations. This study has applied a nondirectorial hypothesis. The hypothesis that is used is based on the modified research hypothesis of Miller and Stoica (2003). Four major hypotheses are tested. Researchers present the framework of hypothesis as follows (see Fig. 30.2).

The research design is experimental and uses a model Pre-Test Post Test Control Group Design. This design involves two groups of subjects, one given the experimental treatment (the experimental group) and others are not given the treatment (the control group). This effect of the treatment design on the dependent variable is tested by way of comparing the state of the dependent variable in the experimental group after treatment through the moderation variable on one hand, with the control group, which is not subject to treatment, on the other hand. The experimental group gets tourist photographs and professional photographs via submissions with a slideshow using a projector, while the control group is the group that gets only the

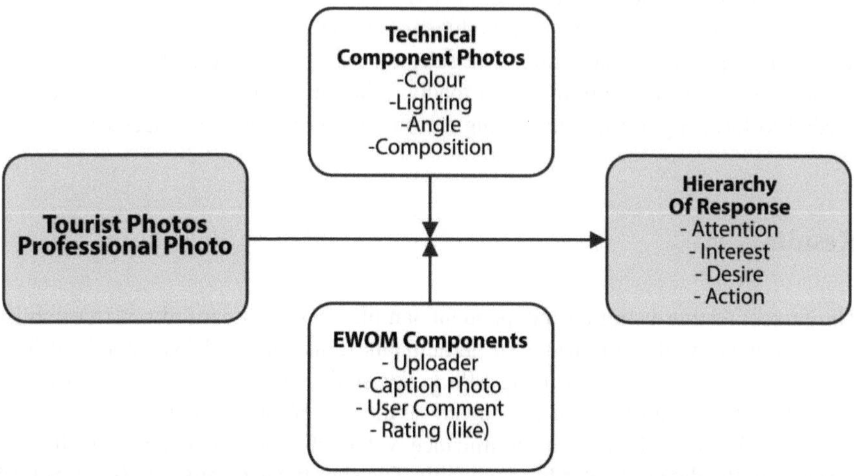


**Fig. 30.1** Theoretical framework

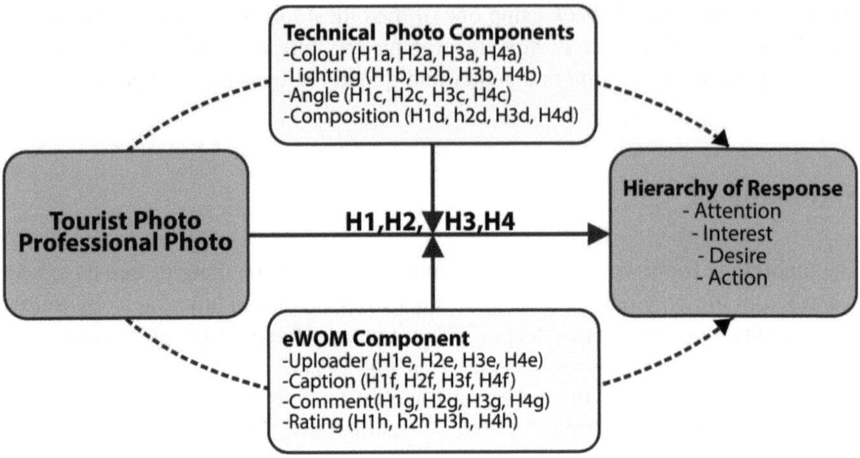


**Fig. 30.2** The hypothesis framework

tourist photographs treatment. This research uses an Independent Test Sample *t*-test on the pretest to test its homogeneity for the two groups and uses a Paired Samples *t*-test to test the difference of response of the experimental group.

## Participants

The sampling method used in this experiment matched the participants in the two groups (control and experimental) necessitating them to have the same characteristics. There should not be differences between the groups: they must be participants in social media and have the same educational background. The participants were recruited by purposive sampling. The selected participants were students majoring in tourism who were users of social media, with an age range of 19–23, and they needed to have a good understanding of tourism activities and processes.

## Results

The pretest testing using an independent sample *t*-test was carried out to establish the homogeneity of both groups of participants. This test result was obtained from processing the primary data using SPSS. For statistical tests (see Table 30.1), the independent sample *t*-test was conducted on the two groups of participants. The test results in Table 30.1 show a significance value of 0.598 which means that the variance in the data indicates homogeneity. The significance value of $t$ 0.356 > 0.05 means there is no significant difference in response between the two groups.

**Table 30.1** The results of the independent sample *T*-test

| Test | The control group—group of experiments |
|---|---|
| *F* | 0.281<br>(0.598) |
| *t* | 0.929<br>(0.356) |

*Description* sign in brackets "(…)" indicates the value significance. *Source* prepared from primary data (2015)

Therefore, the two groups could be used for the experiment. Then the posttest was conducted on the experimental group. Testing the hypothesis in this study was carried out using a paired *t*-test. Based on the results of this research (see Table 30.2), a decision-making process can be identified in social media users regarding their trips when they have been exposed to tourist photographs and professional photographs in social media, starting with when they began to respond to the photographs uploaded to social media up until the decision to visit was made.

The research found a different response in the hierarchy process in so far as professional photographs invoke a better response compared to the tourist photographs. This is in accordance with the research conducted by Edensor (2005) that showed professional photographs to have more influence on interest in sightseeing. To understand this process, a deeper and more thorough effort is required to analyze the difference between the response component of the photograph and the Ewom components at each stage of AIDA. Social media users paying attention to the photographs in social media is the first step in the hierarchy of response. Tourist photographs and professional photographs look different in terms of technical components (e.g., colors and angles). These are visible so users can immediately recognize the differences when they first see the photographs. Color photograph attracts more attention from social media users (Miller and Stoica 2003). In the response stages for professional photograph, there was significant difference in

**Table 30.2** The results of the paired sample *T*-test on the hierarchy process

| Component | Mean | *T* |
|---|---|---|
| *Attention*<br>*Attention* tourist photographs<br>*Attention* pro photo | −0.193**<br>3.703<br>3.896 | −3.056<br>(0.003) |
| *Interest*<br>*Interest* tourist photographs<br>*Interest* pro photo | −0.427*<br>3.307<br>3.734 | −5.827<br>(0.000) |
| *Desire*<br>*Desire* tourist photographs<br>*Desire* pro photo | −0.333**<br>3.442<br>3.776 | −4.575<br>(0.000) |
| *Action*<br>*Action* tourist photographs<br>*Action* pro photo | −0.208**<br>3.276<br>3.484 | −2.823<br>(0.005) |

*Description* sign in brackets "(…)" shows the significance value, sign *, **, and *** indicates significance at the level of the 1 %, 5 %, and 10 %. *Source* prepared from primary data (2015)

terms of a better attention response. In the attention stage for the eWOM component, users could distinguish between tourist and professional photographs. The use of language in captions that is close to that used in the daily lives of social media users used on the tourist photographs was able to get the attention of social media users compared to the captions on professional photographs that use formal language. Then ratings (using 'like') on the professional photographs are able to attract the attention of social media users because the number of 'likes' on the professional photographs was more than that on the tourist photo. That is because the accounts of uploaders of professional photographs have a lot more followers than personal accounts. The number of existing 'likes' on Facebook was able to attract the attention of more users due to the amount of content on Facebook that is rated in this way and which has become a sign of the popularity of such content (Yu et al. 2011). On table 30.3, participants have a tendency to look at the rating first to measure the level of quality of the uploaded photographs. The second AIDA stage is 'interest' which means users have already looked at the photograph more carefully in order to get additional information, meaning that more users can better realize the differences between existing components of both types of photograph. All technical components, namely the color and angle seen in professional photographs, produce a divergence in the interest social media users plus the technical components of lighting and composition.

As for the eWOM component, social media users are starting to realize the difference between uploaders of photographs. Once they become more interested in professional photographs, they will find out who the uploaders of the photograph are and those uploading professional photographs get a better response. Social media users also exhibit a difference in terms of captions for tourist photographs compared to professional ones. Changes in the response showed that professional photographs get a better response than tourist photographs at different stages of attention because social media users want to find information that is more reliable. This can be obtained from the information in captions for the professional photographs which are written by professionals so the response to these captions is better than to those on tourist photographs written by amateurs. The third stage is 'desire' which means social media users have already reached the stage before Action (making trips or visits). The results of the study show that all of the technical components of both types of photograph make a difference in invoking social media users' desire to visit. Professional photographs draw a better response in terms of social media users having a meaningful desire to visit. Participants are more eager to visit an attraction that is depicted by professional photographs than tourists ones. This is because professional photographers can capture and enhance a moment, making it look more interesting when photographed, causing a desire to go to the venue (Edensor 2005), unlike tourist photographs which simply records holiday or traveling activity.

As for the eWOM component, only the photograph's uploader gets a response, with professional photographs getting a better response. This means the uploader of the professional photograph can arouse the desire to visit in social media users. This means the user wants to visit a destination that has already been visited by the

**Table 30.3** Paired samples test results on the technical components and eWOM

| Component | | *Attention* | | *Interest* | | *Desire* | | *Action* | |
|---|---|---|---|---|---|---|---|---|---|
| | | Mean | *t* | Mean | *t* | Mean | *t* | Mean | *t* |
| Technical component photographs | Color<br>Tourist photographs<br>Pro photo | −0.417<br>3.708<br>4.125 | −3.122* | −0.667<br>3.250<br>3.917 | −3.112* | −0.667<br>3.417<br>4.083 | −2.798* | −0.292<br>3.375<br>3.667 | −1.574* |
| | Lighting<br>Tourist photographs<br>Pro photo | −0.209<br>3.750<br>3.958 | −1.225 | −0.708<br>3.208<br>3.917 | −3.205* | −0.500<br>3.500<br>4.00 | −2.937* | −0.417<br>3.083<br>3.500 | −2.198 |
| | Angle<br>Tourist photographs<br>Pro Photo | −0.542<br>3.875<br>4.417 | −3.406* | −0.458<br>3.708<br>4.167 | −1.848*** | −0.500<br>3.667<br>4.167 | −2.220** | −0.375<br>3.458<br>3.833 | −1.519 |
| | Composition<br>Tourist photographs<br>Pro photo | −0.375<br>3.458<br>3.833 | −2.584 | −0.500<br>3.250<br>3.750 | −2.220** | −0.417<br>3.333<br>3.750 | −2.198** | −0.292<br>3.292<br>3.583 | −1.497 |
| EWOM component photographs | Uploader<br>Tourist photographs<br>Pro photo | 0.292<br>3.750<br>3.458 | 1.320 | −0.375<br>2.917<br>3.291 | −2.387** | −0.375<br>3.292<br>3.667 | −1.813*** | −0.292<br>3.042<br>3.333 | −1.232 |
| | Photograph captions<br>Tourist photographs<br>Pro photo | 0.500<br>4.000<br>3.500 | 2.398** | −0.333<br>3.50<br>3.833 | −2.000*** | 0<br>3.750<br>3.750 | 0 | −0.083<br>3.458<br>3.542 | −0.401 |
| | User comments | 0.208<br>3.958 | 1.551 | −0.042<br>3.458 | −0.225 | 0.0833<br>3.375 | 0.385 | 0.167<br>3.333 | 0.749 |

(continued)

**Table 30.3** (continued)

| Component | | *Attention* | | *Interest* | | *Desire* | | *Action* | |
|---|---|---|---|---|---|---|---|---|---|
| | | Mean | *t* | Mean | *t* | Mean | *t* | Mean | *t* |
| | Tourist photographs<br>Pro photo | 3.750 | | 3.500 | | 3.291 | | 3.167 | |
| | Rating<br>Tourist photographs<br>Pro photo | −0.542<br>3.125<br>3.667 | −2.600** | −0.333<br>3.167<br>3.500 | −1.498 | −0.292<br>3.208<br>3.500 | −1.497 | −0.083<br>3.167<br>3.250 | −0.464 |

*Description* sign *, **, and *** indicates significance at the level of the 1 %, 5 %, and 10 %. *Source* prepared from primary data (2015)

uploader of the photographs. A destination that has been visited by a professional photographer, who has produced artistic photographs that can arouse the desire of the user, who viewed the photographs to replicate the image in the same place because photography can inspire a visit (Haldrup and Larsen 2003; Jenkins 2003; Urry and Larsen 2011). The fourth stage is Action. The results of this research show it is only the technical component of color in the photograph that gets the different response from social media users. So the color component draws different responses for both types of photographs from the early stage (Attention) to the final stage (Action). The colors on professional photographs draw a better response in influencing the decisions of users to visit.

## Conclusion

The conclusions of this study are; (1) There is a difference in the responses at the attention stage in the professional photographs in two technical components (color, angle) and one eWOM component (rating), whereas differences are found in responses to tourist photographs in one component eWOM (caption). (2) There are differences in the interest stage responses in the professional photographs in terms of all technical components (color, lighting, angle, composition) and two eWOMc components (uploader, caption), whereas for tourist photographs there is no difference for all components. (3) There is a difference in the responses at the desire stage in terms of professional photographs in all technical components (color, lighting, angle, composition) and one eWOM component (uploader), whereas for tourist photographs there is no difference for all components. (4) There is a difference in the responses at the action stage in the professional photographs in one technical component (color), whereas for tourist photographs there is no difference for all components. (5) Professional photographs get better responses overall compared to tourists' photographs. Technical components and eWOM components of photographs that become mediators in the AIDA response hierarchy, are: (a) color, angle, and rating in terms of Attention; (b) color, lighting, angle, composition, uploader, and caption in terms of Interest; (c) color, lighting, angle, composition and the uploader in terms of Desire; and (d) color in terms of Action. Then the eWOM component that is the sole mediator is the tourist description of the photograph (caption).

## Recommendations

According to the findings of this research on the use of images in social media for tourism marketing strategies, professional photographs are the best option for utilization in tourism promotion through social media because they get better responses with their components that are superior to tourist photographs, and managers of tourist attractions can use the services of a professional photographers to promote those destinations through social media. Futhermore, they should pay attention to how photographs' color accentuates characteristics if they want to promote attractions using social media. Also, managers of tourist attractions need to pay attention to the availability of the Internet at that destination.

## The Limitations of the Research

This study uses Facebook as the social media component, but in a lot of Facebook, eWOM has no difference in terms of the response, and therefore it is possible that other social media are more suitable. The findings in this study indicate that Instagram is more accessible than Facebook which illustrates that social media users have already begun to shift to other social media platforms. The results of this research cannot be generalized to other social media because other social media have different characteristics, components, and the user interfaces. This study did not examine the overall process of eWOM, and just researched social media behavior before taking the decision to travel and therefore cannot assess the behavior after the visit (post purchase behavior). Further research might seek a deeper understanding of the eWOM process using the latest response hierarchy model, AISDALS Love. This is an advanced model whereby Search, Like/Dislike, Share and Love/Hate which describes the post purchase behavior are added to AIDA process.

## References

Boley BB, Magnini VP, Tuten TL (2012) Social media picture posting and souvenir purchasing behavior: some initial findings. Tour Manag 37:27–30

Buhalis D, Law R (2008) Progress in information technology and tourism management: 20 Years on and 10 Years after the internet the state of eTourism research. Tour Manag 29(4):609–623

Chan NL, Guillet BD (2011) Investigation of social media marketing: how does the hotel industry in Hong Kong perform in marketing on social media websites? J Travel Tourism Mark 28 (4):345–368

Edensor T (2005) Waste matter—the debris of industrial ruins and the disordering of the material world. J Mater Cult 10(3):311–332

Gopopcorn (2012) Sixty seconds of social media sharing. Retrieved from www.gopopcorn.ca

Gretzel U (2006) Consumer generated content—trends and implications for branding. E-Rev Tourism Res 4(3):9–11

Haldrup M, Larsen J (2003) The family gaze. Tourist Stud 3(1):23–45

Jenkins OH (2003) Photography and travel brochures; the circle of representation. Tourism Geogr 5:305–328

Kemp S (2015) Digital, social & mobile in 2015. http://wearesocial.sg/blog/2015/01/digital-social-mobile-2015/. Retrieved 19 Jun 2015

Lo IS, McKercher B, Lo A, Cheung C, Law R (2010) Tourism dan online photography. Tourism Manage 31:725–731

McKean M (2013) Internet marketing for tourism demand media. http://smallbusiness.chron.com/internet-marketing-tourism-43292.html. Retrieved 17 Nov 2014

Miller DW, Stoica M (2003) Comparing the effects of a photograph versus artistic renditions of a beach scene in a direct-response print ad for caribbean resort island: a mental imagery perspective. J Vacation Mark 10(1):11–21

Robinson M, Picard D (2009) Moments, magic, and memories: photographing tourist, tourist photographs, and making worlds. In: Robinson M, Picard D (eds) The framed world: tourism, tourist, and photography. Ashgate Publishing, Surrey, pp 1–37

Urry J, Larsen J (2011) The tourist gaze 3.0: vision and photography. Sage Publications, London

Walters G, Sparks B, Herington C (2012) The impact of consumption vision and emotion on the tourism consumer's decision behavior. J Hospitality Tourism Res 36(3):366–389

Xiang Z, Gretzel U (2010) Role of social media in online travel information Search. Tour Manag 31:179–188

Yu B, Chen M, Kwok L (2011) Toward predicting popularity of social marketing messages. Lect Notes Comput Sci 6589:317–324

# Chapter 31
# Tweeting and Retweeting Tourism and Airline Service Attributes

**Arissetyanto Nugroho and Janfry Sihite**

**Abstract** Online sales for business trips in Indonesia have increased during 2008–2013 from 2.55 trillion to 26.54 trillion rupiahs. This development is supported by increasing numbers of social network users who convey their experiences in Indonesia, which includes the rise of low-cost airlines, using social media, such as Twitter, to promote the tourism experience in Indonesia. These developments have not been deeply explored in academic research; in particular, there is a lack of research examining low-cost airlines and their use of social media. Therefore, we take an inductive approach using qualitative and quantitative analysis methods (mixed method) for this research. Gathering primary data from the Twitter account @Citilink from the fourth Quarter of 2014 (October, November, December) and the first quarter of 2015 (January, February, March), we have extracted the data, preparing it for processing using the Mixed Method Proval is Research software. There were 2194 tweets from fourth quarter of 2014. After cleaning the data, 992 tweets were excluded from the analysis. The data were further categorized into 821 tweets and 381 retweets. Of the 1724 tweets from first quarter of 2015, 436 tweets were excluded from the analysis; the remaining tweets were further categorized into 588 tweets and 436 retweets. Exploratory factor analysis was conducted and the findings show that the @Citilink Twitter account delivered five service attributes, including (1) routes, (2) quizzes and ticket promotions, (3) call center information, (4) tourism destination exposure, and (5) Citilink promotional gimmicks for the netizen (Twitter followers). Furthermore, the *n* words paired *t* test to compare the tweet and retweet data found that @Citilink preferred other service attributes to retweet, compared with the service attribute tweet. The retweets (netizen tweets forwarded by the @Citilink twitter administrator) often show the service attributes of on-time airline performance and airline smooth landings. The smooth landing is an innovative and distinctive service attribute that is proposed for the potential customer when compared with the other service attributes from the @Citilink

A. Nugroho (✉) · J. Sihite
Department Management, Mercu Buana University, Jakarta, Indonesia
e-mail: arissoehardjo@yahoo.com

J. Sihite
e-mail: janfry_sihite@yahoo.com

 
A. Saufi et al. (eds.), *Balancing Development and Sustainability in Tourism Destinations*, DOI 10.1007/978-981-10-1718-6_31

account. These findings show the advantage of using exploratory factor analysis to construct and discover the relationships and motives within the unstructured dataset.

**Keywords** Low-cost airline service · Citilink · Qualitative quantitative analysis · Twitter · Retweet preference

## Introduction

Indonesia's current consumption is growing enormously and becoming one of the dominant elements driving economic growth when compared to government spending and investment from private companies. Projections of Indonesia's economy show an increasing trend of consumption along with the growth of both the Indonesian middle-class and the percentage of the population that is under 30 years of age. Until the year 2020, the population of Indonesians that are under 30 years old will reach 47 % of Indonesia's total population (Accentur 2012). Both the under 30 and the middle-class demographics are the target customers for a variety of products and services in Indonesia. Along with the growth of the Indonesian purchasing power, the airline service demand is also growing (Darmawan and Wandebori 2013). Data on the number of trips from 2008 to 2013 show that the number of arrivals (inbound) and departure (outbound) users of airlines in Indonesia is increasing significantly (UNWTO 2015). The online airline sales trend in Indonesia has also increased from 2008 to 2013 (UNWTO 2015). In 2013, the online airline sales in Indonesia had risen to 26.54 trillion rupiah, up from 2.55 trillion rupiah in 2008. The percentage of online airline ticket sales is also the largest proportion of online travel sales when compared to online agency travel sales, tour operator sales, and accommodations sales in Indonesia.

The growth of airline services is supported by mobile Internet technological developments and social network applications. Social network applications such as Facebook, Twitter, and Instagram provide access for users to interact and exchange information, photos, and videos that deliver remarks about the users' respective tour experiences and tourism in Indonesia. The social network users show high enthusiasm for airline service and have put two airlines, which are Air Asia (1.028.686 followers) and Garuda Indonesia (861.847 followers), among the five accounts with the largest number of Twitter followers for Indonesian companies (Socialbakers 2015). As the number of mobile Internet users in Indonesia has increased from only 989.100 in 2008 to 90.831.400 users in 2013 (ITU 2014), the number of Twitter followers for the airline companies will continue to increase with the rapid growth in mobile Internet users.

High demand growth for airline services also increases the intensity of competition, giving birth to a new category of low-cost airlines (low-cost carriers) like Lion Air, Air Asia, and Citilink. These airlines offer reduced services, such as meals, baggage facilities, and seat space, in comparison with full service carriers like Garuda Indonesia. This phenomenon is consistent with the predictions of ICAO (International Civil Aviation Organization) and IATA (International Air Transport Association), which show that airline industry competition in the Asia Pacific region will be more stringent for a low-cost airline in Indonesia because the majority of the airline companies (14 of 17) are low-cost airlines (OECD 2014).

While social network applications will play an important part in the customer decision-making process affecting low-cost airline customers, the research parsing the phenomenon of social networking as a marketing communications channel, online marketing platform for products and services, and tourism and travel is still undeveloped (Virginia Phelan et al. 2013). Given this background, additional research on a low-cost airline service using the social network Twitter as a marketing communications channel for the customer is required. The lack of the literature on low-cost airlines in Indonesia is a reason d'etre to conduct fundamental research exploring tweets that will uncover patterns (Powell and Dodd 2007) of low-cost airline service attributes in Indonesia.

## Review Literature

### *Indonesian Airline Industry*

The Airline Industry is a service industry, which means that service satisfaction is connected with overall marketing activities to meet the expectations of the customer. These marketing activities begin with delivering service offerings to the customer through the process of fulfilling promises. The services consist of the price (cost), the development of new routes, the plane, the queue in the airport, an interesting promotion, and the ease of moving between airlines in cooperation with a network of airlines (Driver 1999). The entire process is a holistic unity of services the customer values in the decision-making process (Aksoy et al. 2003). However, along with the start of low-cost airlines, price competition has become the main attribute that the company delivers.

A meta-analysis of the factors that influence passenger loyalty in the low-cost airline space shows that the low-cost airline focuses on low prices (Akamavi et al. 2015). Furthermore, the word cloud is needed to visualize the 31 dominant words in the empirical research that are selected in the Journal of Tourism Management and discover that the costs, services, and low are the keywords that characterize the main principles of the low-cost airline.

The phenomenon of low-cost airlines is in contention with whether the airline can offer more beneficial service to its customers. Fundamental research is needed to uncover low-cost airline service attributes so that the industry participants, government, and academia can understand the low-cost airline industry.

## *Social Network*

The airline delivers its service offerings to customers via social networking channels. Social networking is becoming an effective and efficient tool in building impressions, creating the presence, and connecting companies with customers (Virginia Phelan et al. 2013). Social networking develops responsive interactions between various parties in the online ecosystem and encourages interaction between prospective customers and companies, between prospective customers, between the candidate and the company's customers, the product and service supplier companies, and between the company and the supporting product and service companies.

While Twitter is a social network application, which has been utilized for marketing activities since 2008 (Bulearca and Bulearca 2010), Twitter microblogging is popular and has been growing since 2006 (Sakaki et al. 2010). Twitter users can create tweets (messages or comments) with a limit of 140 characters, follow activity on other accounts (follow), and receive tweets generated by another account (Kwak et al. 2010). Twitter is growing rapidly, which is reflected in the number of tweets that Indonesian Twitter users generate. Indonesian Twitter users account for 4.9 % or 19 million tweets per day of total tweets throughout the world (Wibisono and Faruqi 2013); therefore, the Citilink tweet research is essential.

## *Research Position*

Exploratory research on the low-cost airline Citilink service attributes is new in Indonesia. Earlier studies such as Open Skies research policy of Citilink (Nurhendiarni et al. 2015), the research on revenue management (Lupiyoadi and Putra 2014) strategy analysis, and research conducted by (Darmawan and Wandebori 2013) propose a strategy to improve the Citilink market share by not focusing the research on the service attributes of the low-cost airline.

This project is an inductive research that examines Citilink marketing activities from the Twitter tweet @Citilink. The tweet data are open access data collected from @Citilink. The corpus (textual documents) of tweets are delivered by Citilink during the period from October, November, December 2014 to January, February, March 2015 and are collected for further processing and analysis.

## Research Method

This is an exploratory research study that extracts data using a collection of tweets from a Twitter social networking account. A sampling of tweets from the fourth quarter 2014 (October, November, December) and first quarter 2015 (January,

February, March) are used. We will be able to show the company's marketing activities through the use of a six-month data interval.

The tweets from the @Citilink are unstructured data, therefore data preparation is needed before analysis (Silver and Lewins 2014). Data preparation consists of four stages of the process, which are the stages of inputting data descriptions, cleansing and transforming the data, stemming and lemmatization, and developing a dictionary.

Next, we use a cluster analysis technique, or the analysis of textual data with great numbers, to complete the EFA. Cluster analysis shows the dimensions of different tweets that can aid researchers in interpreting these dimensions (Campbell et al. 2011). We use mixed method research software in this research because it offers exploratory techniques that show the relationship between keywords with the co-occurrence method (Silver and Lewins 2014).

The values we use in this measurement are based on the degree of similarity. The higher the index of similarity, the narrower the distance of the keywords within the text (Chung and Lee 2001). On the other hand, the lower the index of similarity, the wider the distance of the keywords within the text. The dendrogram offers visualization of an element in the unit of analysis. The coefficient value for this measurement is the Jaccard coefficient; values close to 1.0 indicate the close relationship of the elements in the unit of analysis (Lewis and Maas 2007).

## Analysis

The primary data from the Twitter account @Citilink from the fourth Quarter 2014 (October, November, December) and the first quarter 2015 (January, February, March) are extracted and prepared for processing using the Provalis mixed method research software analysis. There are 2194 tweets from fourth quarter 2014. After cleaning the data and excluding 992 tweets from the analysis, we further categorize these data into 821 tweets and 381 retweets. Of the 1724 tweets from first quarter 2015, after cleaning the data and excluding 700 tweets from the analysis, we further categorize these data into 588 tweets and 436 retweets. Table 31.1 details the data preparation process.

**Table 31.1** Data preparation

| Data interval | Oct, Nov, Dec 2014 (4th Q) | | Jan, Feb, March 2015 (1st Q) | |
|---|---|---|---|---|
| Data extracted | 2194 | | 1724 | |
| Data excluded | 992 | | 700 | |
| Total data processed | 1202 | | 1024 | |
| Type of data | Tweet | Retweet | Tweet | Retweet |
| Data processed | 821 | 381 | 588 | 436 |

*Source* Data process

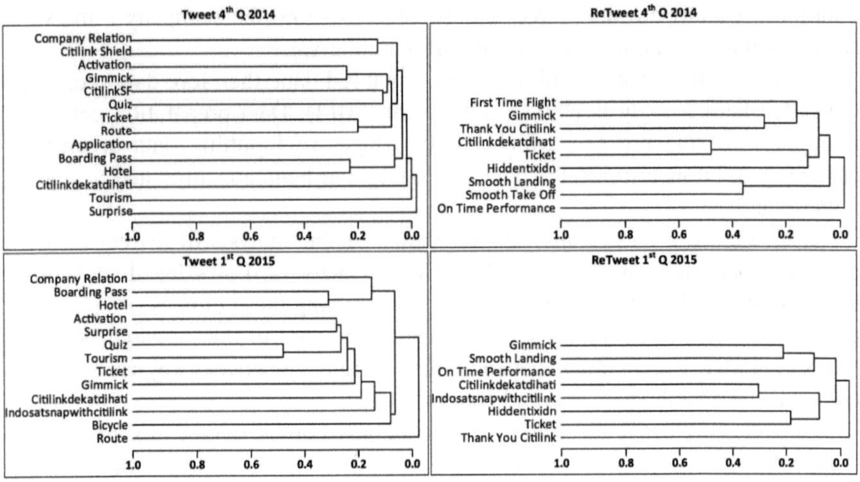


**Fig. 31.1** Dendrogram (Jaccard coefficient of similarity index). *Source* Data process

After preparing the data, exploratory factor analysis is employed to discover the attributes of the tweets and retweets from @Citilink in the given period. This research method will be able to define the brand identity using these attributes (Sevin 2014) and develop relevant measurement in terms of the company's strategic goal. These attributes also support the statement by Kietzmann et al. (2012) and Pitt et al. (2007) that the brand personality and brand identity can be analyzed with the social network application research approach.

We then conduct exploratory factor analysis using the data and categorizing it into four quadrants, which are the tweets for the fourth quarter 2014, the tweets for the first quarter 2015, the retweets for the fourth quarter 2014 and the retweets for the first quarter 2015. The separation based on the interval and the type of tweet will benefit from further analysis; furthermore, the separation will show the distinctive online brand personalities in cyberspace (Opoku and Hinson 2006). The result of the Exploratory Factor Analysis in using the four quadrants is in Fig. 31.1.

The Exploratory Factor Analysis of the @Citilink tweets for the fourth quarter 2014 shows that the tweet attributes are the promotion program Citilink Shield (the travel insurance), Citilink SF (Sounds Fair Concert), Boarding Pass (Hotel discount), Hotel (Hotel Discount), Citilink Dekat Di Hati (which encourages the follower to use the tweet hash tag #citilinkdekatdihati). The @Citilink delivers Quiz, Surprise Activity, Activation and proposes Ticket promotions to engage with the follower. Furthermore, the @Citilink delivers information about the company's Public Relations activities, the Citilink Application for the smartphone user, and the Citilink Route. Finally, the @Citilink account delivers Gimmick and the Tourism experience to the follower.

The quadrant of tweets for the fourth quarter 2014 shows that there are 14 attributes and five direct clusters which are the Activation and Gimmick (similarity index 0.231), Boarding Pass and Hotel (0.212), Ticket and Route (0.186), Company Relation and Citilink Shield (0.136), and Citilink SF and Quiz (0.126). The higher similarity index shows that there are higher similarities between the two cluster attributes in the data. The analysis states that in the tweet data for the fourth quarter 2014, the cluster Activation, and Gimmick (0.231) has a higher similarity when compared to the cluster Citilink SF and Quiz (0.126).

The Exploratory Factor Analysis of the @Citilink retweets for the fourth quarter 2014 reveals that the retweet attributes are Citilink Dekat Di Hati, Hidden TixIdn (a hidden ticket game that encourages the follower to interact and guess the game), and finally the Ticket promotion. The retweets also deliver the Gimmick toward the follower to improve the proximity between the @Citilink account and the follower. Furthermore, the @Citilink account delivers follower satisfaction and endorsement toward the airline industry's main attributes which are Smooth Landing, Smooth Take Off, and On-Time Performance. As three of the main attributes are clustered, the follower also delivers varieties of Thank You Citilink wishes and service satisfaction. Finally, there are many retweets from the First Timer Flight experience of the follower.

The quadrant of retweets for the fourth quarter 2014 show that there are nine attributes and three direct clusters which are the Citilink dekat di hati and Ticket (Similarity index 0.504), Smooth Landing and Smooth Take Off (0.370), and Gimmick and Thank You Citilink (0.233). The analysis states that in the data retweets for the fourth quarter 2014, the cluster Citilink dekat di hati and Ticket (0.504) have higher similarities when compared to the cluster Gimmick and Thank You Citilink (0.233).

The Exploratory Factor Analysis of the @Citilink tweets for the first quarter 2015 shows that the @Citilink data can be used to develop many promotional programs for the potential customer. The attributes are Boarding Pass, Hotel, Citilink Dekat Di Hati, Indosat Snap With Citilink (Indosat Citilink Co Program), Bicycle (proposing a bicycle gift and free luggage benefit). The @Citilink also delivers Quiz, Surprised Activity, Activation, and proposes Ticket promotions to engage the follower. The @Citilink also delivers the varieties of Thank You Citilink wishes and service satisfaction. Furthermore, the @Citilink account delivers information about the company's Public Relations activities and the Citilink Route. Finally, the @Citilink delivers Gimmicks and Tourism experiences that are geared toward the follower.

The quadrant of tweets for the first quarter 2015 show that there are 13 attributes and three direct clusters which are Quiz and Tourism (0.493), Boarding Pass and Hotel (0.304) and Activation and Surprise (0.256). The analysis shows in the tweet data for the first quarter 2015, the cluster Quiz and Tourism (0.493) has a higher similarity when compared with the cluster Activation and Surprise (0.256).

The Exploratory Factor Analysis of the @Citilink retweets for the first quarter 2015 reveals that the retweet attributes are Citilink Dekat Di Hati, Indosat Snap With Citilink, Hidden TixId, and Ticket promotion. The retweet also delivers Gimmicks toward the follower to improve the proximity between the @Citilink account and the follower. The follower also delivers variations of Thank You

Citilink wishes and service satisfaction, follower satisfaction, and endorsement toward the airline industry main attributes of Smooth Landing and On-Time Performance.

The quadrant of retweets for the first quarter 2015 shows that there are eight attributes and three direct clusters which are Citilink dekat di hati and Indosat snap with citilink (0.311), Gimmick and Smooth Landing (0.198), and Hidden tixidn and Ticket (0.184). As in the retweet data for the first quarter 2015, the analysis shows the clusters Citilink dekat di hati and Indosat snap with citilink (0.311) have higher similarities when compared with the cluster Hidden tixidn and Ticket (0.184).

## *Tweet Analysis*

The exploratory factor analysis findings support the statement that route, ticket proposition, and interesting promotion programs are the airline service offerings (Driver 1999). The route and ticket proposition are the main service attributes for the airline. Furthermore, the research discovers varieties of interesting promotion programs targeting the potential market such as the Quiz. The Quiz is a tool that influences customer engagement (Vinerean et al. 2014), delivers incentive to interact with the community of followers, and influences customer purchasing behavior, referral behavior, influencer value, and knowledge value.

The Gimmick the @Citilink Twitter account delivers explains the phenomena of human behavior symbolic interaction in the social network application. This symbolic interaction is necessary to develop intimacy and proximity from the twitter account to the customer. The application itself is a technology that should deliver the human behavior symbolic interaction to extend the self as human (Rader 2009). This is the advantage of the social network application for a brand because human behavior symbolic interaction is necessary in the social network application. Rapid and responsive platforms better support interaction and the development of the human behavior symbolic interaction when compared with conventional media, such as newspapers, magazines, television, and radio.

The Tourism attribute in the tweets of @Citilink explains that the travel industry is closely related to tourism. The tweets deliver exposure to varieties of tourism destinations in Indonesia and encourage inbound travel to Indonesia. The airlines' main service is supporting mobilization from point to point, which is related to geography; therefore, the @Citilink account promotes a variety of point to point airline tickets.

## *Retweet Analysis*

The retweet exploratory factor analysis discovers that Smooth Landing and On-Time Performance analogically compare to aesthetic innovation of service attributes for a high end technology product such as airplane technological innovation (Eisenman 2006). These service attributes can influence the decision-making process of the customer, even though these service attributes are not an instrumental benefit from the airline.

The Thank You Citilink attribute from the retweet exploratory factor analysis is a means to deliver customer feelings; therefore, the company should analyze the context of the comment to better understand customers' feelings (Pullman et al. 2005). The @Citilink retweet of a follower's tweet mentioning the @Citilink account to promote himself uses the airline service attributes in cyberspace. This mode of electronic word of mouth influences the decision-making process of the potential customer.

Finally, the research shows that while the @Citilink has a retweet preference toward select main attributes, the airline does not retweet all the tweets mentioning the @Citilink account. Instead, the @Citilink focuses on the retweet preference, which are Smooth Landing and On-Time Performance. The Smooth Landing is an innovative and distinctive service attribute proposed for the potential customer when compared with the other service attributes from the @Citilink account. The findings show the advantages of conducting text analysis to construct and discover the relationships and motives within the dataset (van Perlo-ten Kleij 2004).

## Conclusion and Implication for Future Research

The tweets and retweets in the @Citilink account have different attributes, and this research shows that the tweets of @Citilink delivers tourism exposure experience while the retweet preferences are the airline service attributes that target the twitter followers. The mixed method research method is used to explore the brand identity developed by the company with exploratory factor analysis, while the four quadrant distribution based on processing the data also develops the reliability of the exploratory factor analysis.

Future research should compare the tweets delivered that mention @Citilink within the same period to be able to define the intended brand identity and the perceived brand image in the eyes of the beholder (Marchi et al. 2011). Furthermore, future research should make a scale development that takes intermediaries such as travel agencies that promote the perceived experience of the destination into consideration (Vicol and Zait 2014). Further research should develop the method to combine the tweet and the photo delivered in the tweet because the photos could deliver experiential and affective qualities of the place (Pan et al. 2014). Finally, when the top companies use social network applications seriously compared with other medium-sized companies (Dzyaloshinsky and Pilgun 2015), the use of social networking applications will be more intense, which will lead to an improvement over the current research methodology.

## References

Accenture A (2012) Ready for Indonesia's digital future? Retrieved from http://www.accenture.com/sitecollectiondocuments/pdf/accenture-asean-ready-indonesias-digital-future.pdf

Akamavi RK, Mohamed E, Pellmann K, Xu Y (2015) Key determinants of passenger loyalty in the low-cost airline business. Tour Manag 46:528–545. doi:10.1016/j.tourman.2014.07.010

Aksoy S, Atilgan E, Akinci S (2003) Airline services marketing by domestic and foreign firms: differences from the customers' viewpoint. J Air Transp Manag 9(6):343–351

Bulearca M, Bulearca S (2010) Twitter: a viable marketing tool for SMEs. Global Bus Manag Res Int J 2(4):296–309

Campbell C, Pitt LF, Parent M, Berthon P (2011) Tracking back-talk in consumer-generated advertising: an analysis of two interpretative approaches. J Advertising Res 51(1):224. doi:10.2501/jar-51-1-224-238

Chung YM, Lee JY (2001) A corpus-based approach to comparative evaluation of statistical term association measures. J Am Soc Inform Sci Technol 52(4):283–296

Darmawan AP, Wandebori H (2013) Proposed strategy improvement for Citilink airline to increase market share. Ind J Bus Adm 2(14):1706–1717

Driver JC (1999) Developments in airline marketing practice. J Mark Pract Appl Mark Sci 5(5):134–150

Dzyaloshinsky IM, Pilgun MA (2015) Social media in Russian business communication: analysis and development prospects. Rus J Commun 7(1):27–39. doi:10.1080/19409419.2015.1008944

Eisenman M (2006) Essays on aesthetic innovation. Graduate School of Business

ITU (2014) The world telecommunication/ICT indicators database

Kietzmann JH, Silvestre BS, McCarthy IP, Pitt LF (2012) Unpacking the social media phenomenon: towards a research agenda. J Publ Aff 12(2):109–119. doi:10.1002/pa.1412

Kwak H, Lee C, Park H, Moon S (2010) What is Twitter, a social network or a news media? Paper presented at the proceedings of the 19th international conference on world wide web

Lewis RB, Maas SM (2007) QDA miner 2.0: mixed-model qualitative data analysis software. Field Methods 19(1):87–108. doi:10.1177/1525822x06296589

Lupiyoadi R, Putra B (2014) The effects of applying revenue management on customer satisfaction in airline industry: an experimental study in Indonesia. ASEAN Mark J 6(1):25–37

Marchi G, Giachetti C, de Gennaro P (2011) Extending lead-user theory to online brand communities: the case of the community Ducati. Technovation 31(8):350–361. doi:10.1016/j.technovation.2011.04.005

Nurhendiarni S, Hidayat NK, Pasasa L (2015) The effect of ASEAN open skies policy 2015 upon opportunities for low cost carriers in Indonesia. A case study of PT, Citilink. S East Asian J Manag 9(1):34

OECD (2014) Airline competition. Retrieved from 121st meeting of OECD competition committee on 19 June 2014

Opoku R, Hinson R (2006) Online brand personalities: an exploratory analysis of selected African countries. Place Branding 2(2):118–129

Pan S, Lee J, Tsai H (2014) Travel photos: motivations, image dimensions, and affective qualities of places. Tour Manag 40:59–69. doi:10.1016/j.tourman.2013.05.007

Pitt LF, Opoku R, Hultman M, Abratt R, Spyropoulou S (2007) What I say about myself: communication of brand personality by African countries. Tour Manag 28(3):835–844. doi:10.1016/j.tourman.2006.06.003

Powell S, Dodd C (2007) Managing vision and the brand within the creative industries. Corp Commun Int J 12(4):394–413. doi:10.1108/13563280710832533

Pullman M, McGuire K, Cleveland C (2005) Let me count the words quantifying open-ended interactions with guests. Cornell Hotel Restaurant Adm Q 46(3):323–343

Rader CS (2009) Toward a theory of consumer interaction with mobile technology devices. Doctoral dissertations, p 104

Sakaki T, Okazaki M, Matsuo Y (2010) Earthquake shakes Twitter users: real-time event detection by social sensors. Paper presented at the proceedings of the 19th international conference on world wide web

Sevin HE (2014) Understanding cities through city brands: city branding as a social and semantic network. Cities. doi:10.1016/j.cities.2014.01.003

Silver C, Lewins A (2014) Using software in qualitative research: a step-by-step guide

Socialbakers S (2015) Social marketing report: Indonesia regional

UNWTO (2015) Compendium of tourism statistics. Data 2009–2013, 2015 edn

van Perlo-ten Kleij F (2004) Contributions to multivariate analysis with applications in marketing. [University Library Groningen] [Host]

Vicol O, Zait A (2014) A country's image as tourist destination for external intermediaries. Manag Mark 9(1):47–74. Retrieved from http://search.proquest.com/docview/1521708615?accountid=17242

Vinerean S, Opreana A, Ţichindelean M (2014) Analyzing consumer engagement programs from the perspective of a qualitative research of marketing executives. Procedia Econ Finance 16:621–630. doi:10.1016/s2212-5671(14)00849-1

Virginia Phelan K, Mills JE, Douglas AC, Brian Aday J (2013) Digital personalities: an examination of the online identity of travel and tourism web sites. J Hosp Tourism Technol 4 (3):248–262. doi:10.1108/jhtt-11-2012-0032

Wibisono Y, Faruqi N (2013) Penentuan Gender Otomatis Berdasarkan Isi Microblog Memanfaatkan Fitur Sosiolinguistik. Jurnal Cybermatika 1(1)

Sakaki T, Okazaki M, Matsuo Y (2010) Earthquake shakes Twitter users: real-time event detection by social sensors. Paper presented at the proceedings of the 19th international conference on world wide web

Sevin HE (2014) Understanding cities through city brands: city branding as a social and semantic network. Cities [illegible] doi:10.1016/j.cities.2014.01.003

[illegible]

[illegible]

# Chapter 32
# Spatial Relationship of Tourist Distribution in Turkey

**Deniz Yüncü, Semra Günay and Yeliz Mert Kantar**

**Abstract** This study aims to investigate the spatial dependence of international and domestic tourists in 81 Turkish provinces for the period 2001–2013 using spatial statistics and mapping analysis. In the study, data relating to the number of domestic and international tourists participating in tourism movements in Turkey between 2001 and 2013 was used. The obtained results of an autocorrelation value of near zero shows that there is no global spatial pattern at province levels in Turkey. Since the study was conducted at city level, differences within cities could not be studied. Therefore, the spatial distribution of tourists was analyzed by mapping analysis. The west of Turkey receives more visitors compared to the east. International tourists travel in Turkey becomes denser in the southwest and west coasts of the country. An increase can be seen from the west towards the central parts of Turkey in terms of both domestic and international tourists. The effects of projects, designed to increase rest areas to improve mass tourism can be observed in the distribution of the number of tourists. Furthermore, it may be stated that the effects of practiced strategies, such as the spread of tourism nationwide, thereby decreasing regional development differences, can be seen.

**Keywords** Tourist flows · Exploratory spatial data analysis · Domestic tourists · International tourists · Turkey

D. Yüncü (✉) · S. Günay
Faculty of Tourism, University of Anadolu, Eskisehir, Turkey
e-mail: dkaragoz@anadolu.edu.tr

S. Günay
e-mail: semragunay@anadolu.edu.tr

Y.M. Kantar
Faculty of Science, University of Anadolu, Eskisehir, Turkey
e-mail: ymert@anadolu.edu.tr



A. Saufi et al. (eds.), *Balancing Development and Sustainability in Tourism Destinations*, DOI 10.1007/978-981-10-1718-6_32

## Introduction

### *Tourism Policies and Tourism Development in Turkey*

Tourism development and tourism policies in Turkey are grouped in five important time periods depending on the general economic development of the country. These are as follows: strong nationalistic policy (1924–1950), protectionist liberalism (1950–1960), planned economy period (1661–1980), introduction to private entrepreneurship (1980–1990) and emphasis on the private sector and liberal economic policy (1990–to the present) (Var 2005: 2–3). Similar to development in Europe, until the 1950s, tourism, especially international tourism, did not receive any real importance or priority from Turkish governments. On the contrary, because of war psychology, international visitors were taken with suspicion by the authorities and their arrival was discouraged (Tarhan 1997: 61). The period up to 1950 saw the birth of the New Republic with efforts to set national policies, so this was a time when tourism policies could not receive much importance. From the 1950s, government interest in Turkish tourism has stated to show a steady rise, parallel to developments in international tourism, and governments have started to handle the economic benefits of the tourism sector in creating currency flow and new employment opportunities as a modern means of development. Such ever-increasing government interest in the tourism industry has shown itself in various tourism organizations and institutions (Tezcan 2004). According to statistics, the 1950–1960 period yielded a 12 % increase in the number of tourists. In 1961, Turkey's total capacity was 15,685 and with only 258 hotels which could barely satisfy the needs of international tourists. During the same year, 129,000 tourists visited Turkey and spent 7.5 million dollars (Var 2005). The Ministry of Tourism was created in 1965 and tourism revenues increased from 7.7 million dollars to 193.7 million by 1974, which was an important achievement that surprised planners. This was the first time the number of tourists reached the one million mark, with the balance of tourism account showing a surplus.

The 1980s witnessed noticeable development in the tourism sector with certain legislative improvements, ongoing physical plans, infrastructure upgrading and so forth. This period can be called the 'First Attack in Tourism Development' (T.C. Turizm Bakanlığı 2002) as such progress depended on a relatively greater comprehensive attitude. The recent entry of Turkey to the tourist industry is mirrored in the fact that, prior to 1980, levels of international tourism were minimal. Thus, growth was highly compressed into a period of a little more than a decade. Between 1982 and 1992, the fact that total arrivals increased by over 400 % in visitor levels in ten years is indicative of the speed with which Turkey entered the international market. Estimates suggest that in 1992, tourism employed around 250,000 people in Turkey and earned nearly US$4 billion (Williams 1998: 63).

Initially, four Five-Year development plans were prepared, which included development of the tourism sector in a period called as the Planned Period (1960–1980) in Turkey. For the first Five-Year Development Plan, the mission of gaining currency for the tourism sector and closing the balance of payment deficit was commenced. The concepts of mass tourism and social tourism became a real issue during those years. In the second Five-Year Development Plan (1968–1972), for the first time, the focus was on condensing the investments on coastal regions, such as the Mediterranean and South Aegean regions which had high tourist potential to increase international tourism income and the number of international tourists (Soyak 2013). The coastal region of 3 km from the Canakkale-Balıkesir province boundary to the Antalya-Icel province boundary were declared a tourism development region to utilize mass tourism development through a decision of the Council of Ministers in 1969 (Tezcan 2004).

With the Third Five-Year Development Plan, it was stated that, in order to provide domestic tourism opportunities, necessary measures would be taken (Tarhan 1997). The first significant government action on domestic tourism was to establish low price and public holiday facilities (Kamu Kampları) for public sector employees. These facilities increased every year and reached a capacity of 60,000 beds. After 1972, the government implemented a credit program to encourage family pensions, aiming to increase low price accommodation capacity at established holiday resorts. Incentive measures were aimed at motivating families to convert rooms to low priced accommodation for middle-income tourists (Tarhan 1997). As mentioned earlier, the aim of the Five-Year Development Plans was to intensify tourism activities in high tourism potential regions in order to reduce infrastructure costs and maximize currency invested.

In compliance with this policy, the South Antalya Tourism Development Project was instituted. The South Antalya Tourism Development Project is the most outstanding integrated tourism development in Turkey, incorporating, from the beginning, planning, programming, finance and operational stages within one project (Inskeep 1991: 1). The South Antalya Tourism Development Project includes Kemer, Beldibi, Çamyuva, Göynük, Tekirova, Adrasan and Çıralıas well as three ancient cities of Phaselis, Olympos and Ldyros. The aim of the project was to create a bed capacity to meet the accommodation needs of domestic and international tourists in order to serve mass tourism (Turizm ve Tanıtma Bakanlığı 1980: 25).

With the third Five-Year Development Plan (1973–1977), mass tourism was much more clearly defined and special attention was paid to ensuring the suitability of all sectors and public supervision to mass tourism (Soyak 2013). For a fifth Five-Year Development Plan (1985–1989), for the first time, as well as, mass tourism, the history of the country, archaeological and cultural aspects, winter, hunting and water sports, festivals, health and youth tourism and also other tourism potential began to be evaluated in line with sustainable tourism principles. In this way, it was stated that, in addition to, giving priority to mass tourism, individual tourism would not be neglected. The sixth Five-Year Development plan (1990–1994) is important too, in the sense that the policies for promoting mass tourism alternatives and new tourism trends would be sustained. Also, in this plan, developing promotional policies, which would make winter, hunting and water sports, festivals, health, youth, conferences, thermal, golf and third year tourism more attractive, were prioritized. In the seventh Five-Year Development Plan

(1996–2000), in the light of new trends in tourism, it was stated that creating new alternative tourism fields would be sustained through consideration of the idea that developing mass tourism was merely related to increasing bed capacity ended. Thus, the concept 'sustainable tourism' was again referred (Yedinci Beş Yıllık Kalkınma Planı 1995: 174–175).

The eighth Five-Year Development Plan (2001–2005) saw the bankruptcy of the idea of promoting bed capacity in mass tourism (Soyak 2013). This plan included precautions to sustain inducement facilities related to alternative tourism in the creation of new potential fields, according to consumer tendencies in international markets (Sekizinci Beş Yıllık Kalkınma Planı 2000: 206). Similarly, the ninth Five-Year Development Plan (2007–2013) included findings which showed there was a need for structural reform of recognition and marketing issues, although there had been important improvements and an instant increase in bed capacity in recent years (DPT, 9. Kalkınma Planı, 2006: 35). In the period of the ninth Five-Year Development Plan, a tourism policies document 'Turkey Tourism Strategy 2023' was published by the Ministry of Culture and Tourism to promote new tourism types other than mass tourism, and to catch up with new trends in international tourism. In the document, it was stated that there was an extensive piling problem in the Mediterranean and Aegean coastal regions as a result of improvements in mass tourism in Turkey and as a result of a piecemeal approach to tourism planning (Türkiye Turizm Stratejisi 2023: 2).

In the study, fields where tourism potentials were high in coastal areas were to be planned and improved on a regional scale as 'Tourism Cities', and the issue of promoting private sector investment to be planned and improved in an understanding of a promotive 'field improvement model' was included in the plan. In this way, it was possible to plan not only the coastal region, but also areas in land and settlements in the region as a whole to increase tourism variety, and to improve and transform existing settlements. These policies were stated as vital in terms of providing the sustainability of the tourism sector. Moreover, it was targeted to contribute to a sustainable development target by avoiding unequal improvement among the regions.

It is thought that the spatial distribution of domestic and international tourists in a country will reflect the tourism policies and strategies of the country. Since the spatial distribution of tourists in Turkey has not been previously studied makes the current study significant. The research subject investigates the existing spatial relationship among tourist distribution differences in Turkish cities. Within this aim, spatial autocorrelation was used to determine the spatial distribution of spatial econometrics focusing on a global and local scale. In the study, data relating to domestic and international tourists participating in tourism movements in Turkey between 2001 and 2013 was used. It is believed that the findings of the study will be helpful in the future design of appropriate tour packages, to improve appropriate tourism programs and activities and to design collaborative cooperation among regions.

## *Spatial Autocorrelation in Tourism Studies*

Studies analyzing spatial models and spatial relationships are uncommon in the literature on tourism. It can be seen that related studies are limited to clustering in the sector and to spatial attitudes of tourists. There have been a number of empirical studies investigating the differences among tourists and tourist categories or estimating tourist numbers for the future in the literature. In most of the studies, tourists were categorized according to travel characteristics. Also, within this aim, based on time sequence data, Summary (1987), Naude and Saayman (2005) and Emir and Kiliç (2014) used different regression or statistical models to estimate tourism demand or tourist arrivals. Cooper (1981) studied the spatial behavior of tourists on the Channel Islands and defined differences of spatial models of tourist behavior according to a lifecycle stage of destination and socio-economic status variables.

Chadefaud (1981) examined tourist activities and time-spatial models of local public, and created a detailed map showing organized tourist groups and individual tourist activity fields. The researcher found that organized group activities were much denser, and that tourist activities showed more scattered and spatially spread characteristics. Depending on Plog's tourist typology, Debbage (1991) studied tourists' spatial behavior at a holiday destination on the Bahamas and found that differences of spatial behavior of tourists arose from individual differences. Montanari and Muscara (1995) identified nine tourist profiles for travelers to Venice based on time-space. The researchers determined the underlying basic aim of the visitors besides travel characteristics, such as previous visits and length of stay. Zhang et al. (2011) examined the spatial distribution of domestic and international tourists in 299 cities in China. Lu et al. (2011) examined the spatial distribution of inbound tourist arrivals for the period 2001–2009 in all cities in the Jiangsu region in China.

The researchers used global and local spatial autocorrelation analyses to measure the differences and significance of tourist distribution among the cities and examined their spatial aggregation. The researchers stated that the distribution of inbound tourist arrivals for the period 2001–2009 in the Jiangsu region indicates an unbalanced tourism development in the region and that cities accepting a high level of tourists showed overcrowding in the south of the region. Yang et al. (2011) based their research on the spatial interaction theory, using Wilson's model and spatial autocorrelation, and studied the spatial differences of tourism in cities in the Anhui region in China. According to the findings of study, the 17 cities studied were grouped into four different categories; developed tourism cities, less developed tourism cities, less undeveloped tourism cities and undeveloped cities. According to the global spatial autocorrelation results, the significance level of tourism global spatial autocorrelation was low. Xing-Zhu and Qun (2014), for the period 2000–2009, studied the tourist flow pattern applied using discovery space-time analysis.

## Method

In the study, data related to the number of tourists in Turkey, according to domestic and international tourist arrivals for the period of 2001–2013 was used. The data was obtained from the database of the Turkish Statistical Institute. Data concerning 81 provinces was reported. We analyzed the number of tourists for international and domestic tourism for the periods of 2001, 2005, 2009 and 2013, in Turkey at the province level by using Global Moran I. Analyses were conducted using GeoDA.

### *Moran's I Spatial Autocorrelation*

It is known that 'everything is related to everything else, but near things are more related than distant things' as stated in the first law of geography. Statistically, this relationship is explained by the concept of spatial autocorrelation. This well-known spatial correlation measure is Moran's *I* statistic. Moran's *I* statistic measures spatial autocorrelation between locations. This statistic determines whether spatial pattern is clustered, dispersed, or random (Al-Ahmadi and Al-Zahrani 2013). To explore the global spatial correlation, global Moran's *I* statistic, as follows, is generally used:

$$I = \frac{n}{\sum_{i=1}^{n} (x_i - \bar{x})^2} \times \frac{\sum_{i=1}^{n} \sum_{j=1}^{n} w_{ij}(x_i - \bar{x})(x_j - \bar{x})}{\sum_{i=1}^{n} \sum_{j=1}^{n} w_{ij}}$$

where $n$ is the total number of spatial observations (cities in this study), $x_i$ is the value measured in city $i$, $x_j$ is the value measured in city $j$. $\bar{x}$ is the mean value of all cities, $w_{ij}$ is the spatial weight between cities $i$ and $j$. While a positive value of Moran's *I* indicates that neighboring values are similar, suggesting spatial dependency or correlation, a negative value of Moran's *I* means that neighboring values are not similar. Also, a zero value of Moran's *I* indicates that there is no spatial pattern. As seen from the formula of Moran's *I*, the spatial weights matrix should be known or calculated before calculating Moran's *I*. In this study, a spatial weights matrix, based on rook measure, was used.

Since spatial autocorrelation was not detected in the study, Local Moran's *I* is not calculated. Having obtained the number of tourists at province level to investigate the distribution of tourists country-wide constitutes the limitation of the study. In one province, many tourist destinations can be found. This situation causes heterogeneity. Therefore, only a limited evaluation of tourism flow can be made. The province-level data set does not enable us to detect tourism hot spots composed of destinations across province boundaries. To examine the spatial distribution of tourism activities considering spatial perspectives, we also used mapping analysis. Moreover, the change in the number of tourists at province level for the period 2001–2013 was calculated using growth rate and mapping.

## *Growth Rate of Tourism*

Specifically, growth rate is expressed as the change in population over a unit time period and is generally expressed as a percentage value. The formula of the growth rate is given as follows:

$$\text{Growth rate} = \frac{P(t_2) - P(t_1)}{P(t_1)}$$

While a positive value of growth rate indicates an increase, a negative growth rate shows a decrease. A zero value of growth rate indicates that there is no change.

## *Drawing the Maps*

The distribution of tourist numbers according to province is explained by maps drawn for the periods (2001, 2005, 2009 and 2013). Tourist numbers for all settlements in each province are displayed within province borders according to the importance of the province. A MapInfo 12.5 Program was used to draw up the maps showing the distribution of domestic and international tourists according to city. The maps were spaced using Natural Break Classification. While drawing the maps, an interpolation method was applied using Inverse Distance Weighted (IDW) algorithm. The maps were bipolar colored. Cities having the highest number of tourists are displayed in dark red and cities with the lowest number of tourists are shown in dark blue.

# Findings

## *Change in the Number of Domestic and International Tourists in Turkey According to Years*

Turkey is an important destination for international tourism movements in terms of its abundance of natural, cultural attractions and its tourism platform of international quality. The tourism industry takes an important place in the Turkish economy (Yurtseven 2014). Turkey ranks sixth of countries receiving the highest number of tourists (Kültür ve Turizm Bakanlığı 2007). The main objective of a long

Table 32.1 The number of domestic and international tourists in Turkey (2000–2013)

| Years | Number of domestic tourists | Number of international tourists |
|---|---|---|
| 2000 | 8,785,905 | 6,784,599 |
| 2001 | 7,729,583 | 8,778,076 |
| 2002 | 7,916,530 | 9,871,594 |
| 2003 | 8,429,868 | 8,991,756 |
| 2004 | 9,723,087 | 10,981,573 |
| 2005 | 10,458,386 | 12,952,616 |
| 2006 | 11,570,101 | 11,896,301 |
| 2007 | 12,038,281 | 14,794,270 |
| 2008 | 11,286,396 | 13,647,336 |
| 2009 | 12,137,822 | 14,385,998 |
| 2010 | 12,435,836 | 17,633,683 |
| 2011 | 14,350,129 | 19,264,688 |
| 2012 | 15,722,753 | 20,477,308 |
| 2013 | 17,100,765 | 21,181,618 |

*Source* (T.C. Kültür ve Turizm Bakanligi 2007)

term Turkey tourism strategy is to be one of the first five tourist attracting and revenue generating countries in terms of tourism by 2023.

In 2000, 8,785,905 domestic tourists and 6,784,599 international tourists traveled to Turkey. By 2013, domestic tourist numbers had increased to 17,100,765, and international tourist numbers to 21,181,618. While domestic tourist numbers were higher in 2000, after 2001 international tourist numbers overtook domestic tourist numbers. The difference between domestic and international tourist numbers reached its highest value in 2010. The years 2011, 2012 and 2013 are those years when the difference was highest after 2010 (see Table 32.1). Although there is a steady increase in the number of tourists visiting Turkey, the rate of increase declined in years with economic difficulties (Günay Aktaş and Yılmaz 2013).

## *The Spatial Autocorrelation of Domestic and International Tourist Numbers in Turkey*

Table 32.2 shows the value of Moran's *I* for distribution of domestic and international tourist numbers. In the period 2001–2013, there is no spatial autocorrelation at province level. The non-existence spatial clustering, in terms of domestic and international tourist numbers, may show that tourists engage in multi-destination travel in Turkey. However, since the study is at province level, differences within the provinces are not investigated. Thus, the spatial distribution of the tourists is examined through mapping analysis.

**Table 32.2** Moran's $I$ value for the spatial autocorrelation of tourist number distributions ($p$-value in parentheses)

| | 2001 | 2005 | 2009 | 2013 |
|---|---|---|---|---|
| Domestic tourist numbers | 0.059784 (0.320) | 0.020735 (0.722) | 0.047555 (0.437) | 0.0548403 (0.388) |
| International tourist numbers | 0.067585 (0.287) | 0.0759485 (0.229) | 0.0480803 (0.452) | 0.0268648 (0.670) |

## *The Distribution of Domestic and International Tourists in Turkey According to Province*

The distribution of domestic and international tourists in Turkey according to provinces was examined using mapping analysis with five-year gaps. The distribution for domestic tourist numbers is shown in Fig. 32.1.

## *When the Distribution for Domestic Tourist Numbers Was Examined According to Province*

As it can be seen in Fig. 32.1, for all years, the most visited cities by domestic tourists were in the west, south and central west parts of Turkey. Since 2005, the are

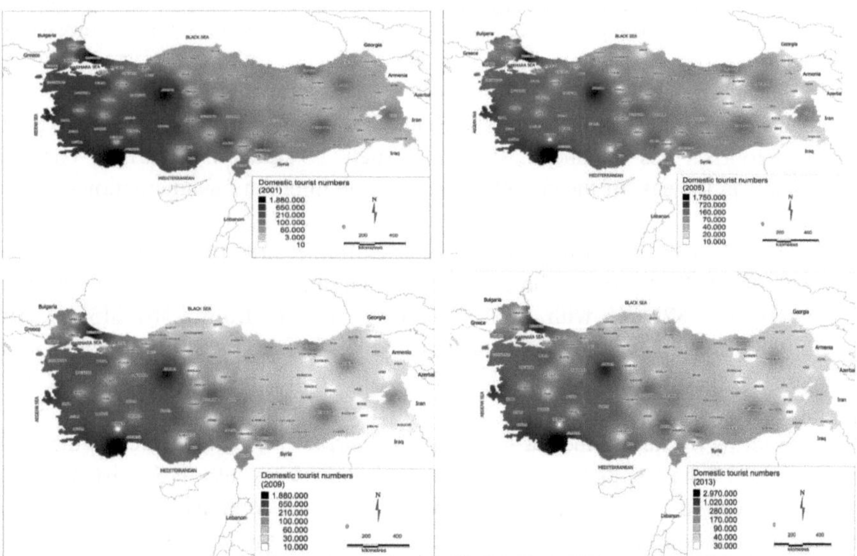


**Fig. 32.1** The distribution of domestic tourist number according to province (2001, 2005, 2009 and 2013)

as visited by most domestic tourists have expanded. In the East of Turkey, the number of domestic tourists is decreasing. Istanbul ranks first among cities which domestic tourists visited in high numbers for the years 2001 and 2005. However, in 2009 and 2013, it ranked behind second Antalya, which was ranked second for previous periods. The number of domestic tourists visiting İstanbul, which was 1,105,804 in 2001, reached 1,839,131 in 2013 which is an increase of 168 % over twelve years. Nonetheless, Antalya showed a 358 % increase for the same period, with the number of domestic tourists growing from 829,762 to 2,966,418. Turkey's capital Ankara was consistently placed third in the rankings.

Muğla, Izmir, Aydın and Balıkesir, known for coastel tourism opportunities and Bursa, birthplace of the Ottoman Empire and prominent with its cultural tourism attractions, are the cities which appeared in the first seven ranked cities, although their ranking positions changed. Afyon, famous for its thermal springs, Denizli known for the Pamukkale travertines, and Adana and Mersin, known for coastal tourism were ranked in tenth places. Nevşehir, famous for geological eroded structures, cultural, natural and rural tourism elements ranked in tenth place 2009.

## *When the Distribution of International Tourists Was Examined According to Province*

The distribution of international tourists according to province for five-year periods is presented in Fig. 32.2. The spatial pattern of international tourists travelling to Turkey becomes dense in the west of the country just as it does for domestic tourists. As observed in Fig. 32.2, for all years, the provinces visited mostly by international tourists become denseston the southwest and west coasts of Turkey and in Istanbul. Towards the centre, Ankara, Nevşehir and Konya are the cities most visited by international tourists. Since 2005, the area most visited by international tourists has expanded. Provinces such as Trabzon, Erzurum, Van and Diyarbakır in the east of the country have become important as international tourist destinations.

Antalya, Istanbul and Mugla were consistently the three provinces most visited for the period 2001–2013. The number of international tourists visiting Antalya increased to 10,827,645 with a 308 % increase in 2013 from 3,507,343 in 2001. 1,875,988 travelers visiting Istanbul reached 4,475,838 in 2013 with a 238 % increase. Furthermore, in Muğla, 1,018,974 visitor numbers increased to 1,661,453, an increase of 63 %. İzmir, Aydın and Denizli are the provinces following the first three in terms of the number of international tourists. Nevşehir, which is the most visited city of the Cappadocia tourism region, ranked seventh in the list. In 2013, Nevşehir overtook Denizli and rose to sixth. The capital Ankara consistently ranked eighth. In addition to Bursa and Çanakkale, the site of the battle of Gallipolli, and Konya, an important center of religion, are cities competing ninth and tenth places.

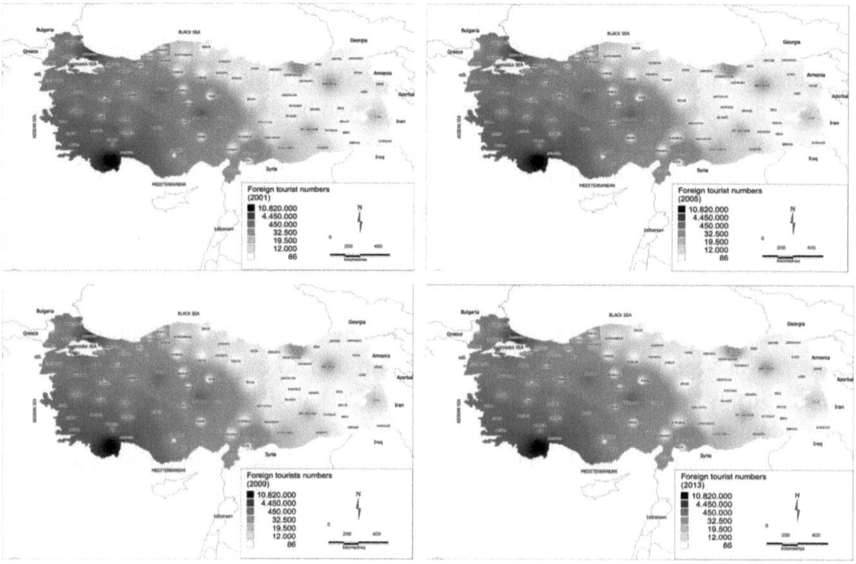


**Fig. 32.2** The distribution of international tourists according to province (2001, 2005, 2009, 2013)

## *The Growth Rate of Domestic and International Tourism in Turkey*

Between 2001 and 2013 the growth rate in the number of international tourists was 141.30 %, and for domestic tourists, it was 94.63 %. The growth rate for the number of domestic and international tourists between 2001 and 2013 is shown in Fig. 32.3.

Ascan be seen in Fig. 32.3, from 2001 to 2013, the highest growth rate in terms of domestic tourist numbers is seen in Manisa (45.73 %), which has cultural

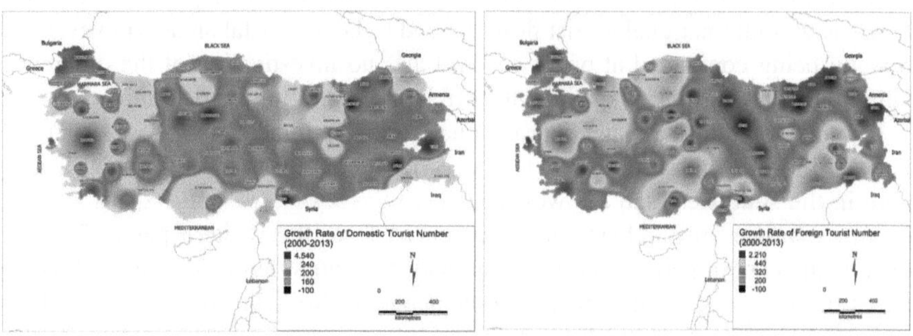


**Fig. 32.3** The distribution of the growth rate of domestic and international tourist number between the years 2001 and 2013 according to province

tourism attractions without a coastline in the west of Turkey. Ardahan (12.48 %), with ecological tourist attractions in the northeast of Turkey, and Hatay, known as an intercultural communication center on the in Syria border, are the cities with the fastest growth rate in terms of domestic tourist numbers. In five cities, the number of domestic tourists decreased.

When the growth rate of tourist numbers was investigated in terms of Istanbul, Antalya and Ankara, which receive the most tourists in Turkey, the tourist numbers grew at the rate of 43 % for Istanbul, 182 % for Antalya and 48 % for Ankara. This indicates that these cities, already having a high number of domestic tourists, continued to grow at a moderate rate.

As shown in Fig. 32.3, the highest growth rate in terms of international tourist numbers was seen in Sırnak, at the southeast tip of Turkey. 2225 % of growth was seen in Sırnak. Kocaeli, on the Blacksea and east of Istanbul was raked second place, Manisa, with no a coastline in the western Anatolia is a cultural tourism city ranked third, having the highest growth rate in terms of international tourist numbers. In 11 cities, international tourist numbers decreased.

When the tourist number growth rate was analyzed for Antalya, Istanbul and Mugla, with the highest number of international tourists in Turkey, the tourist numbers increased 209 % for Antalya, 139 % for İstanbul and 63 % for Mugla. This shows that these cities, already having the highest number of tourists, continue to grow at moderate rate. Şırnak, Kocaeli and Manisa are the cities with the highest growth rates in terms of both domestic and international tourists.

## Conclusions

In Turkey, spatial autocorrelation for spatial distribution of domestic and international tourists is shown at province level. This may indicate that multi-destination travel is conducted in Turkey. However, at province level, differences within provinces cannot be analyzed. Hence, the spatial distribution of the tourists was analyzed using mapping analysis.

As it can be understood from the maps, the most visited places in terms of domestic and international tourist flows tended to be in coastal areas. However, due to study being conducted at province level and no investigation of the differences within cities taking place, spatial autocorrelation could not be determined.

In terms of domestic and international tourist numbers, the west of Turkey receives more visitors compared to the east. International tourists travel Turkey more in the southwest and the west coasts.

An expansion in numbers can be observed from the western parts of Turkey towards the central parts in terms of international tourist numbers. Cities attracting tourist in the east of Turkey have become apparent. Nevsehir, the most visited city in the Cappadocia Tourism Region, is visited more by international tourists than domestic tourists.

When the growth rate for both domestic and international tourist numbers was analyzed, the tourist numbers for the southeast, improved for mass tourism and greatly attractive, and for the west coast and Istanbul, keep growing. On the other hand, the largest growth rates can be observed in provinces which do not have a coastline offering health, culture and ecotourism attractions.

It can be seen that domestic tourist travel has expanded nationwide, although it falls behind international tourist numbers. The findings draw attention to the promotion of policies to increase the participation of Turkish population in domestic tourism.

From the distribution of tourist numbers, the effect of projects to increase accommodation areas to improve mass tourism can be observed. On the other hand, it can be concluded that the effects of the practiced strategies on policies such as expanding tourism nationwide, and decreasing differences between regional developments, have begun to be felt.

## References

Al-Ahmadi K, Al-Zahrani A (2013) Spatial autocorrelation of cancer incidence in Saudi Arabia. Int J Environ Res Public Health 10(12):7207–7228

Chadefaud M (1981) Lourdes: Un P′elerinage (Aix-en-Provence: une Ville Edisud)

Cooper CP (1981) Spatial and temporal patterns of tourist behavior. Reg Stud 15:359–371

Debbage KG (1991) Spatial behaviour in a Bahamian resort. Ann Tourism Res 18:251–268

DPT (1995) Yedinci Beş Yıllık Kalkınma PLanı (1996–2000) (Seventh five development plan), Ankara

DPT (2000) Sekizinci Beş Yıllık Kalkınma Planı (2001–2005) (Eighth five development plan), Ankara

DPT (2006) Dokuzuncu Kalkınma Planı, (2007–2013) (Ninth five development plan), Ankara

Emir O, Kılıç İ (2014) Cok degişkenli analiz yöntemiyle turizm göstergeleri açisindan dünya ülkelerinin incelenmesi. J Yaşar Univ 9(33):5597–5622

Günay Aktaş S, Yılmaz A (2013) The effects of the 2008 global economic crisis on tourist flows to Turkey. The science and education at the beginning of the 21st in Turkey. In: Atasoy E (ed) Kliment Orhidski University Publishing, Sofya, 2013

Inskeep E (1991) General information South Antalya development project. Ministry of Tourism, Ankara

Kültür ve Turizm Bakanlığı (2007) Türkiye Turizm Stratejisi (2023). Ankara

Lu S, Zhang J, Zhang H (2011) Spatial differential features of inbound tourists in Jiangsu, China. In: 19th International Conference on Geoinformatics, Shanghai, China, 24–26 June 2011, pp 1374–1377

Montanari A, Muscara C (1995) Evaluating tourist flows in historic cities: The case of Venice. Tijdschrift voor Economischeen Sociale Geographie, 86:80–87

Naude W, Saayman A (2005) Determinants of tourist arrivals in Africa: A panel data regression analysis. Tour Econ 11(3):365–391

Soyak M (2013) Uluslararası Turizmde Son Egilimlerve Turkiye'de Turizm Politikalarının Evrimi. J Marmara Soc Res 4:1–18

Tarhan C (1997) Tourism policies. Bilkent University School of Tourism and Hotel Management

Tezcan B (2004) Developing alternative modes of tourism in Turkey. Unpublished Master Thesis, Middle East Technical University

Turizm ve Tanıtma Bakanlığı (1980) Dünyave Türkiye Turizm

Turizm Bakanlığı TC (Ministry of Culture and Tourism) (2002). II. Turizm Şurası Alt Komisyon Raporları Genel Kurul Görüşmelerive II. Turizm Şurası Kararları. 12–14 Nisan: 2002, Ankara

Var T (2005) The role of coastal tourism in Turkish tourism development: a historical perspective. In: International congress on coastal & marine tourism: development, impacts and management. Cesme, Turkey

Williams S (1998) Tourism geograph. Routledge

Xing-Zhu Y, Qun W (2014) Exploratory space–time analysis of inbound tourism flows to China cities. Int J Tour Res 16:303–312

Yang Z, Lu S, Jin X (2011) Tourism spatial association analysis based on GIS technology for the cities in Anhui of China. In: 19th International conference on geoinformatics, Shanghai, China 24–26 June 2011, pp 208–212

Yurtseven Ç (2014) International tourism and economic development in Turkey: a vector approach. Afro Eurasian Stud 1(2):37–50

Zhang Y, Xu J, Zhuang P (2011) The spatial relationship of tourist distribution in Chinese cities. Tour Geogr 13(1):75–90

# Chapter 33
# Spatial Distribution of Occupancy Rate in the Hospitality Sector in Turkey According to International and Domestic Tourist Arrivals

**Semra Günay Aktaş, Deniz Yüncü and Yeliz Mert Kantar**

**Abstract** Besides resort tourism, Turkey has a huge potential of alternative tourism sources, such as health and thermal tourism, winter sports and mountain nature tourism, mountain tourism, rural and eco-tourism, congress and exhibition tourism, cruise and yacht tourism, golf tourism, and so forth. However, this potential can hardly said to be used effectively. For example, attractive and suitable alternative destinations, as well as routes for thermal, spring, winter and mountain sports, to be distributed in different regions throughout Turkey could not be constructed. In this study, we research occupancy rates in the hospitality sector for the period 2001–2013, by using spatial analysis techniques in order to demonstrate whether there is any spatial pattern of occupancy rates in Turkey. For this purpose, we calculated Moran's I spatial autocorrelation coefficient for occupancy rates in the hospitality sector for both domestic and international tourists. The results show that, while there is spatial autocorrelation, taking accommodation of international tourists into account, spatial autocorrelation does not arise in terms of domestic tourist's occupancy. Thus, domestic and international tourists' spatial patterns are different in Turkey. This result can be interpreted as a preference of domestic tourists to visit relatives in Turkey, while international tourists prefer to stay in the Mediterranean and Aegean coasts along with Istanbul. Also, climate affects the occupancy rates. In other words, the length of the tourist season affects occupancy rates in the hospitality sector. Moreover, we also aim to detect spatial-dependencies among domestic and international tourists' lines and local correlations using bivariate Moran's I and LISA. The analyses show that coastal provinces that have HH autocorrelation are neighboring provinces that have a low degree of occupancy indicating that strong

S.G. Aktaş (✉) · D. Yüncü
Faculty of Tourism, University of Anadolu, Eskişehir, Turkey
e-mail: semragunay@anadolu.edu.tr

D. Yüncü
e-mail: dkaragoz@anadolu.edu.tr

Y.M. Kantar
Faculty of Science, University of Anadolu, Eskişehir, Turkey
e-mail: ymert@anadolu.edu.tr



A. Saufi et al. (eds.), *Balancing Development and Sustainability in Tourism Destinations*, DOI 10.1007/978-981-10-1718-6_33

tourism routes and regional destinations have not yet been established. For this reason, such destinations should be bolstered with accommodation, culture, hand-crafts, and food and beverage facilities. Therefore, the results of this study may help Republic of Turkey's Development Ministry to rearrange their integrated tourism planning approach.

**Keywords** Spatial autocorrelation · LISA · Regional economic development · Occupancy rates in the hospitality sector · Turkey

## Introduction

Occupancy rate is the basic product of a hotel business and measures success in selling rooms, which forms the most important part of income. Also, occupancy is a vital key in carrying out the functions of planning and auditing by the hotel management. Planning starts with estimating the number of full rooms and the resulting income. Later, depending on occupancy, estimation of staff numbers, estimated heat and lighting expenses and the amount of bed linen is calculated. Thus, finding the entire estimated income and cost for the hotel as a whole becomes easier (Emeksiz 2002: 115). While correct estimation of the occupancy rate helps the hotel managers' plan on strategy, tactics, and operations, it also empowers the decision-making processes (Law 1998: 234). Moreover, occupancy rate in the hospitality sector of a country is an important indicator of tourism performance. In general, when considered on a country-wide basis, the distribution of occupancy rates of tourism businesses throughout the country can be considered an element indicating performance of tourism movements and facilities within the country.

Tourism researchers deal with a number of statistical methods in their research and investigations in order to plan and manage tourism activities. Taking the tourism literature into account, most studies have focused on classical statistical methods, such as descriptive statistics, time series, or regression analysis. Summary (1987) found the estimation of tourism demand by multivariable regression analysis evidence from Kenya. Naude and Saayman (2005) used the panel data regression model to identify the determinants of tourism arrivals in 43 African countries, for the period 1996–2000. Emir and Kiliç (2014) studied Turkish tourism demand with linear spline regression. These studies consider various indicators of tourism. In this study, we use global and local spatial association statistics to explore the spatial distribution of tourists from 2001 to 2013 using occupancy rates in the hospitality sector in Turkey.

The aim is to investigate whether a spatial relationship or spatial clustering exists among domestic and international tourists in these sub-provinces, and also to find local correlations using spatial statistics tools, such Moran's I, bivariate Moran's I and LISA. As far as we know, only a few studies have used spatial statistics tools. The first law of geography (Tobler 1979) states that 'everything is related to everything else, but near things are more related than distant things'. That is, spatial data corresponding to a region or a district contains spatial dependence. Spatial dependence is generally measured by spatial autocorrelation statistics. In this context, Moran's I is a well-known spatial correlation measure (Anselin 1988).

If positive spatial autocorrelation exists, this means that there is a significant similarity between the values of variables measured in regions. Although Moran's I statistics indicates global spatial autocorrelation, it does not provide local clustering or local autocorrelation. Anselin (1995) provides the definition of statistics as a local indicator of spatial association (LISA) for measuring local spatial autocorrelation.

## Exploratory Spatial Data Analysis (ESDA)

ESDA focuses on the spatial characteristics of data (Anselin 1996). In other words, ESDA methods help us to visualize spatial distributions, define spatial outliers, discover spatial clusters, make suggestions on spatial stationary or nonstationary and different spatial situations as well as explain them (Anselin 1998: 258). In this study, spatial correlation analysis, covering global and local spatial autocorrelation and space-time correlation, is conducted to find interesting patterns.

Global spatial autocorrelation is generally measured by Moran's I (Moran 1950) given by:

$$I = \frac{n}{\sum_{i=1}^{n} (x_i - \bar{x})^2} \times \frac{\sum_{i=1}^{n} \sum_{j=1}^{n} w_{ij}(x_i - \bar{x})(x_j - \bar{x})}{\sum_{i=1}^{n} \sum_{j=1}^{n} w_{ij}} \tag{33.1}$$

In Eq. (33.1), $n$ denotes the total number of observations measured in districts and, $x_i$ and $x_j$ are the observation values measured in spatial districts $i$ and $j$, respectively $\bar{x}$ is the mean value of observations and $w_{ij}$ is the spatial weight calculated based on the measure between districts $i$ and $j$. The Moran's $I$ varies from $-1$ to $+1$. If it is greater/less than 0, this shows that there is positive/negative autocorrelation. On the other hand, a zero value of Moran's $I$ indicates no spatial autocorrelation. While positive spatial autocorrelation explains similar value cluster together in the whole map, negative spatial autocorrelation means that dissimilar values arise close to one another. As can be seen from the formula of Moran $I$ in Eq. (33.1), the spatial weights based on the determined distance matrix should be calculated. In this study, a spatial weights matrix based on rook measure is used.

Although the Moran's $I$ statistics provide evidence of global spatial correlation on the sample from a whole region, it does not indicate local autocorrelation. The local measure of spatial autocorrelation gives a spatial association coefficient for a particular location. These local and global statistics are distributed as asymptotic normal under the hypothesis which states 'there is no spatial autocorrelation.'

The local measure of Moran's *I*, proposed by Anselin (1995), is defined as follows:

$$I_j = \frac{n}{\sum_{i=1}^{n} (x_i - \bar{x})^2} \frac{\sum_{i=1}^{n} w_{ij}(x_j - \bar{x})(x_i - \bar{x})}{\sum_{i=1}^{n} \sum_{j=1}^{n} w_{ij}} \tag{33.2}$$

Local Moran's *I* is an important member of 'local indicators of spatial association' (LISA) proposed by Anselin (1995). According to Anselin, the LISA statistic should satisfy two conditions: (a) show the extent of significant spatial clustering for each location; (b) the sum of local statistics for all observations is proportional to a global indicator of spatial association.

By using the local Moran's *I* statistics, a map of significant clusters and spatial outliers, and also a map of significance levels for the cluster map, are drawn. A map of significant clusters shows significant local clusters and spatial outliers. High-high (HH) means that high values are surrounded by high values, while low-low (LL) shows that low values are surrounded by low values. The high-low (HL) and low-high (LH) regions point out spatial outliers. This means that while high value is surrounded by low values, a low value is surrounded by high values.

Moran's *I* spatial autocorrelation statistics is used for one variable. This, also in the case of a bivariate, can be performed in order to obtain more information about the level of spatial dependency among the values of one variable and other variables. Similar to the case of the univariate Moran's *I*, a weights matrix, based on the determined distance measure, is needed.

## Method

Data used in this study was obtained from the Turkish Statistical Institute's data base relating to 2001–2013, according to the arrival of domestic and international tourists related to the occupancy rate of the hospitality sector in Turkey. Data concerning 919 sub-provinces was reported. We analyzed occupancy rates in the hospitality sector for international and domestic tourists for the periods 2001, 2005, 2009, and 2013 for Turkey at the district level (at the level of sub-provinces) using Global Moran I, Local Moran I and bivariate Moran I statistics and LISA. All analyses were conducted using Geo DA.

## Findings

### *Spatial Correlation for Occupancy Rates*

In this section, we analyze occupancy rates in the accommodation sector for international tourists for the periods 2001–2013, in Turkey, at the level of

**Table 33.1** Moran's I value for the spatial autocorrelation of occupancy rates in the accommodation sector distributions (*p*-value in parentheses)

| Statistics | 2001 | 2005 | 2009 | 2013 |
|---|---|---|---|---|
| Moran's I for occupancy rates in the hospitality sector for domestic tourism | −0.0168176 (0.3150) | 0.0532108 (0.00243) | 0.06557142 (0.00017) | −0.00216305 (0.887) |
| Moran's I for occupancy rates in the hospitality sector for the international tourism | 0.374647 (0.0000) | 0.326383 (0.0000) | 0.28067 (0.00000) | 0.458025 (0.0000) |
| Bivariate Moran's I for domestic and the international tourism | −0.000019 (0.9990) | 0.0480218 (0.00625) | 0.0565294 (0.00127) | 0.0300794 (0.0488) |

sub-province using Global Moran I, Domestic Moran I and bivariate Moran I statistics and LISA. The results of the Global Moran I and bivariate Moran I statistics are provided in Table 33.1.

Table 33.1 shows the value of Moran's I for occupancy rates in the accommodation sector distributions for domestic and international tourists. Taking into account domestic tourists for the period 2001–2013, there is no spatial correlation between the numbers of domestic tourists in Turkey. These tourists are randomly assigned to clusters. When the occupancy rate for international tourists was examined, it was found to be in positive spatial autocorrelation at a moderate level. In 2013, in particular, the positive Moran's I indicates a clustering pattern among international tourists. Also, the *p*-values of the test statistics are less than 0.05 and thus, the hypothesis which states 'there is no spatial autocorrelation' is rejected. Taking into account the bivariate global association index (Moran's I), it can be seen that there is no correlation between domestic and international tourists. In other words, international tourist stays are not associated spatially with domestic stays.

Of significance is the fact that Moran's *I* statistics for international tourist occupancy rates indicates that the international tourist flow is denser compared to domestic tourist flow in Turkey's sub-provinces. Based on this finding, it can be stated that international tourists prefer accommodation in specific regions. As a result of this, touristic facilities for international tourists are dense in specific regions rather than nationwide. Also, the preferred destinations of domestic and international tourists in Turkey show differences.

## *Local Spatial Association of Occupancy Rates for Domestic and International Tourists*

Spatial typology findings for occupancy rates in the accommodation sector, created by domestic and international tourists are presented in Table 33.2. Table 33.2 shows that tourism sub-provinces are located in a particular Moran scatter plot

**Table 33.2** Significant LISA at 5 % pseudo-significance for occupancy rates of domestic and international tourists arrivals

| Spatial typology for occupancy rates of domestic tourists | 2001 | 2005 | 2009 | 2013 |
|---|---|---|---|---|
| Not significant | 828 | 804 | 812 | 867 |
| High-high | 9 | 28 | 39 | 20 |
| Low-low | 2 | 7 | 10 | 10 |
| Low–high | 52 | 49 | 34 | 44 |
| High–low | 28 | 31 | 24 | 25 |
| Spatial typology for occupancy rates of international tourists | 2001 | 2005 | 2009 | 2013 |
| Not significant | 850 | 845 | 835 | 862 |
| High-high | 30 | 25 | 36 | 46 |
| Low-low | 2 | 10 | 11 | 32 |
| Low–high | 24 | 26 | 29 | 21 |
| High–low | 13 | 13 | 8 | 5 |

quadrant. HH-LL in the table reveals significant and greater spatial associations in both domestic and international tourist occupancy rates. The association in most sub-provinces is noted to be nonsignificant. During the period 2001–2013, the occupancy rate of domestic tourists in terms of accommodation was between 13 and 6 %, and the occupancy rate for international tourists in terms of accommodation was between 10 and 7 %. The variable is significant at the 5 % significance level, which can be explained by the fact that regional inequities and uneven growth in tourism have increased the most during this period. In terms of accommodation occupancy rates of domestic tourists, in 2001, only 1.1 % of the sub-provinces had positive associations or similar values (0.9 % in quadrant HH and 0.2 % in quadrant LL). This value increased to 5.3 in 2009 (4.2 % in quadrant HH and 1.1 % in quadrant LL). In terms of accommodation occupancy rates of international tourists, in 2001, 3.4 % of the sub-provinces had positive associations or similar values (3.2 % in quadrant HH and 0.2 % in quadrant LL). This value increased to 8.4 in 2013 (5 % in quadrant HH and 3.4 % in quadrant LL). Over the years, this spatially similarity can be clearly seen from Table 33.2, for international tourists.

The results of the LISA statistics for accommodation sector' occupancy rates, related to domestic and international tourists, are displayed in Figs. 33.1 and 33.2. Under the significant level, the maps are dominated by LL-type and HH-type sub-provinces.

## *Spatial Typology for Occupancy Rates of Domestic Tourists*

Significant values of domestic Moran's I indicating clustering are provided in the maps (see Fig. 33.1). There are only two sub-provinces with a coastline which have

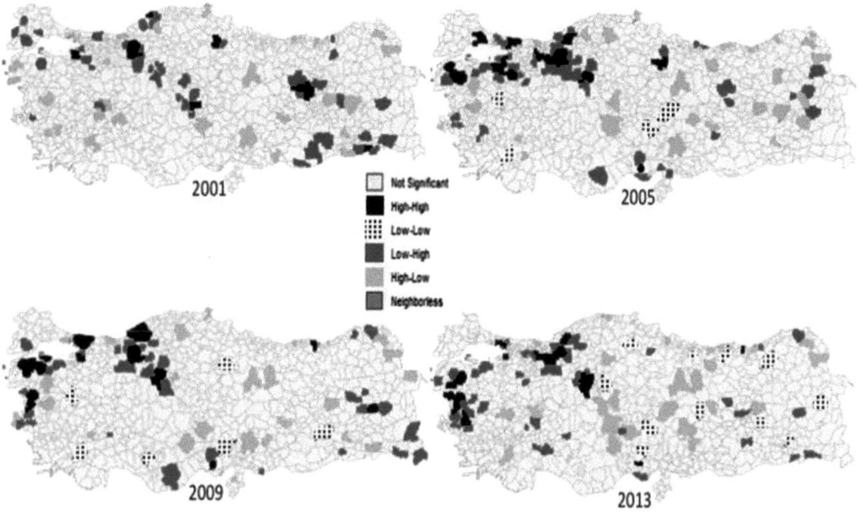


**Fig. 33.1** Local indicators of spatial autocorrelation cluster maps for occupancy rates of domestic tourists (2001, 2005, 2009 and 2013)

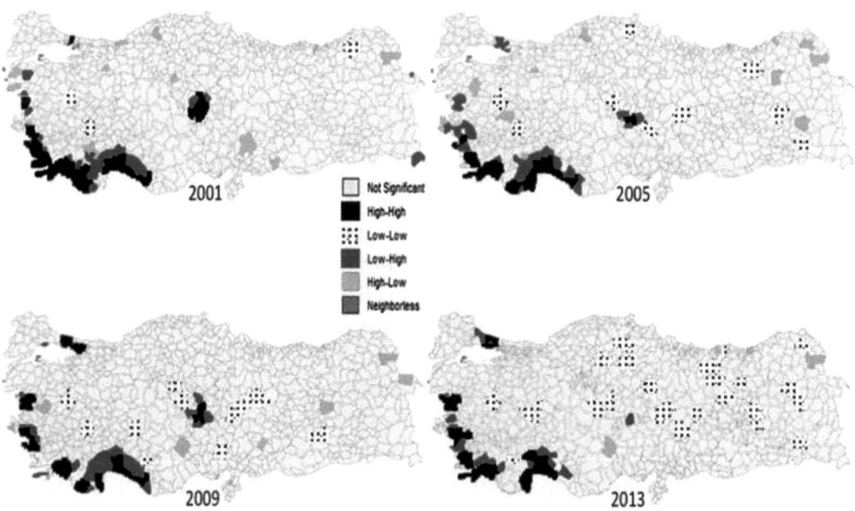


**Fig. 33.2** Local indicator of spatial autocorrelation cluster maps for occupancy rates of international tourists (2001, 2005, 2009 and 2013)

HH occupancy rates in 2001. These are Bursa-Gemlik and Bartın-Amasra. All the other sub-provinces which have HH occupancy rates, apart from these two sub-provinces and Nevşehir-Gülşehir, without a coastline, have natural tourism attractions. However, these sub-provinces do not have common tourist investmentor, a wide range of tourist attractions and facilities. From 2001 to 2013, the sub-provinces and central sub-province of Bolu remained in the HH cluster with decreasing sub-province numbers. Since 2005, the sub-provinces of the capital Ankara, the centre of public services, have consistently formed the largest HH cluster. Since 2009, the sub-provinces of Çanakkale, Balıkesir and İzmir, forming the north part of the Agean Sea Coastline, and the sub-provinces of Istanbul, have formed a wide HH cluster. Again, in 2009, the coast sub-provinces of Zonguldak and Trabzon on the Black Sea coast formed an HH cluster. It can be seen for the accommodation occupancy rate in terms of sea tourism, domestic tourists prefer the north coasts of the Aegean Sea and the Black Sea coastline. Over time, it is observed that LL clusters have increased and expanded to central parts of Turkey.

## *Spatial Typology for Occupancy Rates of International Tourists*

When cluster maps for occupancy rates of international tourists were examined, as in other countries, it can be seen that hot spots tend to take place in narrow and coastal areas (see Fig. 33.2). Until 1980, in terms of tourism policies, Turkey performed development programs to improve mass tourism and especially targeted the international tourist market in settlements of the Mediterranean and Aegean coasts, with high tourism potential, and aimed to promote strategies to increase accommodation facilities' capacity. As can be seen in Fig. 33.2, international tourist occupancy rates in the coastal settlements of Antalya, Muğla, Aydın and İzmir form HH clusters. Istanbul, in the northwest of Turkey, is one of the most important cultural tourism centers. In central Anatolia, the sub-provinces of Nevşehir, which attract most tourists in the Cappadoica Region, form HH clusters in terms of international tourist occupancy rates. From 2001 to 2013, HH cluster spatial patterns showed differences. In 2005, the HH cluster was squeezed into a narrower area, moving back to the south west Anatolian coasts. In Nevşehir, spatial narrowing formed. In 2009 (Günay Aktaş and Yılmaz 2013), following the economic crisis of 2008, while a regression was observed in the number of HH provinces in the Mediterranean and Aegean coasts, prominent for mass tourism, field expansion was again experienced in the cultural tourism areas of Istanbul and Nevşehir. In 2013, the Cappadocia tourism region does not feature in the HH cluster. In Istanbul and the Aegean coasts, the HH cluster continued, while in south Antalya expansion occurred toward the central region, which does not have a coastline. Dalaman, Ortaca, Köyceğiz, Ula, Marmaris, Datça and Milas, all sub-provinces of the city Muğla are the most consistent sub-provinces occurring in the HH cluser in four periods 2001, 2005, 2009, and 2013 period. Although the number of sub-provinces changes between 8 and 12, the sub-provinces of Istanbul consistently occur in HH cluster. In 2013, the sub-provinces occurring in the HH cluster, among the sub-provinces of Antalya, underwent a change. For the first time that year, the sub-provinces Aksu, Kepez, Muratpaşa, and Konyaaltı occured in the HH cluster. With the international occupancy rates, as with the domestic tourist occupancy rates, it can be observed that the number of LL clusters increased and expanded to the central parts of Turkey over the years.

## Conclusions

If Moran's I statistics are positive and significant ($p$-value < 0.05) distributions, it explains that high/low value sub-provinces tend to be clustered close to other high/low value sub-provinces. In the period 2001–2013, there is no clear clustering in the accommodation facilities' occupancy rates in terms of domestic tourists, and it can be seen that there is random distribution. In 2001 and 2013, Moran's I correlation coefficient shows a negative value. This situation indicates that there is no spatial clustering, and there is an expansion close to random. According to international tourist accommodation, when the Moran's I statistics was analyzed, it was detected that a medium level spatial autocorrelation existed, and that international tourists were more densely grouped in particular sub-provinces. In 2009, Moran's I correlation decreased, but reached its highest level in 2013. When the bivariate global association index (Moran's I) is examined, it can be seen that there is a dispersed relationship between the two variables in 2001 and, since 2005, there has been a cluster close to random. Conversely, the low positive value of bivariate Moran's I with regard to the distribution of occupancy rates of the accommodation sector could indicate a non-geographical clustering of similar sub-provinces; that is, the low value of Moran's I indicates a lack of similarity among sub-provinces, with respect to the distribution of occupancy rates of the accommodation sector.

The determinant effect of climate on the occupancy rates of the accommodation sector should not be ignored. During the tourism season, weather conditions, and the number of sunny days in particular, are vital in terms of investment on coastal tourism. On the Mediterranean coastline, daily sunshine duration is between 12 and 13 h on average, and annually, 3000–3200 h. The months when the monthly average temperature is in the year is over 20 °C, are considered convenient for coastal tourism. The average temperature is over 20 °C for two months on the Black Sea coastline in Turkey (June and August), five months on the Aegean coastline (May–September), and six months on the Mediterranean coastline (May–October). The distribution of sea water temperature is within the year has a determinant effect on the coastal tourism season. The number of days when the see temperature is over 18 °C increases from the Black Sea coast to Marmara, toward the Aegean and Mediterranean coasts. Until the 1980s, prioritizing a quantitative increase of the accommodation sector and mass tourism in tourism development policies caused an increase in tourist investment and substructure services in the south and southwest coastlines of Turkey, where the tourism season lasts a long time. Although all the sub-provinces in Turkey are rich in terms of both human and natural resources, when LISA cluster maps are analyzed, it is clear that domestic and international tourist demands show differences, and that domestic and international tourists display different distribution patterns. While international tourists prefer to travel to mass tourism destinations on the southwest and west coasts of Anatolia and İstanbul, where tourism investments are high, domestic tourists prefer to travel to smaller sub-provinces, where tourist investment and tourist attractions are limited, rather than traveling to mass tourism centers with high tourist potential, except for the capital Ankara and Istanbul. Because of developments in mass tourism and a fragmented approach in Turkey, the problem of overaccumulation has occurred on the Mediterranean and Aegean coasts. In 2023, Turkey Tourism Strategy, prepared in 2007, targets planning not only coastlines, but also of inland areas and even of all the settlements as a whole, increasing tourism variety, enhancing existing settlements and providing transformation to contribute to a sustainable development target by eliminating development differences in the region (Türkiye Turizm Stratejisi 2023: 2). The findings of the study show that there is mileage in a target of providing integration of different tourism types, typical to locations, and creating tourism sub-provinces focusing on alternative tourism, which can present great variety of tourism opportunities, such as health and thermal, golf, winter sports, and nature tourism.

## References

Anselin L (1988) Spacial econometrics: methods and models. Kluwer Academic, Dordrecht

Anselin L (1996) The Moran scatter plot as an ESDA tool to assess local instability in spatial association. In: Fischer M, Scholten HJ, Unwin D (eds) Spatial analytical perspectives on GIS. Taylor & Francis, London, pp 111–125

Anselin L (1998) Interactive techniques and exploratory spatial data analysis. In: Longley PA, Goodchild MF, Maguire DJ, Wind DW (eds) Geographical information systems: principles, techniques. Wiley, New York, pp 253–265

Emeksiz M (2002) Beş Yıldızlı Otelİşletmeleriiç in Getiri Yönetimi Uygulama Modeli. Anadolu Üniversitesi Yayınları, Eskişehir, No: 1337

Emir O, Kılıç İ (2014) Cok degişkenli analiz yöntemiyle turizm göstergeleri açisindan dünya ülkelerinin incelenmesi. J Yaşar Univ 9(33):5597–5622

Günay Aktaş S, Yılmaz A (2013) The effects of the 2008 global economic crisis on tourist flows to Turkey. the science and education at the beginning of the 21st in Turkey. In: Atasoy E (ed) Kliment Orhidski University Publishing, Sofya, 2013

Kültürve Turizm Bakanlığı (2007) Türkiye Turizm Stratejisi (2023). Ankara

Law R (1998) Room occupancy rate: a neural network approach. Int J Contemp Hosp Manag 10 (6):234–239

Moran PAP (1950) Notes on continuous stochastic phenomena. Biometrika 37(1):17–23

Naude W, Saayman A (2005) Determinants of tourist arrivals in Africa: a panel data regression analysis. Tourism Economics 11(3):365–391

Summary R (1987) An estimation of tourism demand by multivariate regression analysis: evidence from kenya. Tourism Manage 8(4):317–322

Tobler W (1979) Cellular geography. In: Gale S, Olsson G (eds) Philosophy in geography. Riedel, Dordrecht, pp 379–386

# Chapter 34
# The Influence of Tourism Core Resources on Travel and the Tourism Competitiveness Index and Tourism Performance

**Mohd Hafiz Hanafiah, Mohamad Abdullah Hemdi and Ismail Ahmad**

**Abstract** This paper explores whether tourism core resources promote destination competitiveness and performance, focusing on the travel and tourism (T&T) competitiveness index (TTCI) rankings and tourist arrivals. Utilizing panel data from eight ASEAN countries, this paper examined the statistical relationship between tourism core resources, tourism performance, and TTCI ranking. This empirical research relied on measures of tourism core resources compiled from various tourism organization reports and databanks. The results revealed a mixed correlation effect between these variables, particularly between TTCI ranking and tourism core resources. The lack of TTCI ranking correlation with the tourism core resources arises from the fact that the observed countries are in different economic development stages, which affect the TTCI ranking indirectly.

**Keywords** ASEAN · Core resources · Competitiveness · Performance · Tourism

## Background

Core resources and key attractors are the fundamental reasons why tourists choose one particular destination over another (Dwyer and Kim 2003; Crouch and Ritchie 1999). Natural and cultural resources provide the foundation for a memorable

M.H. Hanafiah (✉) · M.A. Hemdi
Faculty of Hotel and Tourism Management, Universiti Teknologi MARA,
Puncak Alam Campus, 42300 Shah Alam, Selangor, Malaysia
e-mail: hafizhanafiah@salam.uitm.edu.my

M.A. Hemdi
e-mail: moham984@salam.uitm.edu.my

I. Ahmad
Faculty of Business Management, Universiti Teknologi MARA,
40450 Shah Alam, Selangor, Malaysia
e-mail: irnmnaa05@yahoo.com



A. Saufi et al. (eds.), *Balancing Development and Sustainability in Tourism Destinations*, DOI 10.1007/978-981-10-1718-6_34

experience of a destination. Dwyer and Kim (2003) suggested that core resources can be divided into four elements; cultural resources, heritage resources, natural resources, and created resources. Natural resources can be considered as the most valuable resources for a tourism destination (Crouch and Ritchie 1999). Natural sites are restricted to those natural areas that furnish outstanding examples of Earth's record of life or its geologic processes, provide excellent examples of ongoing ecological and biological evolutionary processes, contain natural phenomena that are rare, unique, superlative, or of outstanding beauty, furnish habitats for rare or endangered animals or plants, or are sites of exceptional biodiversity.

There are three types of cultural sites; historical and archaeological sites, artistic and architectural features, and cultural attractors. These cultural heritage sites include hundreds of historic buildings and town sites, important archaeological sites, and works of monumental sculpture or painting. Further, Dwyer and Kim (2003) identified physiographic, climate, flora and fauna, scenery and other physical assets as natural resources. Physiographic and climate are vital tourism elements and dominate other factors in comparative advantage. Physiographic and environment include lakes, rivers, oceans, seas, forests, deserts, rural and urban areas, air quality, scenery, flora, fauna, and wildlife in which these natural resources being out of the control of destination managers. Tourism and natural environment are the two concepts that appear to have become closely related due to the increasing demand among tourists to visit natural environment (Hall and Page 2014; Orams 1996). Meanwhile, tourism product involving the interaction between human and animal appears to be increasing as people continue to seek authentic encounters with wild animals in their natural habitat. The wildlife-based tourism is one of the fastest growing forms of tourism which includes non-consumptive forms of tourism such as viewing, photography, and consumptive forms (Newsome et al. 2012).

As an important element of tourism product, the core resources and attraction sites mutually strengthen the attractiveness and competitiveness of a destination by creating distinctiveness in a crowded global travel marketplace. In line with that, Dwyer and Kim (2003) indicated that destination competitiveness highly depends on the value added to available core resources and attractions. The core resources and attractions are the fundamental reasons for travellers to visit a destination over another which in the end should offer memorable experiences to them. Tourism destinations with one or more favourable and unique physiographic resources tend to compete by employing this strength against others (Ritchie and Crouch 2003). On the other hand, culture and heritage sites are the influential dimensions of destination attractiveness and the primary reasons why visitors choose one particular destination over another (Ritchie and Crouch 2003).

There are much literature on the tourism industry and natural resource management (Ritchie and Crouch 2003). Previous researchers have identified that culture and heritage enhance the attractiveness of tourism destination (Ritchie and Zins 1978; Tang and Rochananond 1990) and provide a significant motivational force for the potential visitor which is represented by the historical and archaeological sites, artistic or architectural features, and cultural attractors (Murphy et al. 2000; Prentice 1993). Furthermore, culture and heritage attractions create a high proportion of cultural tourists influx to a destination (Murphy et al. 2000; Prentice 1993; Ritchie and Zins 1978). Hassan (2000) added that since tourism destinations involve multi-faceted components of natural or cultural resources, a systematic framework for destination planning and development is deemed necessary.

Based on the database from UNESCO World Heritage List, Europe has the most World Heritage natural sites with 54 sites, followed by the Asian continent with 47 sites, North America (34), South America (22), Africa (20), and Oceania (19). In more detail, Italy is the home to the greatest number of World Heritage Sites with 50 sites, followed by China (47), Spain (44), France (39), Germany (39), Mexico (32), and India (32). From the description, it shows that majority of the sites are situated in Europe. In contrast, South America continent has the lowest number of sites. Mexico leads the American continent by hosting 32 sites and is ranked the sixth in the world. On the other hand, the African continent, represented by 21 countries has 20 World Heritage natural sites and 66 World Heritage cultural sites, of which 35 are inscribed for their outstanding qualities. The African nation such as Ethiopia and Morocco are home to the most World Heritage sites with nine sites, followed by Tunisia and South Africa with eight and Egypt with seven sites. In 2014, there are 37 Heritage sites in Southeast Asia. Indonesia and Vietnam lead this list with eight inscribed sites; the Philippines have six sites while Thailand and Malaysia have five sites.

## The Travel and Tourism Competitiveness Index (TTCI)

The World Economic Forum (WEF) has released the Travel and Tourism (T&T) Competitiveness Index (TTCI) ranking by assessing the obstacles and drivers of T&T development. The T&T competitiveness report was developed by the WEF in collaboration with experts from the sector. The TTCI is a comprehensive framework that aims to measure the factors and policies that make it attractive to develop the tourism industry in different countries. From a methodological point of view, TTCI aims to evaluate the elements that ensure the development of tourism sector in different countries through three categories of variables that affect the global tourism competitiveness. The development of TTCI has been supported by major T&T organizations and corporations such as the United Nations World Tourism Organization (UNWTO), International Air Transport Association (IATA), World T&T Council (WTTC), and other tourism enterprises.

## ASEAN Travel and Tourism Competitiveness

In 1976, tourism cooperation was formalized among the Association of Southeast Asian Nations (ASEAN) member countries by setting up the ASEAN Committee on Trade and Tourism to deal exclusively with the sector. ASEAN was established in August 1967 in Bangkok, Thailand by five original member countries; Indonesia,

Malaysia, Philippines, Singapore, and Thailand. Brunei Darussalam joined in January 1984, Vietnam in July 1995, Laos and Myanmar in July 1997, and Cambodia in April 1999. The formation of ASEAN was primarily a response to the threat of communism in South East Asia during the 1960s (Timothy 2000). The primary aims of ASEAN were to ensure peace and stability in the region, to promote and facilitate intra-regional economic development, and to encourage social and cultural progress among members (Hussey 1991: 87, as cited in Timothy 2000).

The T&T industry is an increasing important driver for economic prosperity and social progress in ASEAN. It is estimated that T&T sector accounts for about 9 % of GDP and employment worldwide. For ASEAN, the area has been considered as critical for the creation of ASEAN Economic Community. The potential of T&T in the region is indeed enormous. The region is blessed with a wealth of natural and cultural heritage. It possesses a long tradition of tourism and is strategically located in the heart of Asia. The extraordinary diversity of ASEAN countries further enhances their attractiveness. Despite the many benefits of T&T, numerous obstacles continue to hinder its development.

The TTCI ranking analysis, however, reveals a very mixed picture in terms of the tourism performances of ASEAN countries individually. The TTCI assessment in 2013 showed ASEAN members drew a mixed picture of the region. Singapore (10th), Malaysia (34th), Thailand (43rd), Indonesia (70th), Brunei Darussalam (72nd), Vietnam (80th), the Philippines (82nd), and Cambodia (106th) span almost the entire ranking of 140 economies (Table 34.1).

Based on the TTCI report statistics, the high-rank positions were ruled by advanced economy nations since most of the elements that make up TTCI depend on directly and immediately on the degree of development of the national economy. At the rank of 34th, as a whole, Malaysia is the top-ranked developing country and poses strong performance. As for Singapore, the country is the best performing advanced economy outside Europe and North America. With only three countries featured in the top 50 of the TTCI, the tourism sector in ASEAN undoubtedly faces some challenges. Yet, one must recognize that there is a close relationship between the performance in the TTCI and prosperity, and the rankings continue to be led by advanced economy, of which Singapore is the only representative among ASEAN

**Table 34.1** Selected tourism and economic indicators for ASEAN countries (2013)

| Country | ASEAN rank | TTCI 2013 | ASEAN rank | TTCI 2011 |
|---|---|---|---|---|
| | | Rank | | Rank |
| Singapore | 1 | 10 | 1 | 10 |
| Malaysia | 2 | 34 | 2 | 35 |
| Thailand | 3 | 43 | 3 | 41 |
| Indonesia | 4 | 70 | 5 | 74 |
| Brunei | 5 | 72 | 4 | 67 |
| Vietnam | 6 | 80 | 6 | 80 |
| Philippines | 7 | 82 | 7 | 94 |
| Cambodia | 8 | 106 | 8 | 109 |

members. This is explained by the fact that advanced economy, on average, fare significantly better in most areas of the TTCI, including hard and soft infrastructures, regulatory framework, security, and public health. These factors benefit all sectors, boosting productivity, and economic development.

## Method

This paper used a systematic review approach in reviewing the past studies on tourism core resources and tourism performance, based on the assessment of precedent tourism destination competitiveness (TDC) research and TTCI ranking. The systematic review system guarantees methodological rigidness and eventually will reinforce the study's validity, and later provides knowledge-based evidence that is useful in solving pragmatic management issues. First, this study identified the relevant TDC studies, then focused on the assessment of the core resources and finally analysed the direct effect on tourism performance and TTCI ranking. The study examined eight ASEAN countries namely Brunei, Cambodia, Indonesia, Malaysia, Philippines, Singapore, Thailand, and Vietnam.

Next, the proposition of constructive influence of core resources on tourism performance and TTCI ranking was tested on the basis of WEF and World Bank data on the number of World Heritage natural sites (2013), number of species available (2013), number of World Heritage cultural sites (2013), foreign tourist arrivals (2014), and TTCI ranking (2013). The number of visitors was used to describe the demand conditions of a destination in many studies (Song and Witt 2000). This is supported by Witt and Witt (1995) who found that about 75 out of 118 studies on the demand for tourism used tourist arrivals as a proxy for demand. Similarly, Li (2004) found that 37 out of 45 published articles chose tourist arrivals as the dependent variable (Table 34.2).

**Table 34.2** Selected tourism core resources, performance, and ranking indicators for ASEAN countries

| Country | Number of World Heritage natural sites | Number of species | Number of World Heritage cultural sites | Tourist arrivals (in 1000 million) | TTCI ranking ($n$ = 139) |
|---|---|---|---|---|---|
| Brunei | 0 | 602 | 0 | 209,000 | 20 |
| Cambodia | 0 | 722 | 4 | 3,584,000 | 101 |
| Indonesia | 4 | 2603 | 10 | 8,044,000 | 33 |
| Malaysia | 2 | 1248 | 3 | 25,033,000 | 49 |
| Philippines | 2 | 858 | 5 | 4,273,000 | 66 |
| Singapore | 0 | 457 | 0 | 11,098,000 | 24 |
| Thailand | 2 | 1377 | 3 | 22,354,000 | 85 |
| Vietnam | 2 | 1263 | 11 | 6,848,000 | 45 |

## Findings

The focuses of this paper were macro aspects of tourism core resources, tourism performance, and TTCI ranking. In particular, this paper intended to explore if the tourism core resources promote tourism performance and justify the country ranking in the TTCI. First, the correlation of tourism core resources and tourism performance was tested. The correlation coefficient (value between −1 and +1) states how strong the two variables are related to each other. The results are shown in Table 34.3.

Based on Table 34.3, it is shown that there was a weak uphill (positive) linear relationship between tourist arrivals with the number of world heritage sites and some species. Meanwhile, the analysis results showed a weak downhill (negative) linear relationship between the number of world cultural sites and tourist arrivals. However, on the other hand, the output revealed that there was no linear relationship between TTCI ranking and the number of world heritage natural sites, the number of species, and the number of world heritage cultural sites. Thus, it is clear that the tourism core resources are not the primary attributes affecting the TTCI ranking including another dimension such as destination management, supporting infrastructures, and created resources. However, it is important to note that when all the complexities of destination choice are stripped away, it is an essential core resources and attractors that underlie the fundamental desire to travel to the given destinations.

The result of this study supported the TTCI shortcomings identified by many researchers (Wu et al. 2012). Their criticism regarding TTCI was based on the destination's specific economic conditions rather than the destination competitiveness in offering tourism product. The high ranked destinations in the TTCI were high-end countries such as Singapore and Malaysia, and the developing or less-developed countries (Indonesia and Thailand) were ranked deep down in spite

**Table 34.3** Correlation output between the number of world heritage natural sites, number of species, number of world heritage cultural sites, foreign tourist arrivals, and TTCI ranking

| | Number of World Heritage natural sites | Number of species | Number of World Heritage cultural sites | TTCI ranking | Tourism performance (arrivals) |
|---|---|---|---|---|---|
| Number of World Heritage natural sites | 1 | | | | |
| Number of species | 0.922 | 1 | | | |
| Number of World Heritage cultural sites | 0.738 | 0.711 | 1 | | |
| TTCI ranking | −0.043 | −0.067 | 0.068 | 1 | |
| Tourism performance (arrivals) | 0.304 | 0.237 | −0.127 | 0.169 | 1 |

of their lucrative natural or cultural resources (WEF 2013). As the TTCI report treats all sub-indices/pillars with equal weights, the continuous disparity between the current determinants of tourism competitiveness concerning different economic level may cause a deviation in the indexing result.

## Discussion

The effect of tourism core resources and tourism performance in ASEAN countries was not explained by the TTCI ranking. This is inline with m the arguments by many researchers on the reliability of TTCI ranking in explaining tourism destination competitiveness. Majority of them argued on the strengths and weaknesses of the TTCI, reliability and validity of the index, and suggested limitations based on the fundamental competitiveness theory (Ivanov and Webster 2013; Mazanec and Ring 2011; Wu et al. 2012). The reason behind this impending issues are the T&T Competitiveness Report treats all sub-indices/pillars with equal weights. Thus the TTCI ranking does not represent a country's actual tourism performance.

Furthermore, TTCI seems more as a collection of data rather than a model that shows a clear testable relationship among variables. In contrast, the highest rank destinations in the TTCI were high-end countries, and developing or less-developed countries were listed deep down in spite of their lucrative natural or cultural resources (Mazanec and Ring 2011). Furthermore, the TTCI only assumes the influence of the existing situation, and thus fails to address the effect of futuristic focus of investment decisions. It also fails to delineate between demand and supply focus of competitiveness and assumes a common perspective between domestic and international market.

## Conclusion

The core resources and attractors are the fundamental factors for visitors to visit a destination. The destination's physiographic; culture and history provide the necessary and powerful attraction forces to prospective visitors. Further, creating memorable experiences create competitive advantages and customer experiences, which are in line with the new economic growth trait. However, while inherited resources are paramount to attract visitors, their role in terms of long-term prosperity may be questionable. Destinations that rely solely on their resources may not be able to upgrade real underlying competitiveness and long-term prosperity.

For ASEAN destinations to be successful in the competition for development, they must assess the available internal resources and then convert its qualities into values. On the other hand, ASEAN's tourism competitiveness appears to have much growth prospect. As an example, Malaysia's high competitive score was not achieved by chance, but by formulating creative strategies to sustain destinations' competitiveness in the marketplace over time. In the current competitive tourism market, destination competitiveness is gradually seen as a critical influence on the performance of tourism destinations. Thus, to ensure the long-term success of destination competitiveness in the tourism industry, the policy-making should be decided based on the principles of every essential pillar contributing towards destination competitiveness.

**Acknowledgments** This work was funded by the Ministry of Education Malaysia, Universiti Teknologi MARA, Malaysia (UiTM) under Fundamental Research Grant Scheme: 600-RMI/FRGS 5/3 (129/2014).

## References

Crouch GI, Ritchie JB (1999) Tourism, competitiveness, and societal prosperity. J Bus Res 44 (3):137–152
Dwyer L, Kim C (2003) Destination competitiveness: determinants and indicators. Curr Issues Tour 6(5):369–414. doi:10.1080/13683500308667962
Hall CM, Page SJ (2014) The geography of tourism and recreation: environment, place and space 4th edition: environment place and space. Routledge, Abingdon
Hassan SS (2000) Determinants of market competitiveness in an environmentally sustainable tourism industry. J Travel Res 38(3):239–245
Hussey A (1991) Regional development and cooperation through ASEAN. Geographical Review. 87–98
Ivanov S, Webster C (2013) Globalisation as a driver of destination competitiveness. Ann Tour Res 43:628–633. doi:10.1016/j.annals.2013.07.010
Li Y (2004) Exploring community tourism in China: The case of Nanshan cultural tourism zone. J Sustainable Tour 12(3):175–193
Mazanec JA, Ring A (2011) Tourism destination competitiveness: second thoughts on the World Economic Forum reports. Tour Econ 17(4):725–751. doi:10.5367/te.2011.0065
Murphy P, Pritchard MP, Smith B (2000) The destination product and its impact on traveller perceptions. Tour Manage 21(1):43–52
Newsome D, Moore SA, Dowling RK (2012). Natural area tourism: ecology, impacts and management, vol 58. Channel View Publications, Bristol
Orams MB (1996) A conceptual model of tourist-wildlife interaction: the case for education as a management strategy. Aust Geogr 27(1):39–51
Prentice R (1993) Tourism and heritage attractions. Routledge, Abingdon
Ritchie JRB, Crouch GI (2003) The competitive destination: A sustainable tourism perspective. Cabi, Wallingford, Oxfordshire
Ritchie JRB, Zins M (1978) Culture as determinant of the attractiveness of a tourism region. Ann Tour Res 5(2):252–267
Song H, Witt SF (2000) Tourism demand modelling and forecasting: modern econometric approaches. Routledge
Tang J, Rochananond N (1990) Attractiveness as a tourist destination: a comparative study of Thailand and selected countries. Socio-Economic Plann Sci 24(3):229–236
WEF (2013) The travel and tourism competitiveness report 2013. World Economic Forum, Cologny
Witt SF, Witt CA (1995) Forecasting tourism demand: A review of empirical research. Int J Forecast 11(3):447–475
Wu W-W, Lan LW, Lee Y-T (2012) Critiquing the World Economic Forum's concept of destination competitiveness: a further analysis. Tour Manage Perspect 4:198–206

# Chapter 35
# An Analysis of Travel and Tourism Competitiveness Versus Socio-Political Conditions at the National Level: An Indonesian Case Study

**Gandhi Pawitan, Orpha Jane, Nia Juliawati, Sanerya Hendrawan and Aknolt Kristian Pakpahan**

**Abstract** Recent studies have shown the travel and tourism industry to have positive impacts on the economic development of a country and its global competitiveness. In this research, we have analyzed how the ecosystem of the travel and tourism industry extends beyond the 14 pillars of industrial competitiveness to also include social and political factors. The analysis is based on data at the national level, with a case study discussion of Indonesia's tourism sector. Data which we have obtained illustrate how, the development of Indonesia's travel and tourism industry remains low and lags far behind neighboring ASEAN member states such as Malaysia, Thailand, and Singapore. At present, travel and tourism—from domestic and overseas tourism—contribute 3 % to Indonesia's gross domestic product (GDP). Contributing factors in the country's travel and tourism sector's slow development include unfinished consolidation in the management procedures where by involving a range of factors including bureaucracy involving local governments, the community as well as the central government; the central government has the authority to manage responses for situations affected by macroeconomic, social, and political indicators. We apply the exploratory approach to analyze the correlation between the travel and tourism's competitiveness index and the six

G. Pawitan (✉) · O. Jane · N. Juliawati · S. Hendrawan · A.K. Pakpahan
Department of Business Administration, Faculty of Social and Political Sciences, Universitas Katolik Parahyangan, Bandung, Indonesia
e-mail: gandhi_p@unpar.ac.id

O. Jane
e-mail: janepattiasina@gmail.com

N. Juliawati
e-mail: niajuliawati@gmail.com

S. Hendrawan
e-mail: sanerya@gmail.com

A.K. Pakpahan
e-mail: pakpahan@unpar.ac.id



A. Saufi et al. (eds.), *Balancing Development and Sustainability in Tourism Destinations*, DOI 10.1007/978-981-10-1718-6_35

world governance indicators defined by the World Bank through the provision of a scatter plot and factor analysis. Sources for the secondary data include Indonesia's Central Bureau of Statistics (BPS—*Badan Pusat Statistik Indonesia*), the World Economic Forum (WEF), and the World Bank (WB).The completion of the research proves how there is a positive correlation between the travel and tourism competitiveness index and the six world governance indicators. The effectiveness of governance has the highest correlation to the competitiveness index followed, respectively, by regulatory quality, rule of law, control of corruption, voice and accountability as well as the political stability/absence of violence (the factor with the least effect with the index). Moreover, this paper also highlights the importance of taking into account sociopolitical factors in order to better understand the 14 pillars of the competitiveness index; these factors pinpoint common factors of the 14 pillars. Numerous factors that must be prioritized within sociopolitical indicators include controlling the widespread act of corruption, rule of law, government effectiveness, and regulatory quality.

**Keywords** Travel and tourism competitiveness index · World governance indicators · Fourteen pillars of travel and tourism competitiveness · Exploratory factor analysis

## Introduction

Indonesia ranks behind Malaysia and Thailand in the performance of its travel and tourism industry for wide-ranging contributing factors. According to Indonesia's Central Bureau of Statistics (BPS—*Badan Pusat Statistik*), the sector contributes 3 %to the country's GDP (based on domestic and foreign tourism), whereas the same sectors in Singapore, Thailand, and Malaysia's account more than 7 % on foreign tourism alone Fig. 1 illustrates how Indonesia has a low performance in terms of inbound tourism compared to other ASEAN countries, and China. Based on a World Travel and Tourism Council (WTTC) Report from 2014, global travel and tourism was the fourth-largest contributor to the world's economy after retail, financial services, and mining. The sector contributed an estimated 9.8 % of global GDP in 2014 9.4 % of global employment either directly or indirectly (World Travel & Tourism Council 2015).

In a measure to develop and strengthen the travel and tourism sector, the Indonesian government recently launched a range of strategies, policies, and programs as specified in the National Tourism Development Master Plan for the Year 2010–2025. The master plan itself includes several targets, such as national tourism development areas, strategic areas of national tourism, and a complete guide on tourist attractions (made available in print). Despite their rampant efforts, the plan does not guarantee the delivery of successful results.

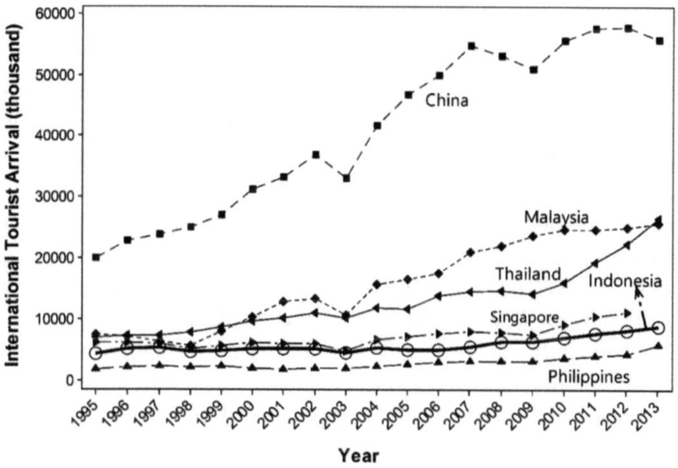


**Fig. 1** International tourist arrival from 1995—2013 ASEAN and China (*Data source* WTTC 2015)

As such, the travel and tourism competitiveness index is an ideal source to further understand why the deliverance of success in the aforementioned regard cannot be guaranteed. According to the ASEAN Travel & Tourism Competitiveness Report—which surveys 140 countries on its competitiveness index (TTCI)—Indonesia's travel and tourism sectors ranked 74th in 2011 and 70th in 2013, whereas ASEAN member states, namely Thailand ranked 41st in 2011 (43rd in 2013) following with Malaysia 35th in 2011 (34th in 2013) and Singapore ranks 10th in 2011 and 2013.

Hampton and Jeyacheya (2015) hosted a discussion examining political and economic factors on small island tourism developments. Using Gili Trawangan–Nusa Tenggara Barat Province as a region of study, the authors discovered several interesting phenomena of how different types of power impact community development, in addition to showcasing the importance of international tourism at the community level. Using sources of economic and political indicators at the macro level, it can be said that social and political factors can impact a country's competitiveness in travel and tourism.

A review of relevant literary sources on the definition of the tourism industry and how it links to economic development has been provided in the following section. In addition, an in-depth analysis on the relationship between travel and tourism competitiveness and sociopolitical dimensions related to economic development is a supplementary information entailed as part of this research. Other sections include details on the research method, sources of data, and will lead to the result and discussion section. This research concludes on presenting findings that have the biggest impact on sociopolitical factors which are necessary for the development of the Indonesian travel and tourism industry.

## Literary Review

The travel and tourism industry can be defined from a supply-side perspective, as explained by Smith (1988) "the aggregate of all businesses that directly provide goods or services to facilitate business, pleasure, and leisure activities away from the home environment." According to Hall et al. (2014), the definition of tourism industry may go beyond the economic aspect, and its needed to take into account the relationship between leisure, recreation, and tourism, as well as other social practices, and behavior (see page 5). The development of the travel and tourism industry itself involves several key stakeholders at the local, regional, national, or global levels (Milne and Ateljevic 2001, p. 371). Analyzing the relationship between travel and tourism competitiveness and its ecosystems are essential in comprehending the wider scope of social, environmental, and political dimensions industry in addition to understanding the suitable role of the state in travel and tourism.

Crotti and Misrahi (2015, p. 3) stated that the travel and tourism sectors contribute a significant amount to the global economy, estimated at approximately 9 % of the global GDP and employment figures with the further increase due to the higher number of international travelers. The World Travel and Tourism Council estimates that the travel and tourism sectors will grow to 4 % annually (World Travel & Tourism Council 2015). Moreover, Crotti and Misrahi (2015) further stated that "understanding the nature and extent of the sector's resilience to shocks is important, as a strong T&T sector is critical for job creation, economic growth and development in advanced and developing economies alike."

Kaufmann et al. (2010) discussed the methodology and analytical issues of the world governance indicators (WGI). The WGI aims to capture the perception of governance from respondents across the globe particularly those in non-governmental organizations, commercial business sectors, and the public sector. The World Governance Indicators—developed by the World Bank Development Research Group—measure perception of governance in six dimensions, namely Voice and Accountability, Political Stability and Absence of Violence/terrorism, Government Effectiveness, Regulatory Quality, Rule of Law, and Control of Corruption; a detailed definition of the dimensions can be found in Kaufmann et al. (2010, p. 4).

### *Overview of the Indonesia's Travel and Tourism Industry*

Based on BPS data collected from 2011–2013, international tourist arrivals occurs mainly in Indonesian airport locations including Jakarta/Cengkareng—Soekarno Hatta, Bali—Ngurah Rai, Medan—Polonia, Batam, Surabaya—Juanda, and Bandung—Husein Sastranegara. As such, the culinary and fashion industries have incurred developments as well as better positioning tourism destinations namely

Jakarta, Bandung, Surabaya, Bali, and Medan. From the tourism figures, individuals from ASEAN countries contribute the most to Indonesia's tourist arrivals. For example, indicate that the number of Malaysian tourists increased from 475,163 tourists to 1,335,531 from 2002–2012 (BPS); an increase of more than 200 % of Malaysian tourists entering Indonesia. Tourists from Singapore take the second position while European and American tourists in total are positioned in third and fourth places, respectively.

Tourists visiting Indonesia come from various regions including Europe, the United States, Africa, Middle East, and Asia (mainly China and ASEAN countries) and the figures have only increased from 2002–2012 from those regions. ASEAN tourists account for 3.37 million individuals in 2012 encompassing the largest group of travelers to Indonesia. Tourists from Europe comprise of 1.11 million arrivals followed with 312,210 arrivals from the United States, and 148,800 arrivals from the Middle East (BPS 2013).

Malaysian and Singaporean travelers opt for activities such as shopping, dining, sightseeing, and leisure as main motivations to visit Indonesia. The BPS's data indicates that dining and shopping are leading factors that result in said visitors' return to Indonesia. On average, tourists are willing to spend up to USD 1300 during their holiday, and show little to hesitance on additional spending for dining, exotic destination visits, and experiencing Indonesia's distinct culture. European and American tourists cite Indonesia's cultural diversity and experiencing Asian culture as motivations for visiting Indonesia.

## *Issues in Developing the Indonesia's Travel and Tourism Industry*

The development of Indonesia's travel and tourism industry is influenced by several factors whereby internal factors are driven by increased purchasing power and external factors are caused by better security and safety in tourist destination locations. For example, Malaysia as a country with consistent patterns of healthy economic growth and whose per capita income is approximately USD 10,000 is considered as one of the countries with the largest contribution to Indonesia's inbound tourism. Hampton and Jeyacheya (2015) indicated that the travel and tourism development in Indonesia must be initiated by the central government, however, there is spread out tourism among different parts of the archipelago where tourism offices are managed at the city or municipality level. As a result from the lack of strong coordination, international tourism has become underdeveloped and incurs a small number of visitors annually. Recently, the central government has issued the National Tourism Development Master Plan 2010–2025 as a measure to diversify the scope of tourism made available in the country.

Indonesia has immense potential in the development of its travel and tourism industry, especially in social, cultural, environmental, and culinary areas.

According to BPS data, there is a total of 88 national tourism strategic areas and 222 national tourism development areas. The natural landscape, warm hospitality, and a wealth of cultural diversity are the main drivers behind Indonesia's tourism industry yet several potential has yet to be explored and developed. Direct contribution from the tourism sector contributed 4 % to the GDP in 2012 while the indirect contribution of the tourism sector contributed 8 %. The low figures in contribution are caused by various factors including poor infrastructure, lack of connectivity across the archipelago, unattractive tourism package, lowly marketed tours promotion, immigration services, and inflexible regulations.

Infrastructure is a key issue to support the growing demand for tourism packages in Indonesia, however, continues to pose as a setback for the country. Inspite of witnessing an annual increase in the number of incoming air passengers, the withstanding capacity of the Soekarno-Hatta and Husein Sastranegara airports are limited. Other problems related to infrastructure additionally include land and port transportation and lack of ICT integration among others. According to the Asian Development Bank (2015) Indonesian government spending is 16.8 % of the country's GDP in 2014, lower compared with other countries such as Laos PDR (26.9 %), Malaysia (23.3 %), Myanmar (33.0 %), China (23.8 %), and Vietnam (25.6 %) over the same period.

Government policies and the implementation of the regulations are essential in bolstering the attractiveness of tourism in Indonesia. The indicators associated with this aspect are the ease in investment and foreign ownership, property rights protection, ease of starting a business, and visa requirements. Analyzing the extent of Policies and Rules related to the policy framework are necessary to build an in-depth understanding of national tourism developments.

Environmental sustainability is an implemented measure in order for the government to control regulations in support of its surroundings while examining the extent of how the regulations are enforced. Forms of environmental sustainability comprise of tourism policies which take into account, there carbon dioxide levels and the number of rare or endangered species remaining.

## Data and Methods

This research uses obtained data from Indonesia's Central Bureau of Statistics (BPS Indonesia), the World Bank (WB), and the World Economic Forum. Statistics on tourism are primarily from BPS while sources from the World Economic Forum are used to define the travel and tourism competitiveness index (TTCI) as well as the 14 pillars of the travel and tourism competitiveness.

The World Bank produced the world governance indicators (WGI) which measures governance perception in six dimensions: Voice and Accountability (VA), Political Stability and Absence of Violence/terrorism (PV), Government

Effectiveness (GE), Regulatory Quality (RQ), Rule of Law (RL), and Control of Corruption (CC). The survey was conducted in 1996 amongst 200 countries and involves hundreds of variables (Kaufmann et al. 2010). For this research, data from 2013 was used to support the research finding.

Crotti and Misrahi (2015) divided the TTCI up into 14 pillars and uses a 7-point scale of measurement (1 as the worst, 7 as the best). The pillars are detailed below:

1) Business Environment
2) Safety and Security
3) Health and Hygiene
4) Human Resources and Labor Market
5) ICT Readiness
6) Prioritization of Travel and Tourism
7) International Openness
8) Price Competitiveness
9) Environmental Sustainability
10) Air Transport Infrastructure
11) Ground and Port Infrastructure
12) Tourist Service Infrastructure
13) Natural Resources
14) Cultural Resources and Business Travel.

Due to the exploratory approach of the research, scattered plots are provided as a visual tool to identify the relationship among numerous variables. In-depth exploration is conducted by applying factor analysis followed by multiple regression of the factors to decide which factor has the greatest impact on the TTCI.

## Result and Discussion

Indonesia's travel and tourism performance is reflected in the TTCI pillars, as represented in Fig. 2. Indonesia has better performance in pillar 6—Prioritization of Travel and Tourism, pillar 8—Price Competitiveness, pillar 13—Natural Resources, and pillar 14—Cultural Resources and Business Travel. The worst performance can be found in pillar 2—Safety and Security, pillar 3—Health and Hygiene, pillar 9—Environmental Sustainability, pillar 11–Ground and Port Infrastructure, and pillar 12—Tourist Service Infrastructure.

Meanwhile, Fig. 3 shows the distribution of the WGI among countries and illustrates Indonesia's positioning compared to ASEAN member nations and China. Results indicate that Indonesia still ranks low in all governance indicators with the worst indicators in Government Effectiveness (GE), Rule of Law (RL), and Control of Corruption (CC). Vietnam and China encounter similar poor performance in similar areas.

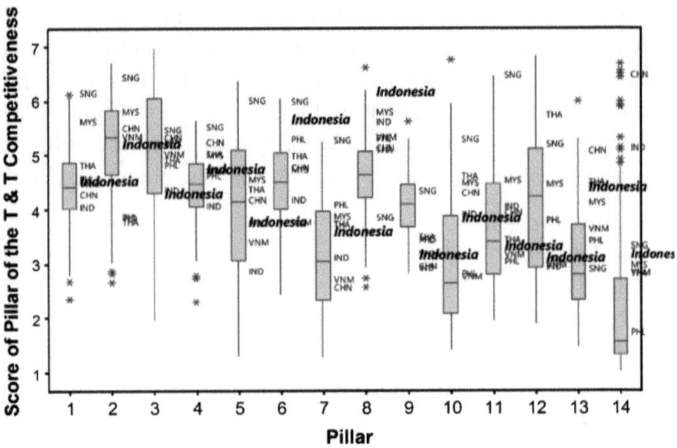


**Fig. 2** The score of Travel & Tourism Competitiveness Index by pillar: Indonesia compared with other countries (*Data source* WEF 2015)

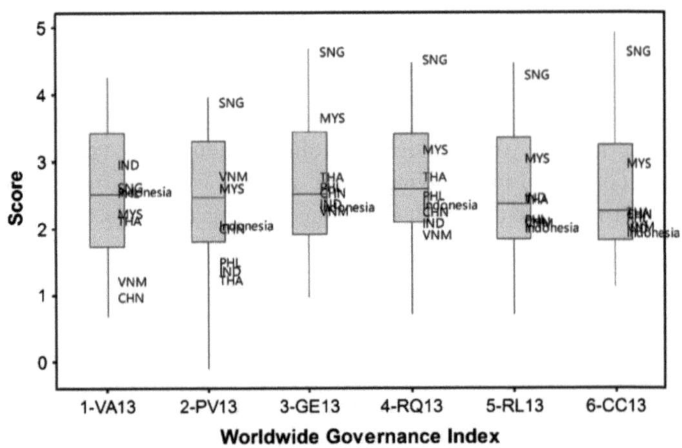


**Fig. 3** Distribution of worldwide governance indicators: Indonesia compared with other countries (*Data source* WB, 2013)

The Voice and Accountability (VA) indicator is used to understand factors that will lead citizens of a country to participate in selecting their government and their priority for freedom of expression, freedom of association as well as freedom of speech. Mietzner (2010) described that since Indonesia's political shift to a democratic system, the country has experienced positive economic growth since 2009. People-based responses reflect optimism to the economic conditions given the Indonesian government's support through measures comprising of direct cash transfer, health support, and school operational support.

## *Relationship Between TTCI and World Governance Indicators*

Figure 4 shows a positive and quadratic relationship and indicates that a better performance in VA indicators can increase the TTCI. Statistically, the magnitude of the correlation indicates a substantial degree of correlation meaning that, if changes/improvements are made to the policy in terms of the freedom voice and accountability measurements, it tends to increase the ranking of tourism competitiveness. Specifically, the policy is used to inform about government policy changes affecting the industry as clear and transparent to companies; giving freedom to the newspapers and the mass media to publish news that is accountable; ensuring equal treatment for businesses (in setting policies and contracts). The increased of VA index (e.g., more freedom expression) has a positive correlation to the increase of competitiveness of the travel and tourism. Policies that can foster a positive perception of Voice and Accountability being crucial for boosting travel and tourism competitiveness ranking.

Improvement of the PV has a positive correlation to the increase of TTCI (see Fig. 5). Stable political conditions (absence of violence) are perceived to improve a country's competitiveness ranking. Policies in relations to security such as law enforcement, protection of business certainty, and the creation of a safe environment whereby guaranteeing tourist's safety contribute to a stable political condition also considered to improve the competitiveness ranking.

Figure 6 indicates that improving the GE has a positive correlation to the increase of TTCI and that the public sector's efficiency is believed to have the biggest capability of increasing the competitiveness ranking. Other examples to

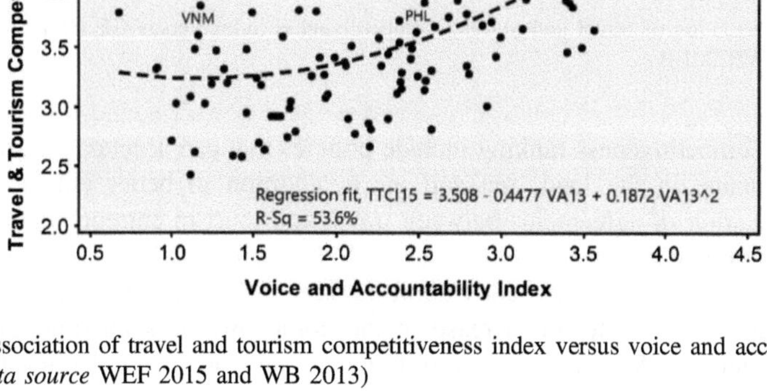


**Fig. 4** Association of travel and tourism competitiveness index versus voice and accountability index (*Data source* WEF 2015 and WB 2013)

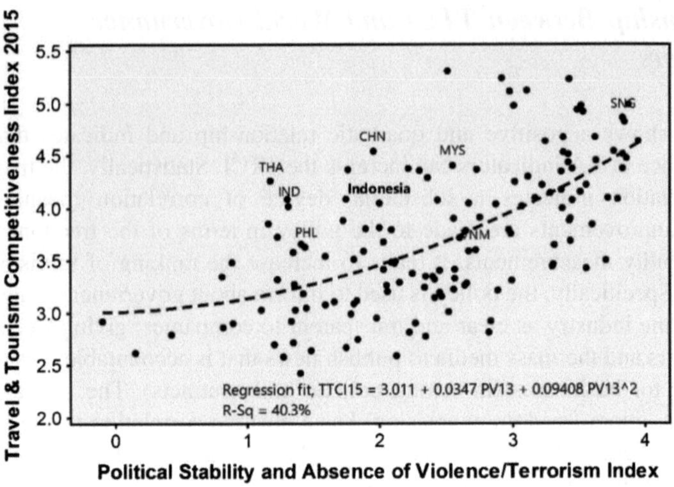


**Fig. 5** Association of travel and tourism competitiveness index versus political stability and absence of violence/terrorism index (*Data source* WEF 2015 and WB 2013)

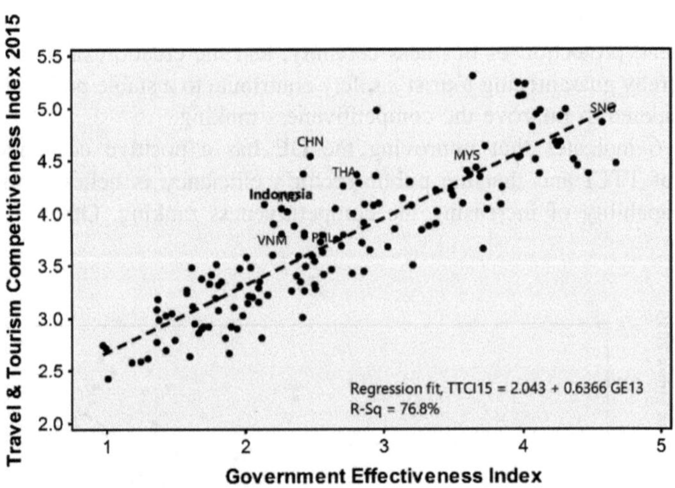


**Fig. 6** Association of travel and tourism competitiveness index versus GE (*Data source* WEF 2015 and WB 2013)

improve competitiveness ranking include policies that can accelerate infrastructure improvements on the land, sea and air in addition to better efficiency in the implementation of agreements between travel and tourism entrepreneurs and relevant government bodies.

Figure 7 shows a positive correlation between TTCI and RQ. Higher quality of regulations will lead to an increase in the competitiveness ranking. The policy change refers to, for example, simplifying administrative regulations through a

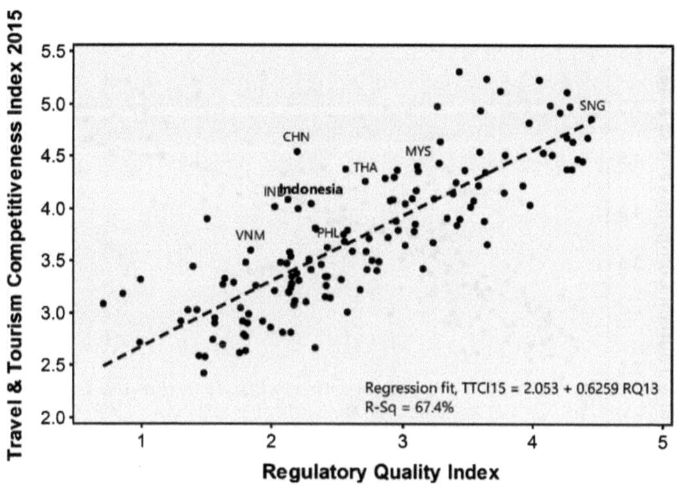


**Fig. 7** Association of travel and tourism competitiveness index versus RQ index (*Data source* WEF 2015 and WB 2013)

one-stop service and reinforcing regulations necessary for establishing environmental preservation.

Figure 8 indicates that the improvement of CC has a positive correlation with the increase of TTCI. A government that is able to demonstrate tighter control over corruption activity is believed to be capable of increasing the competitiveness ranking. Along with other factors, the effort to control corruption would increase travel and tourism competitiveness in an effort to build good governance—mostly flexibility in the attainment of trade licenses, tax payments, and government employment contracts.

Figure 9 shows that improvement in RL has a positive correlation to the increase of the travel and tourism competitiveness ranking. Government regulations supporting economic activities will result in the improvement of the rankings.

## *Factors of the Travel and Tourism Competitiveness Index*

The following section will discuss the factor analysis result of the TTCI pillars. The analysis of the 14 pillars is shown in Tables 1 and 2. Table 1 shows that the 14 pillars can be extracted into three principal components with 72.5 % total variance can be explained. Table 2 shows that the first component can be recognized as travel and tourism governance, the second component is the resources to attract tourism, and the final component relates to environmental sustainability and price competitiveness.

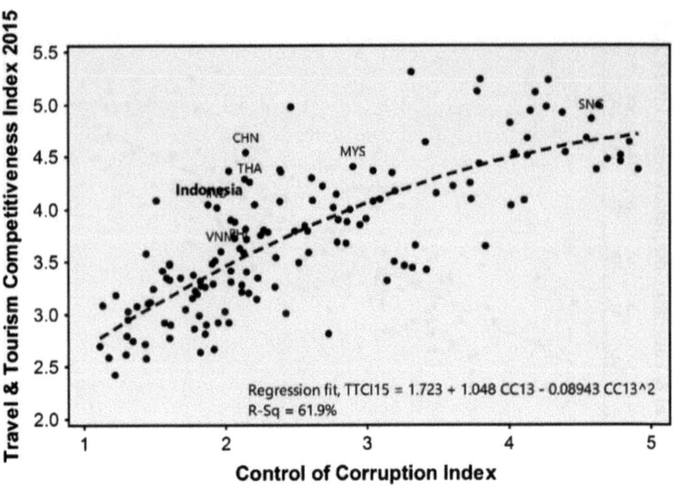


**Fig. 8** Association of travel and tourism competitiveness index versus CC index (*Data source* WEF 2015 and WB 2013)

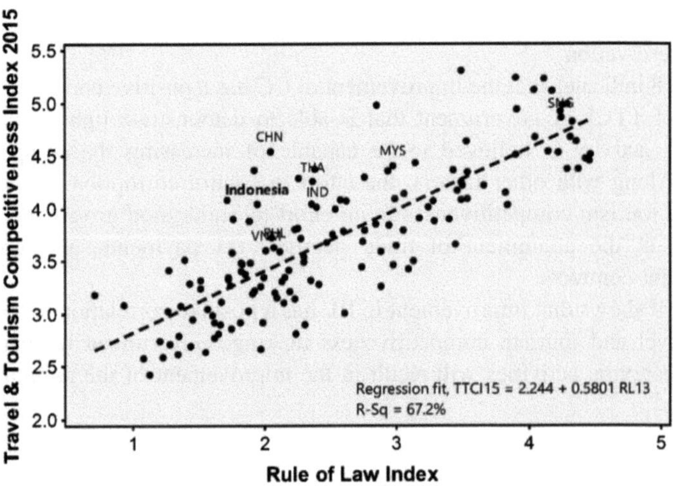


**Fig. 9** Association of travel and tourism competitiveness index versus RL index (*Data source* WEF 2015 and WB 2013)

## *The Effect of World Governance Indicators into TTCI*

The WGI takes into account social and political factors of a country. They play an important role in guiding a policymaker in the betterment of the travel and tourism competitiveness ranking. Exploring their effect on the TTCI, the factor analysis is applied to the 14 pillars combined with the WGI. The results are shown in Table 3

**Table 1** Total variance explained by extraction of the 14 pillars of the TTCI

| Component | Initial eigenvalues | | | Extraction sums of squared loadings | | |
|---|---|---|---|---|---|---|
| | Total | % of Variance | Cumulative % | Total | % of Variance | Cumulative % |
| 1 | 7.512 | 53.658 | 53.658 | 7.512 | 53.658 | 53.658 |
| 2 | 1.625 | 11.608 | 65.266 | 1.625 | 11.608 | 65.266 |
| 3 | 1.017 | 7.261 | 72.527 | 1.017 | 7.261 | 72.527 |
| 4 | 0.754 | 5.389 | 77.916 | | | |
| 5 | 0.721 | 5.150 | 83.066 | | | |
| 6 | 0.493 | 3.523 | 86.590 | | | |
| 7 | 0.441 | 3.150 | 89.740 | | | |
| 8 | 0.399 | 2.850 | 92.590 | | | |
| 9 | 0.312 | 2.232 | 94.822 | | | |
| 10 | 0.222 | 1.585 | 96.407 | | | |
| 11 | 0.177 | 1.267 | 97.674 | | | |
| 12 | 0.129 | 0.923 | 98.597 | | | |
| 13 | 0.117 | 0.838 | 99.435 | | | |
| 14 | 0.079 | 0.565 | 100.000 | | | |

Extraction Method: Principal Component Analysis
*Data source* WEF 2015

**Table 2** Rotated component matrix of the 14 pillars of TTCI

| | Component | | |
|---|---|---|---|
| | 1 | 2 | 3 |
| ICT readiness, 1–7 (best) | 0.872 | | |
| Human resources and labor market, 1–7 (best) | 0.852 | | |
| Health and hygiene, 1–7 (best) | 0.796 | | |
| Ground and port infrastructure, 1–7 (best) | 0.792 | | |
| Business environment, 1–7 (best) | 0.776 | | |
| Tourist service infrastructure, 1–7 (best) | 0.719 | | |
| Air transport infrastructure, 1–7 (best) | 0.701 | | |
| Prioritization of Travel & Tourism, 1–7 (best) | 0.638 | | |
| Safety and security, 1–7 (best) | 0.617 | | |
| International Openness, 1–7 (best) | 0.483 | | |
| Natural resources, 1–7 (best) | | 0.876 | |
| Cultural resources and business travel, 1–7 (best) | | 0.845 | |
| Environmental sustainability, 1–7 (best) | | | 0.848 |
| Price competitiveness, 1–7 (best) | | | −0.780 |

Extraction Method: Principal Component Analysis
Rotation Method: Varimax with Kaiser Normalization
*Data source* WEF 2015

**Table 3** Rotated component matrix TTCI + WGI

| | Component | | | | |
|---|---|---|---|---|---|
| | 1 | 2 | 3 | 4 | 5 |
| Business environment, 1–7 (best) | 0.903 | | | | |
| **Control of Corruption (CC)** | 0.685 | | | | |
| **Rule of Law (RL)** | 0.678 | | | | |
| **Government Effectiveness (GE)** | 0.650 | | | | |
| Safety and security, 1–7 (best) | 0.647 | | | | |
| Ground and port infrastructure, 1–7 (best) | 0.624 | | | | |
| Air transport infrastructure, 1–7 (best) | 0.599 | | | | |
| Human resources and labor market, 1–7 (best) | 0.595 | | | | |
| **Regulatory Quality (RQ)** | 0.577 | | | | |
| *Environmental sustainability, 1–7 (best)* | | *0.782* | | | |
| *Price competitiveness, 1–7 (best)* | | *−0.769* | | | |
| ***Voice and Accountability (VA)*** | | ***0.702*** | | | |
| ***Political Stability and Absence of Violence/Terrorism (PV)*** | | ***0.605*** | | | |
| Health and hygiene, 1–7 (best) | | | 0.884 | | |
| ICT readiness, 1–7 (best) | | | 0.622 | | |
| Tourist service infrastructure, 1–7 (best) | | | 0.539 | | |
| *Prioritization of Travel & Tourism, 1–7 (best)* | | | | *0.751* | |
| *International Openness, 1–7 (best)* | | | | *0.676* | |
| Cultural resources and business travel, 1–7 (best) | | | | | 0.875 |
| Natural resources, 1–7 (best) | | | | | 0.861 |

Extraction Method: Principal Component Analysis
Rotation Method: Varimax with Kaiser Normalization
Data source: WEF 2015 and WB 2013

indicating that the WGI play an important role in the first component of the TTCI, which is the travel and tourism governance. The WGI in the first component is CC, RL, GE, and RQ. Meanwhile, the VA and PV affect the second component of TTCI which deals with environmental sustainability and price competitiveness.

Given the shown interest in uncovering which components have the most impact on the TTCI, the researcher applied a multiple regression analysis to the factors considered as independent variables and TTCI as the dependent variable. The results are shown in Tables 4, 5 and 6.

The regression analysis shows that based on standardized coefficients beta, the fifth component (resources for tourism attraction) have the greatest impact on the TTCI, followed by the first component (business environment), and the third component (infrastructure for health and hygiene, ICT readiness, and tourist services).

**Table 4** Model summary of regression model TTCI + factors

| Model | R | R Square | Adjusted R square | Std. error of the estimate |
|---|---|---|---|---|
| 1 | 0.991[a] | 0.981 | 0.981 | 0.09510 |

[a]Predictors: (Constant), REGR factor score of component 5, REGR factor score of component 4, REGR factor score of component 3, REGR factor score of component 2, REGR factor score of component 1
*Data source* WEF 2015 and WB 2013

**Table 5** ANOVAof regression model TTCI + factors[a]

| Model | | Sum of squares | df | Mean square | F | Sig. |
|---|---|---|---|---|---|---|
| 1 | Regression | 63.734 | 5 | 12.747 | 1409.506 | 0.000[b] |
| | Residual | 1.221 | 135 | 0.009 | | |
| | Total | 64.955 | 140 | | | |

[a]Dependent Variable: travel and tourism competitive index
[b]Predictors: (Constant), REGR factor score of component 5, REGR factor score of component 4, REGR factor score of component 3, REGR factor score of component 2, REGR factor score of component 1
*Data source* WEF 2015 and WB 2013

**Table 6** Coefficients of regression model TTCI + factors[a]

| Model | Unstandardized coefficients | | Standardized coefficients | t | Sig. |
|---|---|---|---|---|---|
| | B | Std. Error | Beta | | |
| (Constant) | 3.740 | 0.008 | | 466.970 | 0.000 |
| REGR factor score of component 1 | 0.343 | 0.008 | 0.503 | 42.625 | 0.000 |
| REGR factor score of component 2 | 0.191 | 0.008 | 0.281 | 23.809 | 0.000 |
| REGR factor score of component 3 | 0.312 | 0.008 | 0.458 | 38.796 | 0.000 |
| REGR factor score of component 4 | 0.257 | 0.008 | 0.378 | 31.996 | 0.000 |
| REGR factor score of component 5 | 0.371 | 0.008 | 0.545 | 46.205 | 0.000 |

[a]Dependent Variable: travel and tourism competitive index
*Data source* WEF 2015 and WB 2013

## Conclusions

Development in travel and tourism competitiveness is primarily concentrated on ICT and infrastructure, however, it is important to note that social and political factors are necessary to take into account to support progress in the aforementioned sectors. Social and political conditions indicate that development thus includes the business environment, human resources factors, and regulations.

In order to increase the level of competitiveness in travel and tourism, sociopolitical factors are necessary to take into considerations in which the highest priority indicators are the CC, RL, GE, and RQ.

## References

Asian Development Bank (2015) Key indicators for Asia and the Pacific 2015. Asian Development Bank, Philippines

Crotti R, Misrahi T (2015) The travel and tourism competitiveness report 2015: growth through shocks. In: Crotti R, Misrahi T (eds) World Economic forum

Hall CM, Williams AM, Lew AA (2014) Tourism conceptualizations, disciplinarity, institutions, and issues. In: Lew AA, Hall CM, Williams AM (eds) The Wiley-Blackwell companion to tourism. Wiley

Hampton MP, Jeyacheya J (2015) Power, ownership and tourism in small islands: evidence from Indonesia. World Dev 70:481–495. doi:10.1016/j.worlddev.2014.12.007

Kaufmann D, Kraay A, Mastruzzi M (2010) The Worldwide Governance indicators: methodology and analytical issues. World Bank policy research working paper No. 5430, pp 31

Mietzner M (2010) Indonesia in 2009: electoral contestation and economic resilience. Asian Surv 50(1):185–194

Milne S, Ateljevic I (2001) Tourism, economic development and the global-local nexus: theory embracing complexity. Tourism Geographies 3(4):369–393. doi:10.1080/14616680110070478

Smith SLJ (1988) Defining tourism a supply-side view. Ann Tourism Res 15(2):179–190. doi:10.1016/0160-7383(88)90081-3

World Travel & Tourism Council (2015). Global Benchmarking Travel & Tourism: How Does Travel & Tourism compare to other sectors?: World Travel & Tourism Council

# Chapter 36
# Marketing Sustainable Beach Holidays: A Tropical Agenda

**K. Thirumaran and Mohit Raghav**

**Abstract** The beach is one of the idyllic tropical attractions requiring and receiving attention from all those who partake in it. As a tourism resource, the issue of preserving its awe and beauty proves challenging for stakeholders in the tropics. Existing academic frameworks of sustainability primarily focus on the triple bottom line, community-based tourism and tourist management strategies to mitigate tourist activities that are detrimental to the beach as a generational resource. This paper proposes a marketing-activities approach to offer policy options when promoting beaches in the tropics. In this approach, we analyse ways natural settings and accompanying activities produce sustainable results if the marketing programmes are in sync with activities at the beach as a perceived valuable resource. The model attempts to understand the beach as a marketable resource commensurate with escalation of activities. A comparative descriptive study was conducted on Singapore's Sentosa Island, Thailand's Phuket and India's Goa beaches to draw a conceptual framework. This framework presents principal stakeholders practical management strategies vis-à-vis marketing and beach activities. Tropical destinations are limited natural resources and face increasing development pressures. A marketing-activities approach can address issues of sustainability alongside experiences of quintessential tropical vacations for tourists.

**Keywords** Tropical agenda · Beach tourism · Marketing-activities framework

K. Thirumaran (✉)
James Cook University (Singapore Campus), Townsville, Australia
e-mail: k.thirumaran@jcu.edu.au

M. Raghav
World Lifestyle Pte. Ltd., Gurgaon, India
e-mail: mohit@elitestyle.net



A. Saufi et al. (eds.), *Balancing Development and Sustainability in Tourism Destinations*, DOI 10.1007/978-981-10-1718-6_36

## Introduction

The tropics are often depicted in a positive light for those intending to savour a holiday under the sun, with exotic landscapes and cultures. This is particularly attractive for those living in temperate zones. In this regard, tropical beaches play a very important part in attracting visitors to a destination. Where the sea meets the sun in picturesque settings of sunrise and sunsets, sandy shores lined with coconut and sea almond trees, beach destinations have created their share of desires and indulgences in the global tourism landscape. In some cases, destinations like Bali, Goa and Phuket among others have developed ancillary urban escapes so that the beach is interlinked with shopping, cultural and other in-land leisure activities.

Here, we adopt a resource-based approach as a priority to be followed in addition to all other balancing objectives of development and attracting tourists. Three tropical beach destinations: Phuket-Thailand, Goa-India and Sentosa–Singapore are selected for our study. Each of these destinations has distinctive and peculiar characteristics that can be useful in a comparative study to the understanding of ways in which a resource-centred approach to sustainability and marketing programmes.

This study unravels the importance of marketing-activities and the intensity level of the use of the beach. These two parameters enable the presentation of marketing-activities model in beach tourism. The outcome of this study allows policy makers and other stakeholders to think about how important marketing is to raising or lowering activities at the beach.

## Studies on Tropical Beaches and Sustainability

In some instances, beach destinations can morph into cities. Smith's (1991) Beach Development Model (BRM) provides stakeholders with an understanding of the development of beach resorts without impacting existing physical and sociocultural ecosystems around the beach itself. The comprehensive BRM approaches destination evolution through developments. The BRM model discusses tourism development that aids this transformation; the model does not examine the specific nature of beach usage as a natural resource marketed on a scale that shows tourism activities concurrent with both the infrastructural development and important activities promoted at the beach itself. However, the BRM model does not provide sufficient balance to show how the beach as a resource in itself can serve to limit or delimit issues of crowding, pollution and urbanisation.

Arias and Martínez (2015), suggest that if the government or authorities do not actively participate in destination promotion or in providing better facilities, tourists may flow to other competitive beach destinations as was the case with Granada's tropical coast. Smith (1991) looked at the development stages and its impact both on the beach and on inland urbanisation. Mycoo and Gobin (2013) found that marine wildlife can be endangered by climate change as well as beach developments that disturb the immediate ecosystem. Therefore, communities have to be mindful of their use of the beach. According to Moscardo and Murphy (2014) the concept of sustainability is questionable since academic understanding has not been translated fruitfully into industry practice.

This paper thus attempts to close the knowledge and policy decision-making gap between resource sustainability and development of a resource for tourism. This paper's entry into the marketing-activities approach contributes to a conceptual framework of the marketing of beaches vis-à-vis activities.

## Approach to Conceptualising Tropical Beach Sustainability

Governments in Asia have capitalised on beaches as resources since the early 1960s when international travel was on the rise. Primarily of interest to the emerging affluent classes of Europeans and Americans, beaches in the tropics, from the Caribbean to Southeast Asia, began to emerge on the tourism scene. Of these, Goa, Phuket, Boracay and many others rode the wave of western interests in tropical holidays. Due to poor governance, climate change and failed strategic management of beaches in many parts of the world, it has become imperative to revisit the beach as an important tourism resource (Smith 1991; Sakolnakorn and Naipinit 2011; Hyman 2014; Marzuki 2012; Moscardo and Murphy 2014). An alternative sustainable way of examining the beach and its promotion needs to be found.

Beach tourism has its own phases of development and popularity crucially propelled by beach activities and marketing. Goa's government does not take an active role in promoting itself as a beach destination. Phuket, on the other hand, was selected because the government assumes an active role in tourism marketing. Sentosa is included because of the private and public partnership when conducting events by the beach and also the heavy government involvement in managing the beach itself.

This method of policy documentation and review of secondary sources allows for a dynamic representation of the beach. The best way to account for the activities was through records and checking through the destinations' calendar of activities. We spent a period of 1–2 weeks in each destination to observe, collect interviews and secondary data to map a descriptive understanding of the beach as a tourism resource in the tropics. In this study, we have developed a marketing-activities model that focuses on the beach itself as a sustainable resource. Beaches were selected as case studies on the basis that they differ from one another and differ in the ways they are being promoted.

## *Phuket, Thailand*

Phuket in the Southwestern Thailand is a popular destination amongst beach tourists. They have a range of high end to budget hotels and resorts. Originally, Phuket was a quiet trading post for rubber and palm oil plantation. In the 1970s, its beach became an attraction and eventually, the interior of Phuket developed in sync with the popularity of the beach. A critical evaluation of Phuket's own experience based on its humble beginnings as an extractive economy to a world-renowned beach destination. In this section, we trace the development and present state of Phuket as a beach destination so that we can explain some of the characteristics as a base use for our marketing-activities model.

Marzuki's (2012) work finds that tourism development in Phuket has benefited local residents in spite of social costs. Sakolnakorn et al. (2013), the Phuket Provincial Administration Organization has thus far promoted the island as marine tourism, healthcare, medical service, spa tourism, and shopping. These are the major themes as part of the promotion of Phuket as a key destination in Southeast Asia. Based on a study of the problems, guidelines of sustainable tourism practices and tourism development policy it was found that tourism had impacted the cost of living and price of commodities (Sakolnakorn et al. 2013).

Phuket's beach destination can be characterised under three categories. First, the beach has been promoted by both the TAT and the Municipal Tourism Authority of Phuket and this has led to the increase in the number of tourists to the island. Further to this development, the interior of the island has benefitted much from the influx of tourists. Banks, residential properties, small manufacturing and services industries have contributed to the growth of Phuket as a leading destination for tourism and an emerging region second to Chiangmai in importance. The military regime means business in securing the sandy beaches of Phuket from wanton beach clubs and restaurants located on the shorefront. The popular beaches of Sarin, Palm and Patong are all targeted to ensure that no permanent structures are erected on the shorefront in hopes of reducing waste and pollution of the seafront.

In a study conducted by Polnyotee and Thadaniti (2015) of tourists' impact on Phuket's Patong Beach in particular, the authors found that visitors were enchanted by the natural beauty of the beach area. Aside from its scenery, beach activities like sunbathing and water activities tourists are primarily for leisure relaxation (Sung and Chi-Ming 2008). Phuket's exoticism was the primary pull factor found in the study by Sung and Chi-Ming (2008). While urban values and the setting of Phuket are incidental attractions for tourists, these studies suggest that Phuket's base value to be the beach itself.

In 2010, tourist arrivals into Phuket were under 2.5 million people and in 2015 arrivals are expected to rise beyond 3.5 million people (Phuket Gazette 2014). This is partly attributable to the popularity of the destination over the years, international air connectivity and Tourism Authority of Thailand's continued active promotion of Thailand and its key attractions internationally. Phuket materials more actively focus on beach. People spend more time specifically on beach engaged in activities such as relaxing, playing sports, indulging in F & B or just hanging out, whereas in Phuket, people specifically go to beach to participate in beach activities. The beach is more popular among couples and families but Phuket is only destination where it makes it popular even to MICE groups. Directly marketed to "Couples" & "Families" and "Groups". So, Phuket tourism promotes beach activities and organise water sports events on Phuket beach which contributes indirectly to the MICE industry.

## *Goa, India*

Goa, a former Portuguese enclave is located in the Western part of India by the Arabian Sea. Since the 1960s, Goa has been attracting a steady flow of visitors—first hippies and returning expatriate Goans, then the charter tourists (starting with Germans in 1987), pilgrims visiting Catholic and Hindu shrines, those opting to settle in Goa as their home, people going for medical treatment, and a growing number of those who attend seminars and conferences in Goa.

People travel to Goa for relaxation diving, kite surfing and paragliding as the setting is an organic one rich with Portuguese heritage. Aguada Beach, Vainginim Beach, Bogmalo Beach are famous for water sports. The infrastructure immediate to the beach is more of a rural setting. Some of the activities at the beach include jet skiing, beach clubs enthusing nightlife, scouting vendors and beach restaurants. Aside from the beach itself, there is very little promotion of this destination. Seldom are Meetings, Incentives, Conventions and Exhibitions (MICE) or other tourist-attracting activities promoted. Therefore, the number of tourists and activities are relatively low. People visit Goa beach just for walking, relaxing, dining, water skiing and beach front night life. Clubs and bars can accommodate more people even during the peak seasons of June–August. All the five star hotels have casinos.

The spillover of a domestic religious festival known as Chaturthi on the 17 and 18 of September is one event that involves the beach and spectating by tourists. In this festival, Ganesha idols are sent off into the sea after a ceremonious march through the city's main streets. Other events more directly related to the tourism industry are the International Travel Mart and the World Tourism Day. The Goa Tourism Ministry also organises the International Film Festival in which movies are played on the beachside.

Only a few events take place like a ZoukOut Party or Film Festival. These few events add to the flavour of Goa as a destination with sporadic events that receives little attention from international tourist circuits as it does not have much government promotional support. In 2014, this popular honeymoon destination for Indians attracted 3.5 million over and half a million internationals (Department of Tourism

2015). Despite best efforts by travel agents and the Goa government, marketing of the beach has not translated into mass international tourist arrivals, a situation that differs greatly from Phuket.

## *Sentosa Island, Singapore*

Sentosa is one of the southern islands off main-island Singapore. Prior to the mid-1970s—Sentosa was untouched and undeveloped. The island was known only to a few people for its beach but otherwise it was thick virgin jungle. Dotted with world war two bunkers, the island was served by a ferry and a double Decker bus was the only public transport able to bring family and other groups to picnic by the beach. During this period, it even lacked modern or basic amenities such as toilets.

In the 1980s, the musical fountain and other attractions were built to draw visitors not so much for the beach but for amusement theme parks. The beach's natural boundary was cleared to make way for a man-configured beach with three lagoons. But the island's overarching selling point was attractions like The Images of Singapore museum, The Cable Car Ride, Fort Siloso, Volcano Land and Underwater World. More recently, Resorts World Sentosa with its gaming and cabaret shows, high-end living with bays for sailing boats and large golf courses have transformed the island into a playground for leisure tourists and wealthy residents. The beach itself remains relatively quiet except on corporate and public event days.

Many local firms periodically organise beach events for their staff and on special occasions, Zookout and an annual foam party coinciding with New Year's Day contribute to activities on the beach. Beach-side parties such as these have become a popular norm on the island. Beach style restaurants and pubs are licensed to operate.

So, while the two islands started off with beaches as spaces of leisure, their present and future trajectories have diverged. The outcome of this divergence is also telling in that, Phuket is world renowned whereas Sentosa remains oblivious in the international name recognition for tropical beaches. Phuket presents a more natural undulating sandy beach with a wild and wonder holiday feeling with night pubs, massage parlours, a red light district and beach shopping walks all encapsulated within the notion of a beach holiday. However, in the Sentosa context, the beach remains just a beach with no street side market and tourists shopping outlets, and with very few beach dining facilities. Sentosa is just simply too clean and artificial that it does not have the pizzazz of Phuket's more internationally favoured beach culture.

## A Framework for Sustainable Tropical Beach Destinations

The study of the three tropical destinations leaves us at a crossroads in marketing and development. Should a beach be preserved with sustainability as a foremost concern or should it be transformed into an international tourist destination? Below are five growth phases through which the natural resource of the tropics such as the beach can be viewed. We are able to draw different types of marketing descriptions for a destination:

1. Holiday Destination

Undisturbed charm because beach goers enjoy sunbathing, a dip in the sea, strolling and/or beachcombing.

2. Water Fun Destination

Informal activities take place with little mechanisation; swimmers abound and boats are non-motorised:

3. Local Beach Festivals

Localised and periodic social events and cultural festivals connected to the beach take place.

4. Mega Event Beach Destination

Regularised International Sea Sports events which may involve construction and motorised boats; waste management requires amplification in this model.

5. Beach Homes and Urbanisation

Significant impact on the beach's physical and social structures. At this stress level, marketing corollary with urban development has to be checked in order to be sync with the character of the natural resource.

Today's growing traveller population and new technological advancements can transform tropical beach destinations into urban developments and conurbations. The marketing-activities model emphasises that policy makers take into account that the beach is susceptible to pollution. If the beach and sea in the vicinity is used for scuba diving, dolphins, or whale-shark viewing, such activities can have a negative impact as does discharge of waste into the sea.

Therefore, a destination has to first decide the position of the beach. If the beach is to be sustained as it is, the first category of marketing should be followed. That means zero marketing effort internationally so that tourists are not attracted in an overwhelming way that demands more development and a significant increase in waste and its management. Likewise, if the policy makers desire to create the beach as a global destination, then marketing efforts would yield large numbers of tourists and increase in infrastructure development to accommodate tourists and the beach itself as an attraction.

## Conclusion

Thus far the model presents an understanding of beach activities and marketing to justify development trajectories. MICE an Urban Beach living can contribute to the value of the beach but it would be subject to further pollution, development and overcrowding. Therefore, Destination Marketing Organisations have to decide on sustainability in terms of the beach as a preservation resource more than the needs of the people and tourists. Managing the beach corollary with tourism and infrastructure development can never secure the beach as a sustainable resource.

The extent to which the beach as a tourism resource is expanded or depleted is very much dependent on the marketing efforts and programmes implemented to stimulate activities. For example, if we know that marine life swim up close to the shore and when too many swimmers are at the beach or requesting a motor boat ride to the sea for a closer look, then it means that marketing efforts have to be reasonably adjusted not to overly promote to bring in more tourists. It is through this types of effort, that resources and demands for them can be sustained. The argument that the visitors can be educated in their approach to marine life and that somehow this can be managed remains inconclusive. Public and visitor education has proved to be far from universally effective. An example is the case of Lovina Beach in Bali.

As this paper proposes, a marketing approach curtailing overwhelming touristic interest might be a way to harness the resource for future generations rather than focusing on just creating tourist behaviour strategies. Measures to keep the beach attractive through crowd control, good signage, proper facilities, and availability of tourist information centres could be a constant challenge.

A study of less-developed beach versus mature beach destinations in terms of marketing differences may be helpful in relating the impact on the beach as a resource. For example, people visiting an underdeveloped beach may not be happy that the beach is too quiet and lacks basic amenities and facilities. Vacationers in a mature destination may find themselves displeased with overdevelopment, crowding and urbanisation. A closer examination of examples of marketing campaigns and the number of tourist arrivals as a result and their impact on the beach as a sustainable resource is needed. Identifying a causal relationship will further this conceptual idea as articulated within a descriptive study of three different tropical beach destinations.

## References

Arias JAE, Martínez FDAR (2015) Trends and opportunities of tourism development in Granada Tropical Coast Local Government visions. Cuadernos De Turismo 35:447–450

Department of Tourism, Government of Goa. 30 January 2015. Goa Welcomes 30 % more tourists in 2014. TravelBizmonitor.com. Retrieved from http://www.travelbizmonitor.com/Ministry/goa-welcomes-30-more-tourists-in-2014-26485

Hyman T-A (2014) Assessing the vulnerability of beach tourism and nonbeach tourism to climate change: a case study from Jamaica. J Sustain Tourism 22(8):1197–1215

Marzuki A (2012) Local resident's perceptions towards economic impacts of tourism development in Phuket. Tourism 60(2):199–212

Moscardo G, Murphy L (2014) There is no such thing as sustainable tourism: re-conceptualizing tourism as a tool for sustainability. Sustainability 6(5):2538–2561

Mycoo MA, Gobin JF (2013) Coastal management, climate change adaptation and sustainability in small coastal communities: Leatherback turtles and beach loss. Sustain Sci 8(3):441–453

Phuket Gazette (2014) International arrivals and hotel occupancy strong in first quarter. Retrieved from http://www.phuketgazette.net/phuket-business/International-arrivals-hotel-occupancy-strong-first-quarter/30785

Polnyotee M, Thadaniti S (2015) Community-based tourism: Is strategy for sustainable tourism development of patong beach, Phuket Island. Thailand. Asian Social Science 11(27):90–98

Sakolnakorn TPN, Naipinit A (2011) The problem and threat in the management of tourism sustainability in Phuket. Int J Manage Inf Syst 15(2):111–115

Sakolnakorn TPN, Naipinit A, Kroeksakul P (2013) Sustainable tourism development and management in the Phuket Province. Thai Asian Soc Sci 9:75–84

Smith RA (1991) Beach resorts: a model of development evolution. Landscape and Urban Plan 21:189–210

Sung HP, C-M Hsieh (2008) Motivations between Island Tourists. Glob Edge Bus Rev 2(8):1–2

MIX
Papier aus verantwortungsvollen Quellen
Paper from responsible sources
FSC® C105338

If you have any concerns about our products,
you can contact us on
**ProductSafety@springernature.com**

In case Publisher is established outside the EU,
the EU authorized representative is:
**Springer Nature Customer Service Center GmbH**
**Europaplatz 3, 69115 Heidelberg, Germany**

Printed by Libri Plureos GmbH
in Hamburg, Germany